Tim Salditt, Timo Aspelmeier, Sebastian Aeffner

Biomedical Imaging

De Gruyter Graduate

Also of Interest

Multiphoton Microscopy and Fluorescence Lifetime Imaging. Applications in Biology and Medicine
Karsten König (Hrsg.), 2017
ISBN 978-3-11-043898-7, e-ISBN (PDF) 978-3-11-042998-5,
e-ISBN (EPUB) 978-3-11-043007-3

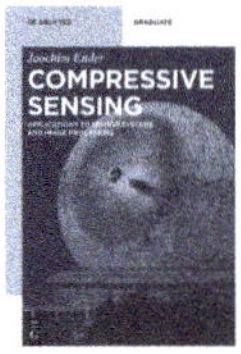

Compressive Sensing. Applications to Sensor Systems and Image Processing
Joachim Ender, 2017
ISBN 978-3-11-033531-6, e-ISBN (PDF) 978-3-11-033539-2,
e-ISBN (EPUB) 978-3-11-039027-8

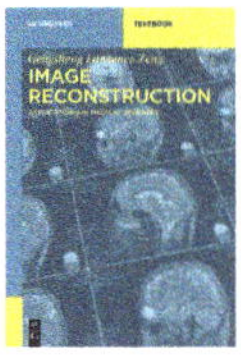

Image Reconstruction. Applications in Medical Sciences
Gengsheng Lawrence Zeng, 2017
ISBN 978-3-11-050048-6, e-ISBN (PDF) 978-3-11-050059-2,
e-ISBN (EPUB) 978-3-11-049802-8

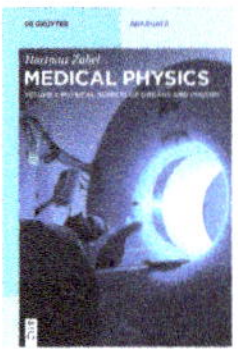

Medical Physics. Volume I
Hartmut Zabel, 2017
ISBN 978-3-11-037281-6, e-ISBN (PDF) 978-3-11-037283-0,
e-ISBN (EPUB) 978-3-11-037285-4

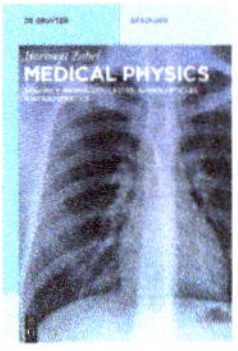

Medical Physics. Volume II
Hartmut Zabel, 2017
ISBN 978-3-11-055310-9, e-ISBN (PDF) 978-3-11-055311-6,
e-ISBN (EPUB) 978-3-11-055317-8

Tim Salditt, Timo Aspelmeier,
Sebastian Aeffner

Biomedical Imaging

Principles of Radiography, Tomography
and Medical Physics

DE GRUYTER

Physics and Astronomy Classification Scheme 2010
87.57.-s, 87.57.C-, 87.57.N-, 87.59.B-, 87.59.-e, 41.50.+h, 32.30.Rj, 41.60.Ap, 87.57.Q-, 78.70.-g, 87.10.Rt, 34.50.Bw, 87.53.-j, 87.19.xj, 87.55.-x, 87.56.-v, 78.20.Ci, 42.25.-p, 42.30.-d

Authors
Prof. Dr. Tim Salditt
Georg-August-Universität
Institut für Röntgenphysik
Friedrich-Hund-Platz 1
37077 Göttingen
Germany
tsaldit@gwdg.de

Dr. Sebastian Aeffner
Georg-August-Universität
Institut für Röntgenphysik
Friedrich-Hund-Platz 1
37077 Göttingen
Germany
Sebastian.Aeffner@stud.uni-goettingen.de

Dr. Timo Aspelmeier
Georg-August-Universität Göttingen
Inst. für Mathematische Stochastik
Goldschmidtstr. 7
37077 Göttingen
Germany
timo.aspelmeier@mathematik.uni-goettingen.de

ISBN 978-3-11-042668-7
e-ISBN (PDF) 978-3-11-042669-4
e-ISBN (EPUB) 978-3-11-042351-8

Library of Congress Cataloging-in-Publication Data
A CIP catalog record for this book has been applied for at the Library of Congress.

Bibliographic information published by the Deutsche Nationalbibliothek
The Deutsche Nationalbibliothek lists this publication in the Deutsche Nationalbibliografie; detailed bibliographic data are available on the Internet at http://dnb.dnb.de.

Cover image: Matthias Bartels, Victor H. Hernandez, Martin Krenkel, Tobias Moser, and Tim Salditt: Phase contrast tomography of the mouse cochlea at microfocus x-ray sources; Appl. Phys. Lett. 103, 083703 (2013); doi: 10.1063/1.4818737
Typesetting: le-tex publishing services GmbH, Leipzig
Printing and binding: CPI books GmbH, Leck
♾ Printed on acid-free paper
Printed in Germany

www.degruyter.com

Contents

Preface and acknowledgements

This book originated from a course at Universität Göttingen for Master's students in physics, first taught by T. Salditt and T. Aspelmeier in 2010, and followed by two iterations, the last in 2015, which then served as an incubation for the book project. As it turned out, the course was a bit wider in terms of techniques covered than the book, since priority and preference was finally given to topics closer to the research of the authors. We will see what extensions any future editions may bring...

As a trio of authors, we bring together expertise from different fields, notably x-ray optics and imaging, statistical physics and applied mathematics, biophysics and medical physics. Merging the different experiences and perspectives into a 'coherent' monograph has been intellectually stimulating. It has also fostered new directions in our research, resulting, for example, in a novel geometry of recording tomographic data. It is a nice example of the synergistic nature of university teaching and research. Because it emerged from a course, this book is far from being complete or free of errors, just as most lecture scripts. However, we do hope that the readers find stimulating insights and, possibly, starting points for further reading, learning, teaching and scientific activity.

Alongside the business of contemporary academia, with its many projects, proposals, publications, beamtime and lectures, the book project has probably not received the time it would have deserved. It did, however, receive a lot of support, help and contributions from the group at Institut für Röntgenphysik. First, and above all, we would like to thank Anna-Lena Robisch, who has been a fantastic tutor for the course and has provided many contributions, in particular concerning graphics, Matlab scripts and careful proofreading. Further, T. S. is grateful for all insights, discussions and ideas developed and shared with former and current Ph.D. students who are active in the different topics of the book. Klaus Giewekemeyer has influenced much work on phase contrast imaging and phase retrieval, and was at the origin of the Matlab toolbox used for phase contrast imaging in the group. Matthias Bartels and Martin Krenkel then extended this to tomography, and provided tutorials and didactic contributions on many levels, in particular graphics and Matlab routines. Finally, we acknowledge the current 'tomo-team' Mareike Töpperwien, Aike Ruhlandt and Malte Vassholz, and also Leon-Marten Lohse, for their outstanding contributions. We are also grateful for the early work on the script of the course, and for valuable tutorials, contributed by Markus Osterhoff and Tobias Reusch.

T. A. would like to thank all people at Scivis GmbH for a very inspiring work atmosphere, bringing together mathematics, physics and computer science. Gernot Ebel and Uwe Engeland introduced me to the field of tomography, without which I could not have contributed to this book. Tilmann Prautzsch and Björn Thiel taught me how to manage and program large scale software projects, which may not seem relevant for this book, but which is, in fact, very important because the mathematical background

https://doi.org/10.1515/9783110426694-201

only comes to life when it can be easily applied. Nils Gladitz's exceptional programming skills continue to be a shining example for me, although I'm afraid he would be appalled at my own feeble attempts.

S. A. would like to thank everyone at the Department of Radiation Oncology at Universitätsklinikum Würzburg, especially Otto Sauer and Klaus Bratengeier, as well as everyone at the Department of Radiotherapy at Klinikum Kassel, especially the former medical physics team around Michaela Ritter and Heike Mögling, for introducing me to the field of radiotherapy physics. Yizhu Li is acknowledged for some very helpful comments on DNA repair mechanisms, and Jan-David Nicolas for careful proofreading of Chapter 5.

We also want to thank the students of the course. Valuable graphical and numerical work evolved from the student projects of the 2015/2016 course. To this end, we particularly thank Marius Herr for illustrations on Fourier filtering and Vitali Telezki for Monte Carlo simulations of depth–dose curves, as well as Gerrit Brehm for proofreading of several chapters.

Last but not least, we thank our families, in particular Nicole and Stella, for their support!

1 Introduction

Physics provides enabling tools for physicians. This is not meant to detract from the importance of biochemistry or genetics, but stresses the unique role of physical methods in modern medicine and biomedical research, in particular for diagnostic purposes. Beginning with simple measurements of body temperature or blood pressure, method development has led to state of the art imaging and microscopy techniques, and is not likely to stop here. In this monograph we treat modern biomedical imaging techniques, not only in the context of clinical radiology, but also in preclinical research. The text is based on a course of medical physics taught in the physics Master's program. Hence, it is primarily intended for a readership with a background in physics, mathematics or computer science. In this course, we show how progress in measurement principles and data recording, as a topic of experimental physics, is intertwined with progress in analysis and algorithms. The design of new methods and instruments requires knowledge and skills both in experimental and theoretical fields of activity. Hence, a textbook covering the state of the art techniques as a starting point for further research must provide a suitable combination, which is attempted here.

1.1 X-ray radiography as an example

Imaging requires thoughtful considerations on image formation, recording, reconstruction and display. X-ray radiography as the oldest non-invasive technique to image the interior of bodies is a perfect example. The principle is seemingly of utmost simplicity – a shadow projection image is recorded with the transmitted intensity reflecting the different absorption coefficients of x-ray radiation in tissues, ranging from strong absorption by heavier atoms such as calcium in bone, to weak absorption in low density soft tissues such as the lung. The concept is entirely based on line integrals and geometric optics. However, when considering the entire wealth of interaction processes between photons and matter, more advanced levels of treatment are required, as illustrated by Figure 1.1, and the issue becomes significantly more complex. In such complexity lies a source of further method development. For example, treating the propagation of x-rays in the body beyond the framework of geometric optics has opened up new contrast mechanisms, see the sketch in Figure 1.2. Phase contrast and scattering (darkfield) contrast, which offer a novel way to visualize also non-absorbing or weakly absorbing tissues, are right now at the transition to clinical applications. In view of further developments, diffraction and wave optical properties hold the promise of increased spatial resolution, at least for preclinical applications.

However, also the relatively simple and conventional form of radiography in absorption contrast requires thoughtful reflection. For example, x-ray film, which prevailed the first hundred years of radiography since Röntgen's discovery, is character-

DOI 10.1515/9783110426694-001

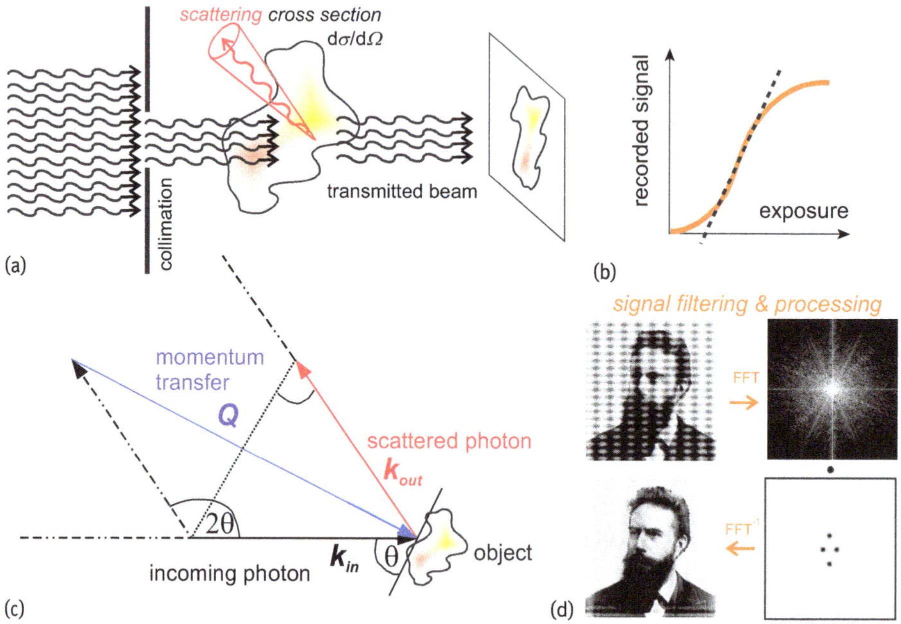

Fig. 1.1: (a) X-ray radiography is based on recording the transmitted x-ray beam behind an object. Absorption contrast can be mostly described in terms of simple geometric optics and differential absorption of x-ray radiation in tissues. (b) In image recording, experimental issues such as the response curve of the detector require careful consideration. The non-linearity arising from exponential intensity decrease in the object and eventually also from detection, can be accounted for by rescaling the signal. (c) To understand the interaction of radiation and matter in more complete terms, wave optical and quantum optical treatments are required, for example, to describe the phase shift or the scattering of radiation from the object. Cross sections can be calculated for different interaction channels. (d) Once the signal is digitalized, numerical processing and filtering, and reconstruction are often indispensable to unravel the information contained in the images.

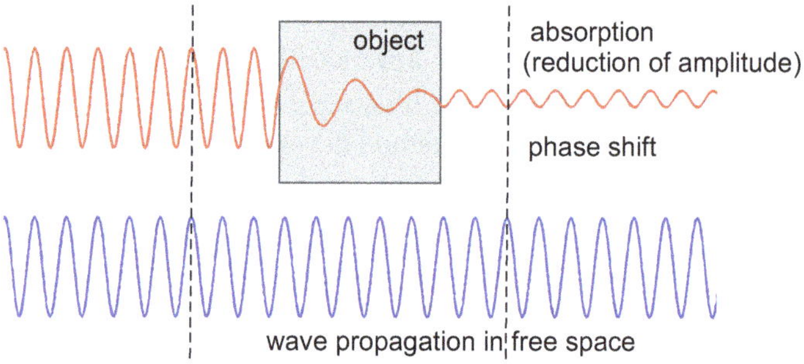

Fig. 1.2: Illustration of two main effects when a wave traverses an object or body: Absorption, i.e., attenuation of the wave's amplitude by matter, and a phase shift resulting from a difference in the wave's phase velocity, with respect to a reference wave propagating in free space. Both effects can be exploited in advanced radiography.

ized by a highly non-linear response curve in optical density, see Figure 1.1 (b). The photon energy and exposure time (and thus the absorbed dose) must be precisely tuned such that the medically relevant region of interest is recorded at the optimum working point of the response curve. Otherwise the image cannot be interpreted by the physician. Modern radiography uses so called flat panel detectors equipped with scintillator films to convert x-rays into visible light for subsequent detection in an adjacent photo diode array. These or other digital detectors offer a sufficiently linear response curve over a much broader signal range of input intensity. Hence, the working point does not have to be precisely tuned, reducing flawed recordings and unnecessary dose. At the same time, their dynamic range covers more orders of magnitude. A typical dynamic range of $2^{12} = 4096$ gray values (for a 12-bit camera) exceeds the number of gray values that can be differentiated by the human eye by about two orders of magnitude. This mismatch alone already creates a need for image processing and filtering, including algorithms that cleverly mine the data for features.

Further points requiring our attention are image resolution and noise. Photons may create a signal in more than a single detector pixel, and a single detector pixel will receive signals from more than one point of the (extended) source. Collimators can be used to select the photon directions and hence to increase spatial resolution, but only at the expense of reduced signals (Figure 1.3). Signal-to-noise, finally, is always a concern, since even with all instrumental noise and perturbation sources suppressed as much as possible, the intrinsic photon shot noise prevails. Within the context of classical sources relevant here, the photon count data is always characterized by a Poissonian distribution, creating a need for a high dose in order to reduce the relative uncertainty of the signal. To obtain optimum image quality and resolution, one has to treat and – if possible – correct for the effects of blurring and noise. This poses a classical inverse problem, one of many that we will encounter throughout this course. The simple example of x-ray radiography has already outlined some challenges of biomedical imaging. Some important issues, which are also often intertwined, relate to the quan-

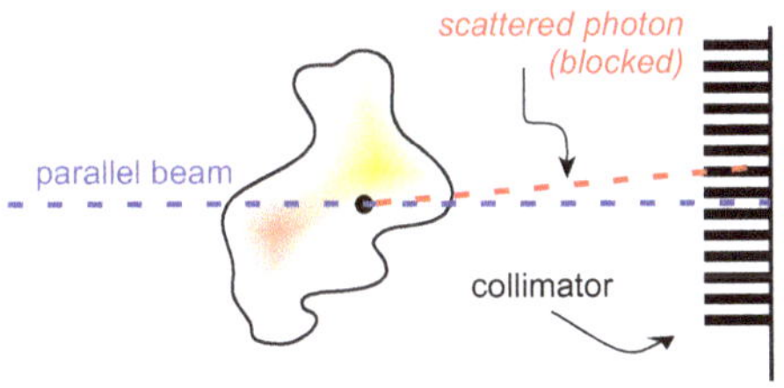

Fig. 1.3: Photons scattered in the body reach the detector at a “wrong” pixel position. This can be suppressed by use of collimators. However, these always involve a reduction of the signal, since also a certain fraction of non-scattered photons is absorbed. In designing experiments and instruments the question arises as to how a maximum of signal for a minimum of photons and dose can be obtained.

tum nature of light, photon shot noise, blurring/resolution, data throughput and discretization, mathematics of reconstruction, inverse problems and human perception.

1.2 Summary of biomedical techniques covered herein

Radiography (x-ray imaging)
In its conventional form, radiography records projection images (Figure 1.4a) using hard x-rays with maximum photon energies ≥ 100 keV for full body exposures, corresponding to wavelengths $\lambda \leq 12.4$ pm. Contrast is formed by differential absorption of mineralized and soft tissues. Since x-rays are ionizing, radiation damage is a significant concern.

Tomography
comprises methods to retrieve the internal three-dimensional structure of an object or body, from its projection integrals. Using tomographic methods, we can inspect virtual slices of the object. Tomographic recordings and image reconstruction (computerized tomography, CT) is not limited to radiographic CT, but also applies to nuclear imaging modalities such as SPECT and PET (Figure 1.4b), which are introduced below.

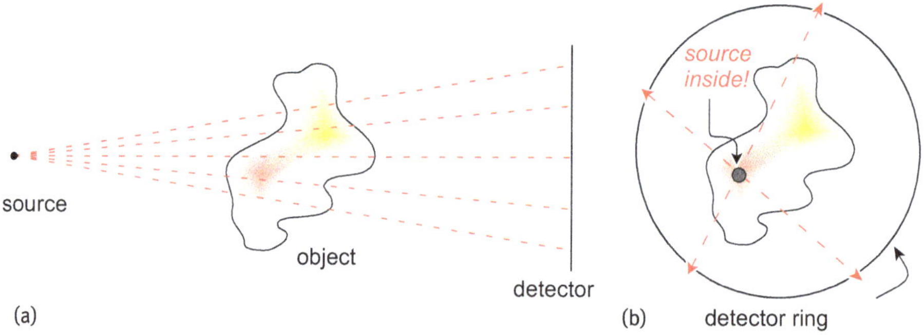

Fig. 1.4: (a) Radiographic and (b) tomographic setup. Tomography requires projections from all angles. In the case of nuclear diagnostic tomography such as PET and SPECT, the sources of radiation are inside the body, and the signal can be recorded by a detector ring.

Beyond absorption contrast
Soft tissues are characterized by weak absorption and small differences in the absorption coefficient μ in the hard x-ray range. Hence, image contrast vanishes easily when the structures are not extended or contrast in μ is too small. A comparatively stronger effect than absorption is in this case the relative difference of the x-ray phase velocity

as a function of tissue density. Unfortunately, the phase retardation of an x-ray wave in matter is not directly accessible to a measurement. However, phase differences in the exit wavefront lead to diffraction, resulting in measurable interference effects.

Positron emission tomography (PET) and single photon emission tomography (SPECT)
A metabolic compound with a radioactive tracer of short decay time is entered into the blood stream and, after transport to the desired organ or region-of-interest, γ-radiation emitted from the tracer isotopes is recorded. The information on the emission direction, as required for tomographic reconstruction, is based on collimators in front of the detector ring around the body, for the case of SPECT. Such (flux-reducing) collimators are not needed, if the isotope exhibits a β^+ decay. In this case, the positron emitted by the tracer and an electron annihilate emitting two γ-quanta. Conservation of momentum warrants emission in the opposite direction, at 180°, so that the line of emission can be inferred from a simultaneous detection of two γ-quanta (coincidence measurement) in the opposite modules of a detection ring. From many such events, the tracer distribution in the body can be reconstructed.

Biomedical imaging techniques not covered in this volume
Unfortunately, we do not (yet) cover the important medical imaging techniques of magnetic resonance imaging (MRI), nor of ultrasound imaging (US). MRI uses strong static magnetic fields and high-frequency electromagnetic radiation to resonantly excite the spin of atomic nuclei, in particular the proton spin. The resulting spinning magnetization induces a signal in recording coils. Image contrast is formed by tissue-dependent relaxation times of excited protons. US imaging is based on the different propagation velocities and reflection and transmission coefficients of tissues and tissue boundaries in the body. Both in MRI and US, information is mostly acquired from a temporal sequence of a signal. Apart from radiological practice, it seems that these techniques have little in common with x-ray imaging, nuclear diagnostic imaging and CT. A closer look, however, reveals many interesting aspects shared between these experimentally very different approaches. For example, the Fourier methods that are important in the design of MRI echo sequences and in mapping the reciprocal space, are closely related to sampling issues in radiography and CT. The full wave optical treatment of US in the body, which could also go far beyond current capabilities, exhibits strong similarities to the approaches used to describe propagation of x-rays beyond the simple projection approximation, which is warranted only for relatively thin objects, high photon energy and low resolution.

2 Digital image processing

Before turning to physical and algorithmic principles of biomedical imaging in particular, we first consider images and image processing from a very general point of view. The scope of the present chapter is to provide definitions and preliminaries of image representation and processing. We define an image as a *discretely sampled* version of a physical observable $f(x, y)$, defined over some domain $\subset \mathbb{R}^2$, the field of view (FOV). In most cases, the image represents a scalar real-valued function $f(x, y) \in \mathbb{R}$, e.g., the x-ray attenuation coefficient, the proton density, the concentration of some radioactive element or the ultrasound impedance, etc. In other cases, we may have $f(x, y) \in \mathbb{C}$; consider, for example, a quasi-monochromatic wavefield, e.g., an ultrasonic wavefield or a component of an electromagnetic field. In biomedical imaging, the phase and amplitude of the wave behind the object can, for example, represent important object properties. In some cases, $\underline{f}$ is a vectorial quantity, even if recoding a vector field certainly involves more effort than recording a scalar field. Representation of vectorial and multi-valued functions could take advantage of sophisticated color representations.

In this chapter, we tacitly imply that images are represented in the prevailing basis system of image pixels. Images can be transformed to many other representations, writing the image as a linear combination of components, such as geometric shapes, Gaussian 'blobs', or more suitably by a shearlet frame. The storing of the corresponding coefficients is very different from storing an image as pixel-by-pixel values. Shearlet transformations can be useful, for example, to implement concepts of sparsity in image processing or to exploit a priori information in reconstruction problems.

The literature on image processing is vast, and many excellent tutorials and resources are available online, see this chapter's bibliography for some examples. The principal references for the topics covered in this chapter are [1, 7, 11, 13].

2.1 Image representation and point operations

Here, we restrict ourselves to scalar images, representing a real-valued physical observable $f(x, y) \in \mathbb{R}$. The pixel values can be displayed as *grayscale* images, or for better visibility eventually in color, mapping pixel value to color according to a predefined colormap. The grayscale or colormap maps the physical observable f for each pixel (n, m) onto a single value representing grayscale or a triplet of values coding for color, respectively,

$$f(x, y) \mapsto g(n, m) \tag{2.1}$$

Creating the image has thus already involved two steps of discretization: A continuous signal is sampled at discrete points, and the signal $\in \mathbb{R}$ in each sample point is quantized into a gray value, e.g., by an analog-to-digital converter (ADC). In total, a

DOI 10.1515/9783110426694-002

grayscale image is a function g on a finite rectangular grid, mapping each point (x, y) to a finite and integer gray value

$$g : \underline{N} \times \underline{M} \longrightarrow \underline{G} \tag{2.2}$$

where

$$\underline{N} = \{0, 1, \ldots, N-1\} \qquad \underline{M} = \{0, 1, \ldots, M-1\} \qquad \underline{G} = \{0, 1, \ldots, G-1\}\,. \tag{2.3}$$

For numerical performance, binary friendly formats are often preferred, for example, $N = M = 2^n$ (square images) and $G = 2^b$, where $n, b \in \mathbb{N}$. Written in matrix form, such a grayscale image with $n = 16$ and $b = 8$ (i.e., $2^8 = 256$ possible gray values) may look as follows (gray value 0 represents black, 255 represents white):

184	208	216	219	193	134	125	157	154	154	134	114	95	68	56	73
199	216	223	177	125	140	159	166	173	162	190	176	148	125	95	68
215	219	157	111	139	140	147	153	154	111	153	177	154	140	131	106
220	130	88	109	120	115	119	117	123	81	77	154	107	103	91	117
106	78	96	80	81	75	60	51	49	43	43	58	43	35	45	91
62	78	67	49	28	37	26	31	33	33	31	26	24	39	24	20
53	45	28	28	35	28	26	28	22	26	28	131	117	47	22	33
65	39	28	28	43	22	41	15	33	62	81	215	224	199	95	18
28	26	28	39	86	28	75	39	18	83	20	190	229	229	212	111
60	51	39	70	112	78	51	75	68	58	78	237	238	236	231	180
125	86	39	65	111	123	80	41	45	96	216	231	239	235	215	170
147	133	91	90	104	140	164	163	176	212	222	233	236	227	199	163
163	162	147	120	107	125	119	145	156	166	183	193	189	207	213	201
166	164	162	156	160	130	122	117	104	142	134	114	104	159	162	189
167	166	157	167	166	151	159	147	145	144	160	142	159	169	169	167
180	179	173	172	164	160	167	159	164	164	170	182	183	190	190	182

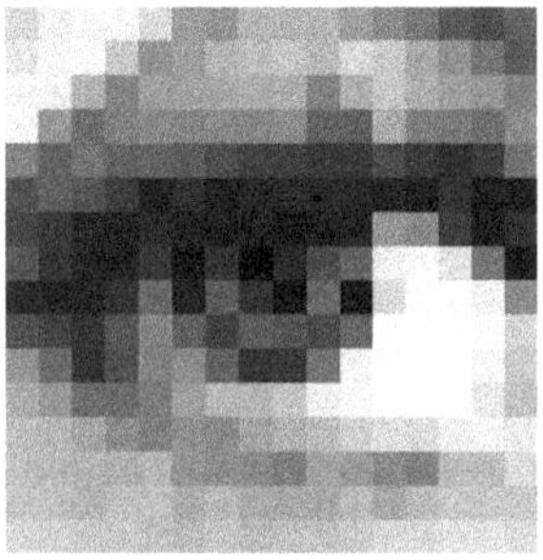

In many applications of image display and processing, the following definitions are used:

- the black point B/white point W of an image is the lowest/highest occurring gray value;
- brightness can be defined as $\mathrm{Br} = \frac{W+B}{2}$;
- contrast can be defined as $K = W - B$.

There are at least three different meanings of contrast. Firstly, it can denote the type of physical observable being imaged by an imaging apparatus, such as darkfield or differential phase contrast in microscopy. Secondly, it can simply mean the difference between black and white points as above, as is often the case in image display and processing software. Thirdly, it can designate how a particular feature or the overall image is 'contrasted', i.e., concerning the range of pixel values. For the third case, contrast can be quantified by the relative modulation of the gray values in the feature $K = (g_{\max} - g_{\min})/(g_{\max} + g_{\min})$ or alternatively by $K = (g_{\max} - g_{\min})/g_{\min}$, as further detailed in Section 2.6.1. The histogram of an image is a function h that yields the number of pixels with a certain gray value (or gray values in a certain range)

$$h(g) = \sum_{n=0}^{N-1} \sum_{m=0}^{M-1} \delta(g - g(n, m)) \tag{2.4}$$

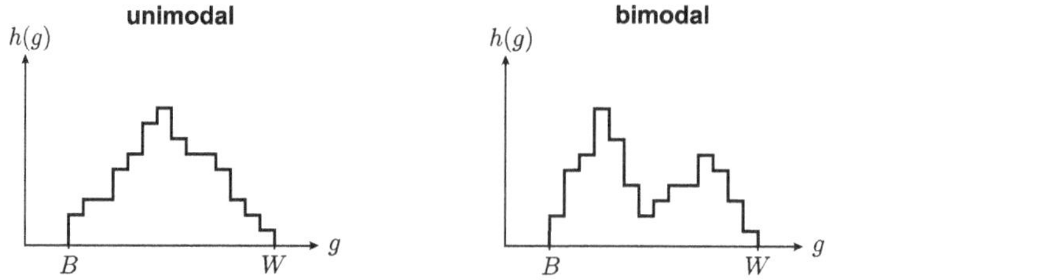

Fig. 2.1: Sketches of a unimodal (*left*) and a bimodal histogram (*right*) with minimum gray value $g_{\min} = B$ and maximum gray value $g_{\max} = W$.

Examples of histograms are shown in Figure 2.1. The sum of all histogram values is the number of pixels NM. Division of the histogram by NM, therefore, yields the probability of an arbitrary pixel having gray value g

$$P(g) = \frac{1}{NM} h(g) . \tag{2.5}$$

The mean gray value is

$$\overline{g} = \frac{1}{NM} \sum_{n=0}^{N-1} \sum_{m=0}^{M-1} g(n, m) , \tag{2.6}$$

the variance is also called the global contrast

$$s^2 = \frac{1}{NM} \sum_{n=0}^{N-1} \sum_{m=0}^{N-1} (g(n, m) - \overline{g})^2 . \tag{2.7}$$

The entropy $\mathcal{E}$ of an image histogram can be defined as

$$\mathcal{E} = -\sum_{i} P(g_i) \log_2 P(g_i) . \tag{2.8}$$

Take the example of an image representing a slice through a tomographic reconstruction of an object with just a few chemical components. If the reconstruction is quantitative with regard to the gray values, the histogram will have a few sharp peaks, and $\mathcal{E}$ will be low. Contrarily, if the histogram is broadened by noise and artifacts, it will be high.

Homogeneous point transformations replace each gray value g of a pixel by a new gray value g', the location of the pixel and the gray values of other pixels in its environment are irrelevant:

$$g'(n, m) = f(g(n, m)) \qquad \text{("pixel-by-pixel"-operation)} \tag{2.9}$$

This class of transformations can be used, e.g., for contrast enhancement (Figure 2.2). A logarithmic transform may help to distinguish differences in regions of low g, while a stepwise linear transform can be used to exploit the full available range of gray values. Another important case is thresholding, resulting in a binary image.

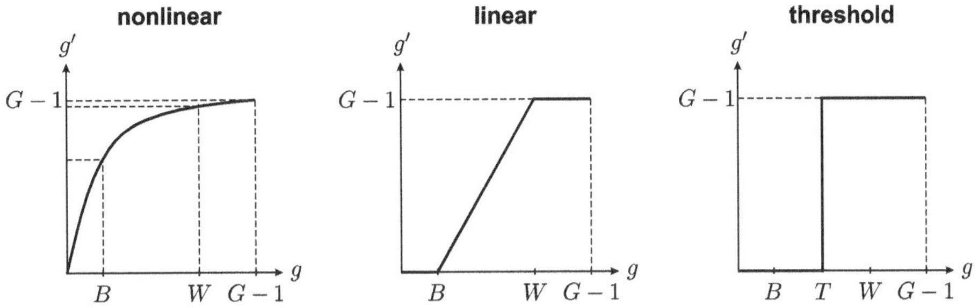

Fig. 2.2: Different homogeneous point transformations $g \rightarrow g'$. *Left*: A nonlinear (e.g., logarithmic) transform can increase the visibility of weak features. *Center*: Adjustment of black and white points to the full available range of gray values. *Right*: Thresholding yields a binary image, i.e., $g' = 0$ or 1, which is sometimes required for further image processing (see, e.g., Section 2.4).

Colormaps

Mapping the pixel values of the image matrix to a single brightness value, yields a grayscale image with a certain bit depth. An 8 bit grayscale image, for example, allows for the display of $2^8 = 256$ brightness values, a 16 bit image for $2^{16} = 65\,536$ different shades of gray.

For color images, each pixel needs brightness and color values. Different schemes exist to represent color in digital format. In physical terms of the visible light spectrum, any color can be represented by a superposition of spectral lines. This would, however, correspond to a basis set with an intractably large number of elements. It is, therefore, necessary to reduce the dimensionality of the representation and to realize a set of colors as a superposition of a few elements. The color space which can be realized (reached by the basis) is called the *gamut*. A few common choices are as follows:

- *RGB colors*: Each pixel has three individual brightness values for red, green and blue, with image brightness defined as the mean. The color impression (to the human eye) emerges by additive color mixing on the monitor.
- *CMYK colors*: Color scheme based on the four colors cyan, magenta, yellow and black, designed for printed graphics. Instead of the radiating monitor pixels, color emerges by filtering of the white impinging light by the color pigments printed on paper, i.e., by subtractive mixing. Despite a digital representation of (in most cases) 4×2^8 color values, the gamut is smaller than for RGB. CMYK also has problems representing some bright colors.
- *HSV colors* (hue, saturation, value), or HSB (hue, saturation, brightness), or HSL (hue, saturation, lightness): As in RGB, color is again represented by a three-dimensional color space, but in cylinder geometry, coding for the spectral line *hue*, the saturation, i.e., the transition from a pure color to a gray value of the same brightness, and finally the brightness *value*.

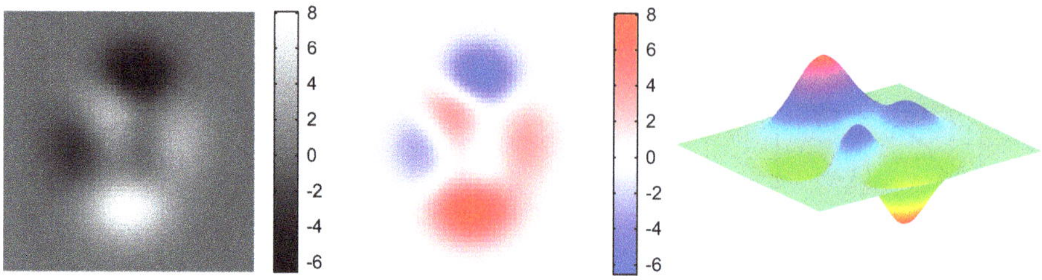

Fig. 2.3: The same image data (image *peaks*, Matlab, Inc.) in three different representations: (*left*) False color image in gray scale, (*center*) false color image in a special red–blue colormap designed to highlight the sign of the pixel values with white representing zero, (*right*) surface plot. For biomedical images, surface plots are rarely used, since they highlight more the range of the pixel values than the shape or geometry of the objects. In the present example, this representation is actually helpful to visualize the height and width of the peaks.

Figures 2.3 and 2.4 illustrate different choices of colormaps, which can be selected in view of the type of information to be highlighted. In practice, it is a good idea to spend sufficient time in trying different types of display, including variations of colormaps, scaling (e.g., linear vs. logarithmic), color limits (lowest and highest pixel value to be represented) and so forth. After few iterations, one will have become sufficiently familiarized with the information content.

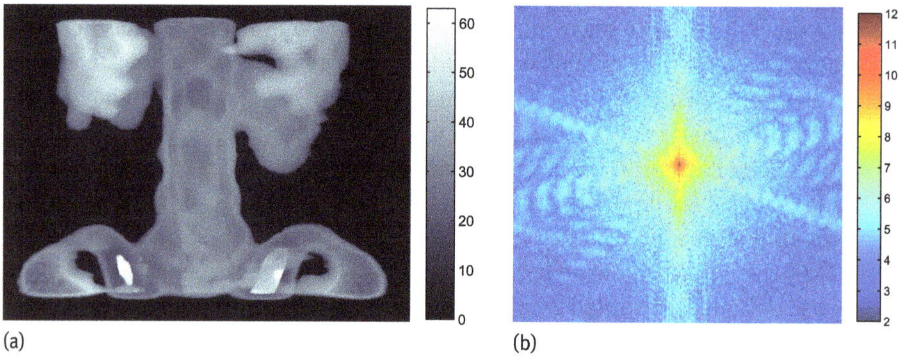

Fig. 2.4: (a) Image of the spine in colormap *bone* (Matlab, Inc.). (b) Corresponding PSD (logarithmic plot) in colormap *jet* (Matlab, Inc.). Features in a typical radiographic image may be well represented by a colormap with continuous shading, such as grayscale, or in this case the *bone* colormap. If the range of the values is more important than the shape (patterns), for example, to highlight the decay of a PSD on a logarithmic scale, the use of color can be helpful.

2.2 Geometric transformations and interpolation

Above we have discussed transformation of pixel values. Next we address transformation of pixel coordinates. In many experimental situations, the value at coordinates (x, y) has to be mapped to new coordinates (x', y'). A simple example could be an image recorded by a digital camera, which is rotated numerically afterwards. Further, it may be required to display the data not on the rectangular spatial grid on which it was recorded (i.e., the sensor chip), but, for example, as a function of angle, photon energy, or momentum transfer. This requires geometric transformations [7], followed by regridding and interpolation. In some cases, it may be advantageous to expand the transformation equations in the form of polynomials

$$\begin{aligned} x' &= f_1(x, y) = a_0 + a_1 x + a_2 y + a_3 x^2 + a_4 xy + \cdots \\ y' &= f_2(x, y) = b_0 + b_1 x + b_2 y + b_3 x^2 + b_4 xy + \cdots . \end{aligned} \tag{2.10}$$

Simple operations such as translation, reflection, rotation and scaling, and combinations of these, belong to the class of affine transformations

$$\begin{aligned} x' &= f_1(x, y) = a_1 x + a_2 y + a_0 \\ y' &= f_2(x, y) = b_1 x + b_2 y + b_0 . \end{aligned} \tag{2.11}$$

Affine transformations preserve collinearity (i.e., points on a straight line are mapped onto another straight line) and parallelism (i.e., a pair of parallel lines is mapped to another pair of parallel lines). Further, they leave the ratios of distances constant. Written in the form of matrix multiplication, we have

$$\begin{pmatrix} x' \\ y' \end{pmatrix} = \begin{pmatrix} a_1 & a_2 \\ b_1 & b_2 \end{pmatrix} \cdot \begin{pmatrix} x \\ y \end{pmatrix} + \begin{pmatrix} a_0 \\ b_0 \end{pmatrix} , \tag{2.12}$$

and in augmented matrix notation, introducing a third coordinate to get rid of the inhomogeneous term, affine transformations can be written as

$$\begin{pmatrix} x' \\ y' \\ 1 \end{pmatrix} = \begin{pmatrix} a_1 & a_2 & a_0 \\ b_1 & b_2 & b_0 \\ 0 & 0 & 1 \end{pmatrix} \cdot \begin{pmatrix} x \\ y \\ 1 \end{pmatrix} , \tag{2.13}$$

which is sometimes preferred for reasons of numerical simplicity. Non-affine transformations, in contrast, represent more complex image deformations. An interesting example is so called image warping [7]. In the original image, N anchoring points (x_i, y_i) are selected and the corresponding locations (x_i', y_i') in the transformed image are defined. For these points, it is required that the transform $\mathcal{T}$ fulfills

$$\mathcal{T}\begin{pmatrix} x_i \\ y_i \end{pmatrix} = \begin{pmatrix} x_i' \\ y_i' \end{pmatrix} . \tag{2.14}$$

Around these points, the deformation field of the image is required to decrease with the distance from the anchoring points. It can be shown that image warping can be decomposed into an affine and a non-affine transformation

$$\begin{pmatrix} x' \\ y' \end{pmatrix} = \mathcal{T}\begin{pmatrix} x \\ y \end{pmatrix} = \underbrace{\begin{pmatrix} a_1 & a_2 \\ b_1 & b_2 \end{pmatrix} \cdot \begin{pmatrix} x \\ y \end{pmatrix} + \begin{pmatrix} a_0 \\ b_0 \end{pmatrix}}_{\text{affine}} + \underbrace{\begin{pmatrix} R_1(x,y) \\ R_2(x,y) \end{pmatrix}}_{\text{non-affine}} \tag{2.15}$$

with functions R_1 and R_2 given by

$$R_k = \sum_{i=1}^{N} c_i\, g(r_i) \qquad \text{where} \qquad r_i = \sqrt{(x - x_i)^2 + (y - y_i)^2}\,. \tag{2.16}$$

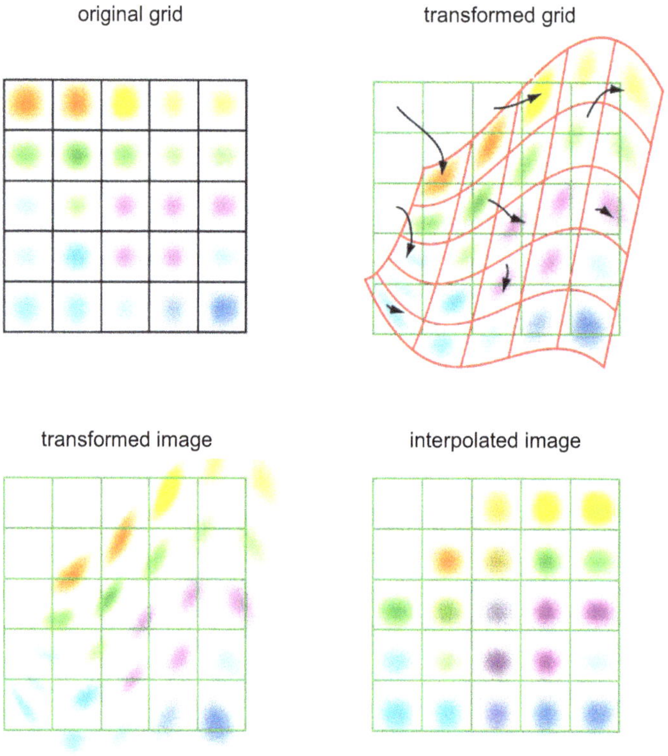

Fig. 2.5: Illustration of regridding and interpolation resulting from a geometric transformation of an image. (*Upper left*) Original image with pixel values defined over an equidistant grid. (*Upper right*) The pixel positions (pixel centers) are mapped onto new values (x', y'), prescribed by the transformation equations. The new values are, in general, not on an equidistant grid. (*Lower left*) The transformed image has values attributed to more than a single pixel, and some pixels may not receive any data at all. A reasonable way to proceed is to use the inverse transformation for each of the new pixels, in order to locate their source points in the old coordinate system. Then the closest data points in the old system are looked up and attributed to the pixel of the new system, resulting in the interpolated image (*lower right*).

Here, $g(r_i)$ is a smooth function that decays monotonously to zero with increasing r_i, for example, $g(r_i) = \exp(-r_i^2/2\sigma^2)$. The coefficients a_j, b_j and c_i are determined in such a way that equation (2.14) is fulfilled. In this way, the anchoring points (x_i, y_i) are mapped to the desired positions (x_i', y_i'), while the deformation of the image decreases for points further away from these anchoring points.

In summary, geometric transformations map the pixels to positions that are, in general, not on a rectilinear grid in the new variables, see the illustration in Figure 2.5. The only exceptions are rotations by 90, 180 or 270 degrees or translations by an integer number of pixels. Therefore, interpolation of pixel values is required, i.e., a prescription how the data is distributed over the new variable grid. In general, the old data points have to be distributed over neighboring pixels of the new coordinate (x_i', y_i'). Most numerical platforms and packages offer regridding routines with different options for interpolation, e.g., (bi)linear interpolation, nearest-neighbor interpolation or spline interpolation. They may all come to their limits when pixels of the new grid do not receive measured data, e.g., at domain boundaries.

2.3 Convolution and filtering

Let us now consider transformations again, in which the gray values are changed and the location of pixels in the image remain constant. In contrast to Section 2.1, however, new values g' depend not only on the pixel's old values, but on the gray values of other pixels as well, say in their immediate neighborhood. Two cases can be distinguished:

- *Linear* filters can be written as a convolution (denoted by the symbol $*$) with a convolution kernel or filter mask f

$$g' = g * f\,. \tag{2.17}$$

- *Nonlinear* filters, in contrast, can not be written as a convolution.

In both cases, special attention has to be paid to pixels at the edges and in the corners, requiring replication, zero-padding, etc., as addressed below.

2.3.1 Linear filters

In the discrete setting of a pixel image, the convolution associated with a linear filter operation can be represented by a matrix w_{ij}, in practice often 3×3 or 5×5 matrices. The filtering or discrete convolution then consists of the following steps (Figure 2.6):

- place filter mask over a certain pixel;
- multiply mask entries w_{ij} with the underlying pixel entries g_{nm};
- compute the sum and write the result into the new image g';
- move the center of filter mask to the next pixel and repeat.

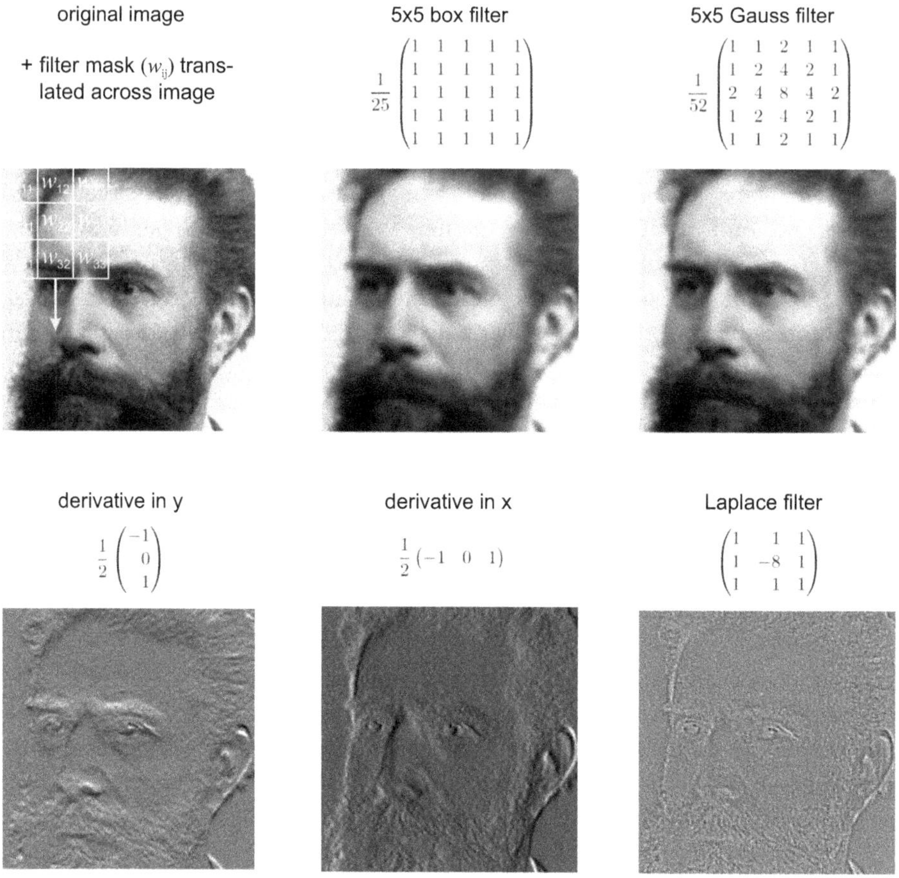
original image
+ filter mask (w_{ij}) translated across image
5x5 box filter
$\frac{1}{25}\begin{pmatrix}1&1&1&1&1\\1&1&1&1&1\\1&1&1&1&1\\1&1&1&1&1\\1&1&1&1&1\end{pmatrix}$
5x5 Gauss filter
$\frac{1}{52}\begin{pmatrix}1&1&2&1&1\\1&2&4&2&1\\2&4&8&4&2\\1&2&4&2&1\\1&1&2&1&1\end{pmatrix}$
derivative in y
$\frac{1}{2}\begin{pmatrix}-1\\0\\1\end{pmatrix}$
derivative in x
$\frac{1}{2}\begin{pmatrix}-1&0&1\end{pmatrix}$
Laplace filter
$\begin{pmatrix}1&1&1\\1&-8&1\\1&1&1\end{pmatrix}$

Fig. 2.6: Photography of W. C. Röntgen before and after convolution with several filter masks. For each mask position, the mask entries w_{ij} are multiplied with the underlying gray values of the original. The sum yields the gray value of the central pixel of this mask position in the filtered image. The mask is then shifted to the next position, and the process is repeated.

For a 3×3 convolution kernel, this can be written as

$$g'(n,m) = \sum_{i=1}^{3}\sum_{j=1}^{3} w_{ij}\, g(n-2+i, m-2+j) \quad \text{with} \quad w_{ij} = \begin{pmatrix} w_{11} & w_{12} & w_{13} \\ w_{21} & w_{22} & w_{23} \\ w_{31} & w_{32} & w_{33} \end{pmatrix}. \tag{2.18}$$

In the following, we present an overview of important linear filters. The mean (or box) filter takes the average of the central pixel and its surrounding pixels. Noise is reduced, and steep edges are smoothed. In frequency space (see Section 2.7.2), this yields the multiplication with a sinc function $\sin(x)/x$, i.e., higher frequencies are damped and some are completely removed,

$$M_{55} = \frac{1}{25}\begin{pmatrix} 1 & 1 & 1 & 1 & 1 \\ 1 & 1 & 1 & 1 & 1 \\ 1 & 1 & 1 & 1 & 1 \\ 1 & 1 & 1 & 1 & 1 \\ 1 & 1 & 1 & 1 & 1 \end{pmatrix} \tag{2.19}$$

A Gaussian filter mask approximates the convolution with a Gaussian:

$$G_{55} = \frac{1}{52}\begin{pmatrix} 1 & 1 & 2 & 1 & 1 \\ 1 & 2 & 4 & 2 & 1 \\ 2 & 4 & 8 & 4 & 2 \\ 1 & 2 & 4 & 2 & 1 \\ 1 & 1 & 2 & 1 & 1 \end{pmatrix} \tag{2.20}$$

A binomial filter has similar properties. In addition, it is separable into two one-dimensional filters. This allows a more efficient numerical implementation: With a 5×5 filter, each pixel requires 25 multiplications and 24 additions, whereas separation reduces this to 10 multiplications and 8 additions [11]:

$$\begin{aligned} B_{55} &= \frac{1}{256}\begin{pmatrix} 1 & 4 & 6 & 4 & 1 \\ 4 & 16 & 24 & 16 & 4 \\ 6 & 24 & 36 & 24 & 6 \\ 4 & 16 & 24 & 16 & 4 \\ 1 & 4 & 6 & 4 & 1 \end{pmatrix} \\ &= \frac{1}{256}\begin{pmatrix} 1 \\ 4 \\ 6 \\ 4 \\ 1 \end{pmatrix} \cdot \begin{pmatrix} 1 & 4 & 6 & 4 & 1 \end{pmatrix} \qquad \text{(separable)} \end{aligned} \tag{2.21}$$

In image processing, derivatives of a function in x or y direction can be approximated by the so called central difference

$$\begin{aligned} \frac{d}{dx} g(x,y) &\longrightarrow \frac{1}{2}[\underbrace{g(i+1,j) - g(i,j)}_{\text{forward difference}} + \underbrace{g(i,j) - g(i-1,j)}_{\text{backward difference}}] \\ &= \frac{1}{2}[g(i+1,j) - g(i-1,j)] \ . \end{aligned} \tag{2.22}$$

Analogously, in y direction

$$\frac{d}{dy} g(x,y) \longrightarrow \frac{1}{2}[g(i,j+1) - g(i,j-1)] \ . \tag{2.23}$$

The corresponding differentiation filter kernels are

$$D_x = \frac{1}{2}\begin{pmatrix} -1 & 0 & 1 \end{pmatrix} \quad \text{and} \quad D_y = \frac{1}{2}\begin{pmatrix} -1 & 0 & 1 \end{pmatrix}^{\mathrm{T}} , \tag{2.24}$$

resulting in large output values for image regions with steeply varying gray values and zero output for regions of constant gray value. The magnitude of the gradient of a grayscale image is given by

$$\left[(D_x * g(x,y))^2 + (D_y * g(x,y))^2\right]^{\frac{1}{2}}, \tag{2.25}$$

which can be used for edge detection. However, these kernels are very noise-sensitive and should only be applied to smoothed images.

Sobel filters combine two operations: smoothing by a Gaussian and differentiation. They are widely used for edge detection. For x and y directions, they are represented by the kernels

$$S_x = \frac{1}{8}\begin{pmatrix} -1 & 0 & 1 \\ -2 & 0 & 2 \\ -1 & 0 & 1 \end{pmatrix} \quad \text{and} \quad S_y = \frac{1}{8}\begin{pmatrix} -1 & -2 & -1 \\ 0 & 0 & 0 \\ 1 & 2 & 1 \end{pmatrix}. \tag{2.26}$$

In addition, edges along the image diagonals can be particularly well detected by

$$S_/ = \frac{1}{8}\begin{pmatrix} 0 & -1 & -2 \\ 1 & 0 & -1 \\ 2 & 1 & 0 \end{pmatrix} \quad \text{and} \quad S_\backslash = \frac{1}{8}\begin{pmatrix} -2 & -1 & 0 \\ -1 & 0 & 1 \\ 0 & 1 & 2 \end{pmatrix}. \tag{2.27}$$

Using these four filter masks, one can define the combined Sobel filter [13]

$$S^* = \max\{|S_x|, |S_y|, |S_/|, |S_\backslash|\}. \tag{2.28}$$

Further combinations of filters leads to the so called Kirsch filter

$$K^* = \max\{|K^{(1)}|, |K^{(2)}|, \ldots, |K^{(8)}|\}, \tag{2.29}$$

with the $K^{(i)}$ filter matrix representing differentiation along a direction $i \cdot 45°$ $(i \in \mathbb{N})$, i.e., the differentiation step is rotated through 8 directions. The first three filter matrices are given by [12]

$$K^{(1)} = \begin{pmatrix} 5 & 5 & 5 \\ -3 & 0 & -3 \\ -3 & -3 & -3 \end{pmatrix} \quad K^{(2)} = \begin{pmatrix} 5 & 5 & -3 \\ 5 & 0 & -3 \\ -3 & -3 & -3 \end{pmatrix} \quad K^{(3)} = \begin{pmatrix} 5 & -3 & -3 \\ 5 & 0 & -3 \\ 5 & -3 & -3 \end{pmatrix}, \tag{2.30}$$

and the other directions follow analogously. Even more advanced edge detection is possible if differentiation filtering is combined with further (nonlinear) steps, such as in the Canny algorithm [5]:

1. The image is smoothed, e.g., by a Gaussian filter.
2. The edges are detected by a combination of Sobel filters $\sqrt{D_x^2 + D_y^2}$.
3. The direction of the gradient is determined for each pixel by $\arctan(D_y/D_x)$.

Fig. 2.7: Different filters for edge detection: The Sobel filter (*right*, equation 2.28) produces a somewhat less noisy result than the Laplace filter (equation 2.35), since it combines edge detection and smoothing.

4. The edges are thinned to a single pixel by putting pixels which are not maxima in the direction of the gradient to zero.
5. Finally, a thresholding procedure is carried out. All pixels with modulus of the gradient $g < T_1$ are put to zero, all pixels larger than T_2 are kept as edge pixels. Then edges are searched in the vicinity of the pixels with $g > T_2$, and kept as edges, as long as they fulfill $g > T_1$.

In the end, a binary image of edges is obtained, which for a suitable parameter choice outperforms the simpler edge detection based on a single differentiation filter. The Canny algorithm can be tested online or be implemented into the open-source image processing software ImageJ.[1,2]

Of course, edge detection algorithms can also include interrogation of filters representing second derivatives. In analogy to equation (2.22), the second derivatives of the gray values in the x direction of an image can be written as

$$\begin{aligned} \frac{d^2}{dx^2} g(x,y) &\longrightarrow [g(i+1,j) - g(i,j)] - [g(i,j) - g(i-1,j)] \\ &= g(i+1,j) - 2g(i,j) + g(i-1,j) \,. \end{aligned} \tag{2.31}$$

This yields the following kernels for the second derivatives in x and y directions

$$D_{xx} = \begin{pmatrix} 1 & -2 & 1 \end{pmatrix} \quad \text{and} \quad D_{yy} = \begin{pmatrix} 1 & -2 & 1 \end{pmatrix}^{\mathrm{T}} . \tag{2.32}$$

Addition yields the Laplace filter, which approximates the sum of second derivatives $\Delta = (\partial_x^2 + \partial_y^2)$

$$L = \begin{pmatrix} 0 & 1 & 0 \\ 1 & -4 & 1 \\ 0 & 1 & 0 \end{pmatrix} . \tag{2.33}$$

1 http://matlabserver.cs.rug.nl/cannyedgedetectionweb/web/index.html
2 http://www.imagescience.org/meijering/software/featurej/

Both regions of constant and linearly varying gray values are set to zero, only regions where the rate of change in gray values is large are amplified. Superposition of original and Laplace filtered images can be used for edge enhancement, leading to

$$\begin{pmatrix} 0 & 0 & 0 \\ 0 & 1 & 0 \\ 0 & 0 & 0 \end{pmatrix} - \epsilon \begin{pmatrix} 0 & 1 & 0 \\ 1 & -4 & 1 \\ 0 & 1 & 0 \end{pmatrix} = \begin{pmatrix} 0 & -\epsilon & 0 \\ -\epsilon & 1+4\epsilon & -\epsilon \\ 0 & -\epsilon & 0 \end{pmatrix}, \tag{2.34}$$

where ϵ is a free parameter. A different version of the Laplace filter mask also detects edges along the image diagonals,

$$L = \begin{pmatrix} 1 & 1 & 1 \\ 1 & -8 & 1 \\ 1 & 1 & 1 \end{pmatrix}. \tag{2.35}$$

2.3.2 Filtering in Fourier space

Above, we saw how the linear image operations as given by convolutions can be approximated by moving discrete filter masks over the image. In practice, these operations are more efficiently implemented in the Fourier domain, just as their counterparts in the continuous setting. With the fast Fourier transformation (FFT) algorithm presented in Section 2.8.3, one can easily obtain the matrix representing the image in Fourier space, on which different filter operations can be applied, by simple multiplication of suitable functions, designed in particular in view of weighting high or low spatial frequencies. The function that is multiplied onto the Fourier matrix is also denoted as the modulation function, since it modulates how the spatial frequencies are transferred between original and processed (filtered) images. Figure 2.8 illustrates the important example of bandpass filtering, where the filter functions in Fourier space are decreasing with spatial frequency in order to implement a low pass, or increasing to implement a high pass filter. In addition, Figure 2.9 presents an example of a sharpening operation, and Figure 2.10 an example of the removal of a periodic artifact.

2.3.3 Nonlinear or rank filters

This class of filters also operates by translating a mask across an image and redefining the central pixel. However, theses filters can *not* be written as a convolution, and there does not exist any modulation transfer function (MTF; see also Section 2.5.1). Gray values within the filter mask are not multiplied by filter mask entries w_{ij}, but ranked

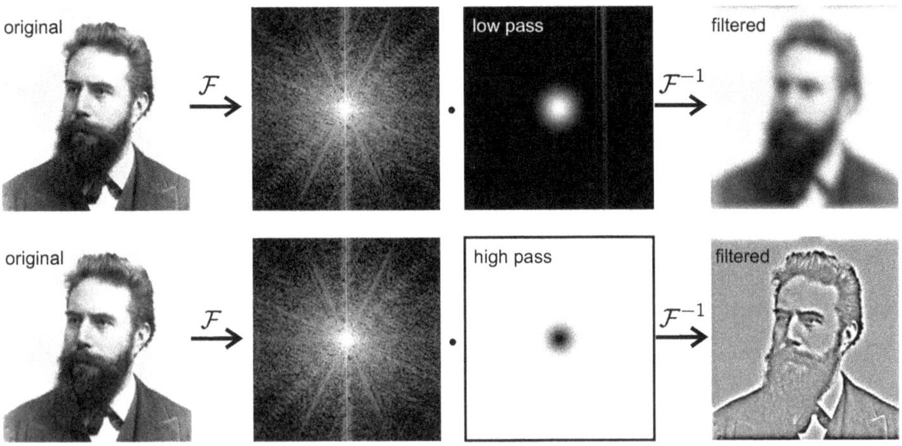

Fig. 2.8: Illustration of low pass and high pass filters. A discrete Fourier transformation is applied to the original image, yielding a matrix of the same dimensions as the original, but representing the image in Fourier space. A filter function is multiplied (here circular, symmetric), designed to pass low (*upper row*) or high (*lower row*) spatial frequencies. After multiplication in Fourier space, the filtered image is obtained by inverse Fourier transformation.

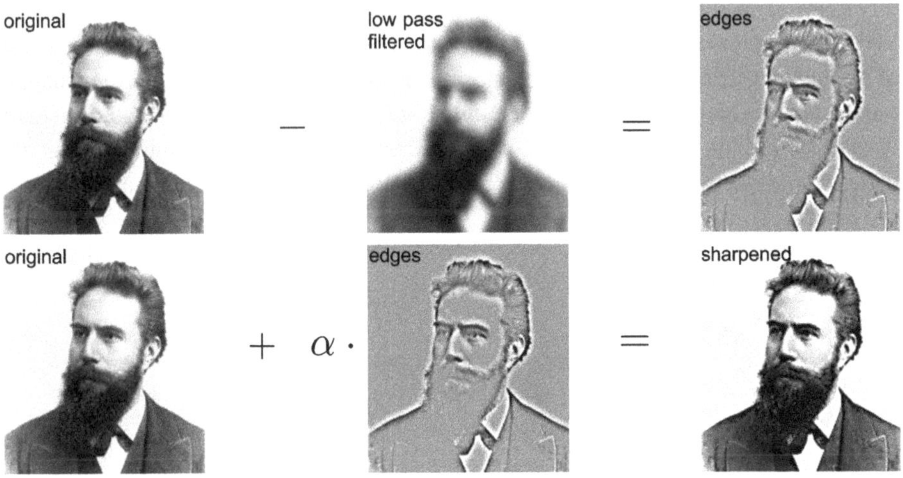

Fig. 2.9: Elementary filtering functions can be combined to achieve slightly more complex operations, such as image sharpening. The low pass filtered image is subtracted from the original, resulting in an output which highlights the edges. This edge enhanced image is then added again to the original (with adjustable parameter α), yielding a sharpened version of the 'Roentgen image'.

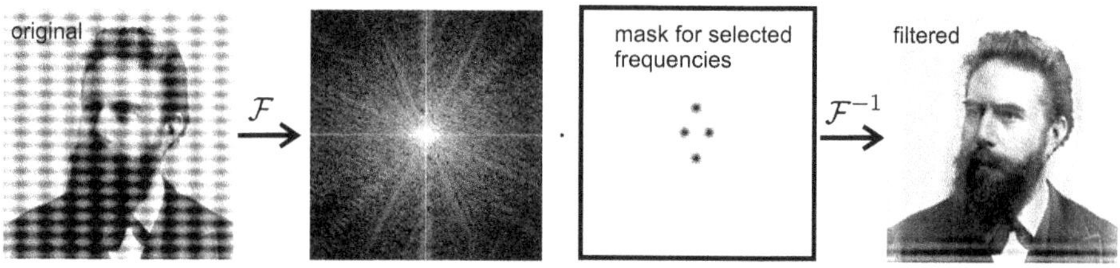

Fig. 2.10: Example of a 'Roentgen image' with a periodic artifact, which can be removed by identifying the corresponding spatial frequencies in the Fourier image, and subsequent masking. Inverse Fourier transformation yields a recovered image without the periodic perturbation.

by their gray value

$$g_1 \leq g_2 \leq \cdots \leq g_{m-1} \leq g_m \qquad (\text{rank } r = 1, \ldots, m)\,. \tag{2.36}$$

The gray value of the central pixel is then replaced by the gray value of a certain rank r. The most important example is the median filter

$$g_{\text{median}} = \begin{cases} g_{\frac{m+1}{2}} & m \text{ odd} \\ \frac{1}{2}\left(g_{\frac{m}{2}} + g_{\frac{m}{2}+1}\right) & m \text{ even} \end{cases}\,. \tag{2.37}$$

For example, the 3×3 median filter ranks the gray values of the central pixel and its 8-connected neighborhood and replaces the gray value of the central pixel by the fifth highest value of the rank list (equation 2.36). The median filter is used for the removal of outliers, e.g., due to defective detector pixels or high energy photons from cosmic radiation. For this purpose, it is much better suited than a mean (box) filter. Instead of distributing the corrupted value over all surrounding pixels, the median filter replaces the 'defective' gray value by a 'common' value in its immediate environment, see Figure 2.11.

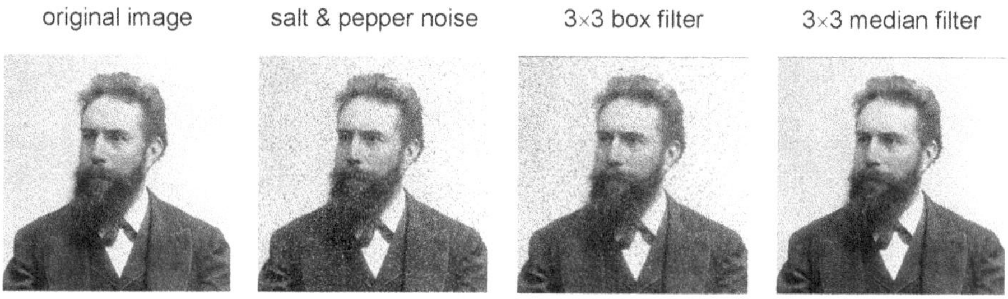

Fig. 2.11: Noise removal by different filters. So called 'salt and pepper noise' has been added to the image, i.e., the gray values of some randomly chosen pixels are set to zero, others to the maximum value. A 3×3 mean (or box) filter (equation 2.19) gives a considerably poorer result than the 3×3 median filter (equation 2.37).

Another interesting class of nonlinear filters is related to the field of image morphology. We start by considering a binary image represented by a matrix that contains only the entries 0 and 1. A structuring element, e.g., a 3×3 box as used in the examples below, is translated across the image. The two basic morphological operations are

- *Erosion* (symbol $\ominus$): If all pixel values of the image within the structuring element are 1, the central pixel remains 1, otherwise it is set to 0.

$$\underbrace{\begin{pmatrix} 0&0&0&0&0&0&0&0&0\\ 0&1&1&1&1&1&1&1&0\\ 0&1&1&1&1&1&1&1&0\\ 0&1&1&1&1&1&1&1&0\\ 0&1&1&1&0&1&1&1&0\\ 0&1&1&1&1&1&1&1&0\\ 0&1&1&1&1&1&1&1&0\\ 0&1&1&1&1&1&1&1&0\\ 0&0&0&0&0&0&0&0&0 \end{pmatrix}}_{\text{original binary image}} \ominus \underbrace{\begin{pmatrix} 1&1&1\\ 1&1&1\\ 1&1&1 \end{pmatrix}}_{\text{struct. element}} = \underbrace{\begin{pmatrix} 0&0&0&0&0&0&0&0&0\\ 0&0&0&0&0&0&0&0&0\\ 0&0&1&1&1&1&1&0&0\\ 0&0&1&0&0&0&1&0&0\\ 0&0&1&0&0&0&1&0&0\\ 0&0&1&0&0&0&1&0&0\\ 0&0&1&1&1&1&1&0&0\\ 0&0&0&0&0&0&0&0&0\\ 0&0&0&0&0&0&0&0&0 \end{pmatrix}}_{\text{eroded binary image}} \tag{2.38}$$

- *Dilation* (symbol $\oplus$): If at least one pixel value within the structuring element is 1, the central pixel is set to (or remains) 1. Only if all values within the structuring element are 0, does the central pixel remain 0.

$$\underbrace{\begin{pmatrix} 1&0&0&0&0&0&0&0&0\\ 0&1&0&0&0&0&0&0&0\\ 0&0&1&0&0&0&0&0&0\\ 0&0&0&1&0&0&0&0&0\\ 0&0&0&0&0&0&0&0&0\\ 0&0&0&0&0&1&0&0&0\\ 0&0&0&0&0&0&1&0&0\\ 0&0&0&0&0&0&0&1&0\\ 0&0&0&0&0&0&0&0&1 \end{pmatrix}}_{\text{original binary image}} \oplus \underbrace{\begin{pmatrix} 1&1&1\\ 1&1&1\\ 1&1&1 \end{pmatrix}}_{\text{struct. element}} = \underbrace{\begin{pmatrix} 1&1&1&0&0&0&0&0&0\\ 1&1&1&1&0&0&0&0&0\\ 1&1&1&1&1&0&0&0&0\\ 0&1&1&1&1&0&0&0&0\\ 0&0&1&1&1&1&1&0&0\\ 0&0&0&0&1&1&1&1&0\\ 0&0&0&0&1&1&1&1&1\\ 0&0&0&0&0&1&1&1&1\\ 0&0&0&0&0&0&1&1&1 \end{pmatrix}}_{\text{dilated binary image}} \tag{2.39}$$

Erosion leads to a shrinking, dilation to an enlargement of an object. This may mostly not be desired. Combination of erosion and dilation, however, has quite interesting properties. The size of objects is left approximately constant, but their appearance and morphology can be changed in ways which facilitate further processing and automated rendering. In this respect, two important operations are

- *Opening* (symbol $\circ$): Erosion and subsequent dilation, i.e., $A \circ S = (A \ominus S) \oplus S$. Thin connections between two objects or isolated noisy pixels are removed:

$$\underbrace{\begin{pmatrix} 1&1&1&0&0&0&0&0&0\\ 1&1&1&0&0&0&0&1&0\\ 1&1&1&0&0&0&0&0&0\\ 0&0&0&1&0&0&0&0&0\\ 0&0&0&0&1&0&0&0&0\\ 0&0&0&0&0&1&0&0&0\\ 0&0&0&0&0&0&1&1&1\\ 0&1&0&0&0&0&1&1&1\\ 0&0&0&0&0&0&1&1&1 \end{pmatrix}}_{\text{original binary image}} \circ \underbrace{\begin{pmatrix} 1&1&1\\ 1&1&1\\ 1&1&1 \end{pmatrix}}_{\text{struct. element}} = \underbrace{\begin{pmatrix} 1&1&1&0&0&0&0&0&0\\ 1&1&1&0&0&0&0&0&0\\ 1&1&1&0&0&0&0&0&0\\ 0&0&0&0&0&0&0&0&0\\ 0&0&0&0&0&0&0&0&0\\ 0&0&0&0&0&0&0&0&0\\ 0&0&0&0&0&0&1&1&1\\ 0&0&0&0&0&0&1&1&1\\ 0&0&0&0&0&0&1&1&1 \end{pmatrix}}_{\text{opened binary image}} \tag{2.40}$$

- *Closing* (symbol $\bullet$): Dilation and subsequent erosion, i.e., $A \bullet S = (A \oplus S) \ominus S$. Gaps within an object are filled, closely adjacent objects are connected:

$$\underbrace{\begin{pmatrix} 1&1&1&1&0&0&0&0&0\\ 1&0&1&1&0&0&0&0&0\\ 1&1&1&1&0&0&0&0&0\\ 1&1&1&1&0&1&0&0&0\\ 0&0&0&0&1&0&0&0&0\\ 0&0&0&1&0&1&1&1&1\\ 0&0&0&0&0&1&1&1&1\\ 0&0&0&0&0&1&1&0&1\\ 0&0&0&0&0&1&1&1&1 \end{pmatrix}}_{\text{original binary image}} \bullet \underbrace{\begin{pmatrix} 1&1&1\\ 1&1&1\\ 1&1&1 \end{pmatrix}}_{\text{struct. element}} = \underbrace{\begin{pmatrix} 1&1&1&1&0&0&0&0&0\\ 1&1&1&1&0&0&0&0&0\\ 1&1&1&1&0&0&0&0&0\\ 1&1&1&1&1&1&0&0&0\\ 0&0&0&1&1&1&0&0&0\\ 0&0&0&1&1&1&1&1&1\\ 0&0&0&0&0&1&1&1&1\\ 0&0&0&0&0&1&1&1&1\\ 0&0&0&0&0&1&1&1&1 \end{pmatrix}}_{\text{closed binary image}} \tag{2.41}$$

Biomedical images are usually not binary, but grayscale images. The principles of image morphology can be extended to grayscale images using the concept of rank filters: A grayscale erosion filter replaces the gray value of the central pixel by the gray value of the pixel with rank $r = 1$, a grayscale dilation filter is a rank filter with rank $r = m$ (Figure 2.12). This is often used in image segmentation, when objects have to be separated and identified.[3]

Problems are encountered in the application of the rank filter mask at the edges of the image if the central pixel is too close to an image boundary: There are not enough pixels to fill the mask. Possible solutions to these boundary issues are: no processing of boundary pixels, replication of boundary columns and rows or shrinking of the mask to a suitable size.

3 Further insightful illustrations can be found at http://homepages.inf.ed.ac.uk/rbf/HIPR2/morops.htm.

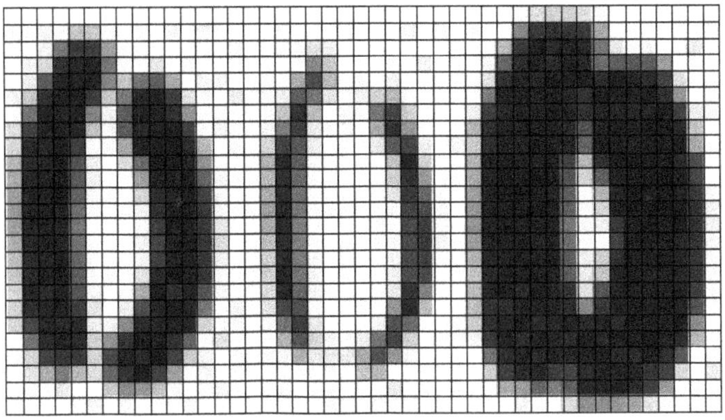

Fig. 2.12: Illustration of opening and closing filters, as applied to a grayscale image. The opening filter leads to a further separation of slightly separated objects, the closing filter merges them into one.

2.4 The Hough transform

The Hough transform is an elegant tool to detect parameterized geometric structures such as lines, circles or ellipses in a binarized digital image. Let us consider the case of detecting lines. We have already discussed edge detection in the section on linear image filters. However, disconnected points falling on a line, or lines in noisy images are often not properly identified by just gradient operators. The method proposed by P. V. C. HOUGH in 1962 [10] is much more robust and tolerant to noise or gaps. The essential prerequisite is that the geometric structure to be detected can be represented by a parametric curve. The parameters span the so called dual space or parameter space. The basic idea is to map each point on an edge in the image to the possible parameter combinations that would generate a curve through this point. All points of a parametric curve have one common point in parameter space. Therefore, one obtains local maxima in parameter space for parameter combinations that indicate likely structures contained in the image. In other words, the Hough transform can also be interpreted as a "voting process", the discretized form of the parameter space is sometimes called the voting matrix. Each edge pixel votes for all parameter combinations which would keep it in the parameterized structure, for example, all parameters which belong to lines going through the pixel. The votes accumulate for parameter combinations of the parameterized structures that are actually present in the image.[4]

We illustrate this procedure for the simplest case of the Hough transform, the detection of lines in an image (Figure 2.13 (1)):

4 An instructive video on the Hough transform can be found at http://www.youtube.com/watch?v=kMK8DjdGtZo.

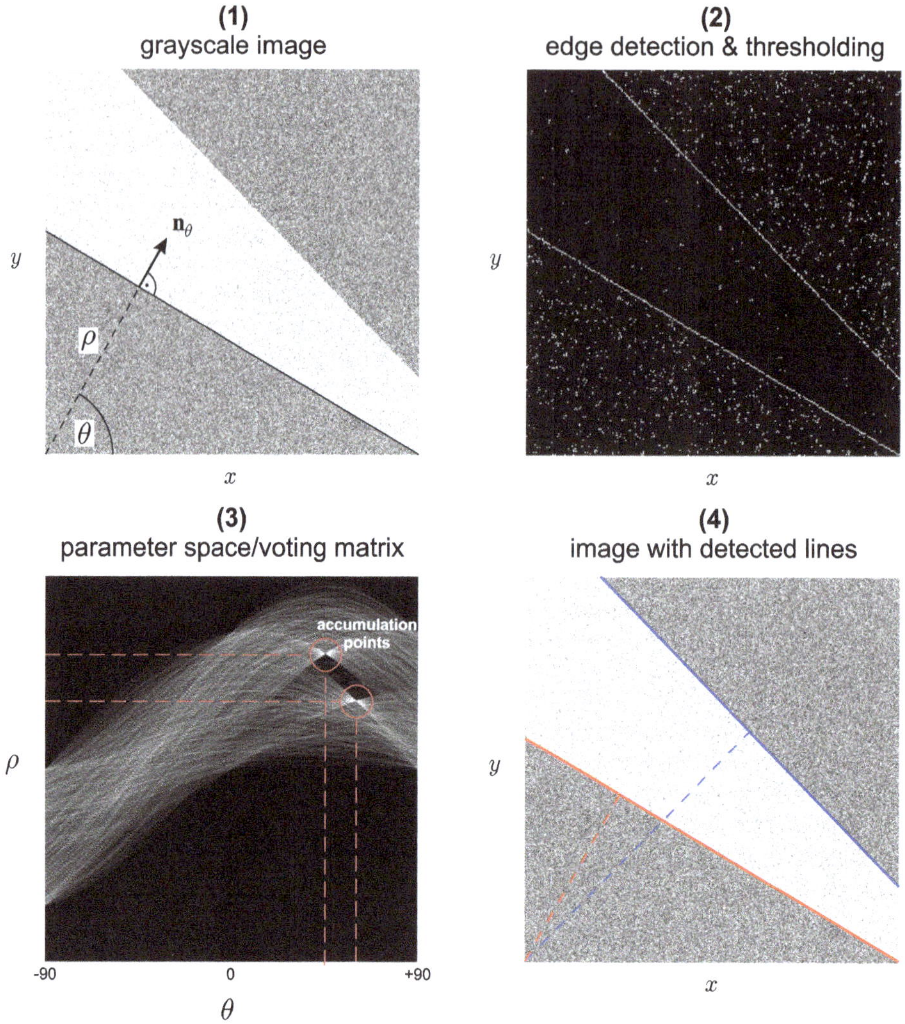

Fig. 2.13: Illustration of line detection based on the Hough transform. (1) All possible lines in an image are parameterized by a pair of values θ, ρ. (2) The original image is transformed to a binary image by edge detection (e.g., by Sobel filters) followed by thresholding. (3) The set of all straight lines going through a candidate for an edge point in the binarized image corresponds to a sinusoidal curve in parameter space, which is unique to that point. A set of points lying on the same straight line results in sinusoidal curves which cross at the parameter pair θ, ρ (accumulation point) representing this common line. (4) The two accumulation points with the highest number of 'votes' correspond to edges in the original image.

1. As the first step, an edge detection filter, e.g., the Sobel filter (equation 2.28) is applied to the grayscale image. Defining a suitable threshold results in a binary image with the assignment "1 $\rightarrow$ edge pixel" and "0 $\rightarrow$ no edge pixel" (Figure 2.13 (2)).

2. Second, a parameterization of the geometric shape to be detected is required. A line is usually parameterized by its slope m and intercept b, i.e., $y = mx + b$. However, in view of the singularity for (almost) vertical lines, it is more appropriate to use the Hesse normal form instead (Figure 2.13 (1)):

$$g = \{\underline{x} \in \mathbb{R}^2 | \underline{x} \cdot \underline{n}_0 = \rho\} \quad \Leftrightarrow \quad \rho = x\cos\theta + y\sin\theta \tag{2.42}$$

   The two parameters ρ and θ are bounded and can only adopt finite values in the intervals $\rho \in [0, \sqrt{x_{\max}^2 + y_{\max}^2}]$ and $\theta \in [0, \pi]$. ρ and θ span the dual space or parameter space.
3. For a single point of an edge, the set of all straight lines that go through it yields a sinusoidal curve in parameter space (θ, ρ). For many points on the same straight line, these curves cross in a single point in parameter space (Figure 2.13 (3)). Numerically, the parameter space is represented by a matrix, since the parameters are discretized. This accumulation process of mapping edge points in the image to matrix entries can be interpreted as a voting scheme with a certain number of votes mapped to a given parameter pair. Each edge pixel in the image can 'vote' for all parameters to which it 'belongs'. The voting scheme is carried out for all edge points.
4. Local maxima in parameter space (with a suitable threshold) yield parameter pairs (θ, ρ) that are most likely associated with edges in the binary image, and the corresponding parameters are regarded as the set of detected lines in the original image.

For a binarized image $f(x, y)$ representing all points on edges with $f = 1$, the mapping to the parameter space of lines (θ, ρ) can be written as

$$H(\theta, \rho) = \iint dx\,dy\,f(x, y)\,\delta(\rho - x\cos\theta - y\sin\theta)\,. \tag{2.43}$$

In the corresponding digitalized version $H(\theta, \rho)$ simply counts the number of pixels which fall on the line of the given parameters. As we will see in Chapter 4, the formula above corresponds to a 2d Radon transformation of a binarized image. In fact, the Radon transformation can be regarded as a special case of the Hough transform, which can be defined in a more general form for arbitrary curves, both in a continuous and in a discrete setting [13]. Applications in medical image processing include, e.g., iris detection in ophthalmology, where the parametric curves are circles [13].

2.5 Theory of linear imaging systems

2.5.1 Introduction and definitions

In the previous sections, we used filters as a way to process and harvest the information contained in images. To this end, we had the situation in mind that the image g

was already recorded, and that numerical processing is applied afterwards by a researcher or user. It turns out that the imaging apparatus or system itself can be understood as an operator processing the physical input to an output. This is the point of view we adopt in the present section. We consider imaging in the framework of linear systems theory, a field originating in the signal processing of radio communication and heavily based on Fourier theory. In this approach, the input data is often called the signal, be it in 1d, 2d or 3d. Here, we primarily have the 2d case in mind, say some imaging modality converting a physical signal f to some image g:

input f (physical property)	$\Rightarrow$	imaging system $\mathcal{H}$	$\Rightarrow$	output g (digitalized) image

The imaging process is represented by an operator $\mathcal{H}$. In most current applications, the final result will be a digital 2d image $g[n, m]$, but may be 3d or even 4d as well. We use the continuous notation $g(x, y)$ here for simplicity. In the most general form, the imaging system is described by the operator equation

$$g = \mathcal{H}[f] \,. \tag{2.44}$$

If the imaging system represented by $\mathcal{H}$ fulfills two fundamental conditions, a very general and powerful description of the imaging process is possible:

1. *Linearity*: Superposition of several arbitrary input signals f_i yields the superposition of the individual outputs $g_i = \mathcal{H}[f_i]$ (α_i: coefficients):

$$\mathcal{H}\left[\sum_i \alpha_i \cdot f_i\right] = \sum_i \mathcal{H}[\alpha_i \cdot f_i] = \sum_i \alpha_i \cdot \mathcal{H}[f_i] = \sum_i \alpha_i \cdot g_i \tag{2.45}$$

   This also implies that imaging of one object is not affected by simultaneous imaging of another object.
2. *Shift invariance*: Translation of the input leads to translation of its image but leaves the image itself unchanged. Hence, for 2d we have

$$\mathcal{H}\left[f(x - x_0, y - y_0)\right] = g(x - x_0, y - y_0) \qquad \forall\, x_0, y_0 \,. \tag{2.46}$$

   For example, in radiography, we would require the gray values and relative distances in the projection image to be constant, if we move the object or body with respect to the center of the x-ray beam.

If these two conditions are met, the imaging system can be described using the theory of linear, shift invariant systems (LSI), where the input-output-relation is given by a convolution with a single function h, the PSF, which characterizes the system

$$g(x, y) = f(x, y) * h(x, y) = \iint dx'\, dy'\, f(x', y')\, h(x - x', y - y') \,. \tag{2.47}$$

2.5.2 Point spread function (PSF)

Next, we sketch how equation (2.47) can be obtained, following the derivation presented in [13]. For the sake of simplicity, we consider a 1d input signal $f(x)$. Generalization to higher dimensions is straightforward. As sketched in Figure 2.14, the signal can be approximated by a sum of shifted rectangular functions:

$$\mathrm{rect}_{\Delta x}(x) = \begin{cases} \frac{1}{\Delta x} & |x| \leq \frac{\Delta x}{2} \\ 0 & \text{else} \end{cases} \tag{2.48}$$

$$\Rightarrow \qquad f(x) \approx \sum_{n=-\infty}^{\infty} f(n\Delta x)\,\mathrm{rect}_{\Delta x}(x - n\Delta x)\,\Delta x \tag{2.49}$$

In the limit $\Delta x \to 0$, $\mathrm{rect}_{\Delta x}(x)$ approaches the Dirac delta function

$$\lim_{\Delta x \to 0} \mathrm{rect}_{\Delta x}(x) = \delta(x)\,. \tag{2.50}$$

From equation (2.49), we obtain

$$\begin{aligned} f(x) &= \lim_{\Delta x \to 0} \sum_{n=-\infty}^{\infty} f(n\Delta x)\,\mathrm{rect}_{\Delta x}(x - n\Delta x)\,\Delta x \\ &= \int_{-\infty}^{\infty} f(x')\,\delta(x - x')\,dx' \\ &= f(x) * \delta(x)\,. \end{aligned} \tag{2.51}$$

We obtain the convolution integral defined in Section 2.7 on Fourier theory. The result simply states the fact that $\delta(x)$ is the neutral element of the convolution, i.e., convolution with $\delta(x)$ returns the original function. If the system is linear, the image of equation (2.49) can be expressed as a superposition of the images of infinitely many rect functions, weighted by the function values $f(n\Delta x)$:

$$g(x) \approx \sum_{n=-\infty}^{+\infty} f(n\Delta x)\,\mathcal{H}[\mathrm{rect}_{\Delta x}(x - n\Delta x)]\,\Delta x\,. \tag{2.52}$$

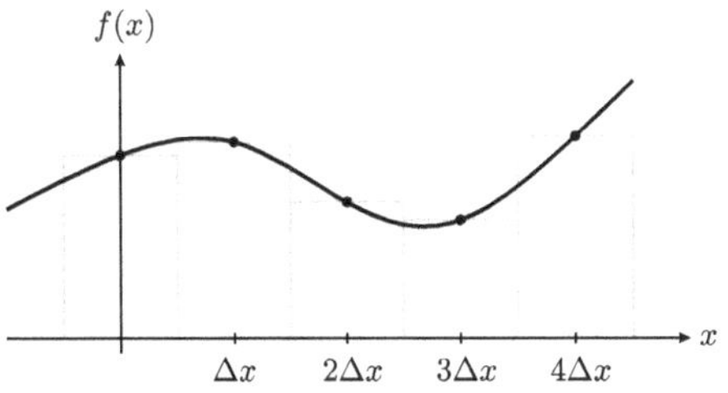

Fig. 2.14: Decomposition of a continuous signal into a sum of rectangular functions. If an imaging system is linear, the image $g(x)$ of $f(x)$ can be approximated by superposition of the images of the individual rectangular functions.

In the limit $\Delta x \longrightarrow 0$, this yields

$$g(x) = \lim_{\Delta x \to 0} \sum_{n=-\infty}^{\infty} f(n\Delta x)\, \mathcal{H}[\delta(x - n\Delta x)]\, \Delta x = \int_{-\infty}^{\infty} f(x')\, \mathcal{H}[\delta(x - x')]\, dx' \,. \tag{2.53}$$

Now, if the system $\mathcal{H}$ is also shift-invariant, the image of a point source (represented by a delta function) will always look the same, no matter where the point source is located. We can thus introduce a single function $h(x)$, the PSF, which allows us to write the input–output relation as

$$g(x) = \int_{-\infty}^{\infty} f(x')\, h(x - x')\, dx' = f(x) * h(x) \,. \tag{2.54}$$

Again, we recognize the convolution integral. We have obtained the following important result:

The relation between input $f(x)$ and output $g(x)$ of an LSI is completely described by a convolution with the PSF $h(x)$ of the system

$$g(x) = f(x) * h(x) = h(x) * f(x) \,. \tag{2.55}$$

The name is due to the fact that the PSF is obtained as the output for a point like input $f(x) = \delta(x)$,

$$g_{\text{point}}(x) = \delta(x) * h(x) = h(x) \,. \tag{2.56}$$

The explicit form in two dimensions applicable to many imaging processes is given by equation (2.47), where $h(x, y)$ is obtained as the image g if a point like object such as a pinhole (radiography), a small radioactive source (SPECT) or a single scatterer (ultrasound) is imaged. Measurement of the PSF is described in Section 2.5.5.

2.5.3 Modulation transfer function (MTF)

According to the convolution theorem (Section 2.7), a convolution in direct space corresponds to a multiplication in frequency space, and vice versa

$$g(x, y) = f(x, y) * h(x, y) \quad \underset{\mathcal{F}^{-1}}{\overset{\mathcal{F}}{\rightleftharpoons}} \quad \tilde{g}(k_x, k_y) = \tilde{f}(k_x, k_y) \cdot \tilde{h}(k_x, k_y) \,. \tag{2.57}$$

The normalized Fourier transform of the PSF is called the optical transfer function (OTF)

$$\text{OTF}(k_x, k_y) = \frac{\tilde{h}(k_x, k_y)}{\tilde{h}(0, 0)} \in \mathbb{C} \,. \tag{2.58}$$

In general, OTF is complex valued, since the Fourier transforms $\tilde{f}, \tilde{g}, \tilde{h}$ are complex valued. If we further separate modulus and phase, we have

$$\mathrm{OTF}(k_x, k_y) = \mathrm{MTF}(k_x, k_y) \cdot \exp[i \cdot \mathrm{PhTF}(k_x, k_y)] \tag{2.59}$$

with the phase transfer function (PhTF) and the modulation transfer function (MTF)

$$\mathrm{MTF}(k_x, k_y) = \left| \frac{\tilde{h}(k_x, k_y)}{\tilde{h}(0,0)} \right| \in \mathbb{R} \,. \tag{2.60}$$

Next, we show that the MTF defined in this way yields the ratio of the modulation in the output to the modulation in the input, following [1]. For simplicity, we limit ourselves to the 1d case. Consider a complex exponential as the input $f(x) = \exp(ik_0 x)$. Since the convolution is commutative, we can write

$$\begin{aligned} g(x) = f(x) * h(x) = h(x) * f(x) &= h(x) * \exp(ik_0 x) \\ &= \int_{-\infty}^{+\infty} dx'\, h(x')\, \exp(ik_0(x - x')) \\ &= \exp(ik_0 x) \int_{-\infty}^{+\infty} dx'\, h(x')\, \exp(-ik_0 x') \\ &= \exp(ik_0 x)\, \tilde{h}(k_0) \,. \end{aligned} \tag{2.61}$$

In other words: A complex exponential $\exp(ik_0 x)$ is transferred through a LSI without changes, except for a multiplication with the Fourier component $\tilde{h}(k_0)$ of its PSF. Now consider a cosine signal with constant background, which can be written as a sum of complex exponentials

$$\begin{aligned} f(x) &= A + B\cos(k_0 x) \\ &= A\exp(ik_0 0) + \frac{B}{2}\exp(ik_0 x) + \frac{B}{2}\exp(-ik_0 x) \,. \end{aligned} \tag{2.62}$$

Using equation (2.61), the image is given by

$$g(x) = A\tilde{h}(0) + \frac{B}{2}\tilde{h}(k_0)\exp(ik_0 x) + \frac{B}{2}\tilde{h}(-k_0)\exp(-ik_0 x) \,. \tag{2.63}$$

If we assume $h(x) \in \mathbb{R}$ and $h(x) = h(-x)$, we have $\tilde{h}(-k_0) = \tilde{h}(k_0)$ and thus

$$\begin{aligned} g(x) &= A\tilde{h}(0) + \frac{B}{2}\tilde{h}(k_0)\,[\exp(ik_0 x) + \exp(-ik_0 x)] \\ &= A\tilde{h}(0) + B\tilde{h}(k_0)\cos(k_0 x) \,. \end{aligned} \tag{2.64}$$

The Michelson (or modulation) contrast (see Section 2.6.1) of a signal g is defined by

$$K_\mathrm{M} = \frac{g_\mathrm{max} - g_\mathrm{min}}{g_\mathrm{max} + g_\mathrm{min}} \,. \tag{2.65}$$

For input and output signals in our example, this yields

$$K_{M,f} = \frac{(A+B)-(A-B)}{(A+B)+(A-B)} = \frac{B}{A} \tag{2.66}$$

$$K_{M,g} = \frac{(A\tilde{h}(0)+B\tilde{h}(k_0))-(A\tilde{h}(0)-B\tilde{h}(k_0))}{(A\tilde{h}(0)+B\tilde{h}(k_0))+(A\tilde{h}(0)-B\tilde{h}(k_0))} = \frac{B}{A}\frac{\tilde{h}(k_0)}{\tilde{h}(0)} \tag{2.67}$$

$$\Rightarrow \qquad \mathrm{MTF}(k_0) = \left|\frac{\tilde{h}(k_0)}{\tilde{h}(0)}\right| = \left|\frac{K_{M,g}}{K_{M,f}}\right| = \frac{\text{modulation contrast in output}}{\text{modulation contrast in input}} \tag{2.68}$$

Thus, we see that the MTF indeed quantifies the degradation of modulation (Michelson) contrast with increasing spatial frequency k. In an ideal system, if a point were imaged as a point, $h(x) = \delta(x) \Rightarrow \tilde{h}(k) = \text{const.}$ and thus $f(x) = g(x)$ and $\mathrm{MTF}(k) = 1$ for all modulation frequencies k. Both uniform regions (low k) and very fine structures (high k) would be precisely reproduced in the image. In a real imaging system, MTF always approaches zero for $k \to \infty$ (but not necessarily in a monotonous way), thus limiting the resolution of the image. Table 2.1 gives a summary of the functions introduced in this section.

Table 2.1: Important functions characterizing a linear, shift-invariant imaging system in 1d. Generalization to higher dimensions is straightforward.

function	description	formula
PSF	image of a point $\delta(x)$	$h(x)$
OTF	Fourier transform of the PSF	$\tilde{h}(k)$
MTF	normalized absolute value of the OTF	$\left\lvert\frac{\tilde{h}(k)}{\tilde{h}(0}\right\rvert$

2.5.4 Example: cosine grating and Gaussian PSF

The considerations so far can be illustrated by the 1d example sketched in Figure 2.15. Let the input be given by a cosine grating

$$f(x) = \frac{1}{2}\left[1+\cos(2\pi x)\right] , \tag{2.69}$$

and the PSF of the system by a normalized Gaussian

$$h(x) = \frac{1}{\sigma\sqrt{2\pi}} \exp\left(-\frac{x^2}{2\sigma^2}\right) . \tag{2.70}$$

The width of the Gaussian is described by the standard deviation σ or equivalently by the full width at half maximum (FWHM)

$$\mathrm{FWHM} = 2\sqrt{2\ln 2}\sigma . \tag{2.71}$$

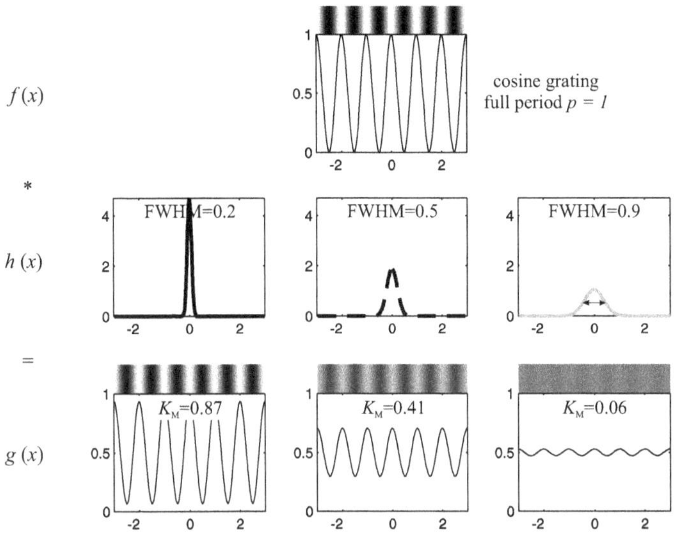

Fig. 2.15: Illustration of the imaging process of a cosine pattern (period $p = 1$) by three different Gaussian PSFs $h(x)$. Convolution of the input signal $f(x)$ with the PSF leads to a decrease in (Michelson) contrast $K_M = (g_{max} - g_{min})/(g_{max} + g_{min})$, while the background signal g_{min} increases. Since the human eye can only distinguish about 20–50 gray values ranging from black to white, it is difficult to resolve the differences in the grayscale representation (0 corresponds to black, 1 to white) for the case $K_M = 0.06$ shown on the right hand side.

Here we use the FWHM, since it is more readily obtained from the plots, in particular for non-Gaussian PSF. If the PSF was a delta function, the image $g(x)$ would be identical to the input $f(x)$. For increasing FWHM, we can observe that the modulation in the image decreases successively. This is quantitatively captured by the MTF

$$\mathrm{MTF}(k) = \left| \frac{\mathcal{F}[h(x)](k)}{\mathcal{F}[h(x)](0)} \right| = \exp\left(-\frac{\sigma^2 k^2}{2} \right) , \tag{2.72}$$

sketched in Figure 2.16. The poorer the quality of the imaging system, the broader its PSF and the stronger the damping of high spatial frequencies in the image. This finally leads to a reduction of the resolution.

2.5.5 Measurement of PSF and MTF

Since the PSF and the OTF/MTF constitute a Fourier pair, the measurement of one of them automatically allows computation of the other. This can be achieved in several ways, see also [6, 8]

- The most intuitive way is to present a point-like input to the imaging system and directly record the PSF (equation 2.56). In practice, this can be achieved, e.g., by

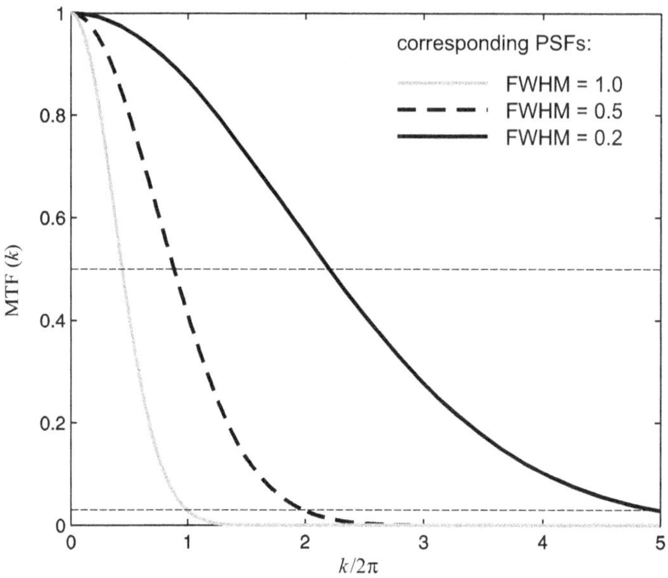

Fig. 2.16: Modulation transfer functions (MTF) corresponding to the three Gaussian PSF shown in Figure 2.15. Higher spatial frequencies in the image are more and more damped if the width of the PSF increases.

imaging a pinhole (radiography), a wire spanned in axial direction (CT), a small radioactive source (SPECT) or a single scatterer (ultrasound).

- The MTF of a radiography system can be measured by placing lead foils with cosine like thickness variations of different periods in the beam and recording the modulation in the resulting images (equation 2.68).
- Another method for radiography systems is to place a narrow slit in vertical direction in the beam and measure the one-dimensional line spread function (LSF)

$$f(x,y) = \delta(x) \quad \Rightarrow \quad g(x,y) = \mathrm{LSF}(x)\,, \tag{2.73}$$

which is an integral representation of the 2d PSF

$$\mathrm{LSF}(x) = \int_{-\infty}^{\infty} dy\, \mathrm{PSF}(x,y)\,. \tag{2.74}$$

The normalized absolute value of the Fourier transform then yields the MTF in the x direction.

- In practice, it is often easier to fabricate a sharp edge than a line, i.e., a structure parameterized by the Heaviside function

$$\Theta(-x) = \begin{cases} 1 & x \leq 0 \\ 0 & \text{else} \end{cases}\,. \tag{2.75}$$

This leads to the edge spread function (ESF)

$$f(x,y) = \Theta(x) \quad \Rightarrow \quad g(x,y) = \mathrm{ESF}(x)\,, \tag{2.76}$$

with

$$\mathrm{LSF}(x) = \frac{d}{dx}\mathrm{ESF}(x)\,. \tag{2.77}$$

The total MTF of an imaging system composed of several components is given by the product of the MTFs of its components

$$\mathrm{MTF}_{\mathrm{system}} = \prod_k \mathrm{MTF}_{\mathrm{component}\ k} \tag{2.78}$$

Therefore, the component of poorest quality determines the performance of the entire system. For example, in a measurement of the MTF of an x-ray system, one records the LSF or ESF using an x-ray film, which is subsequently digitalized by a scanner for further analysis. The measured MTF is then given by

$$\mathrm{MTF} = \mathrm{MTF_a} \cdot \mathrm{MTF_d}\,, \tag{2.79}$$

where $\mathrm{MTF_a}$ denotes the 'analog' MTF of the x-ray system and $\mathrm{MTF_d}$ the MTF of the digitalization process [6].

2.6 Measures of image quality

An image is never a full representation of the original object. In particular, small components in the object may no longer be contained in the image – i.e., the resolution of the imaging process is limited. The ability of the human eye to detect some feature in an image depends on the contrast. Further, two recordings under the same conditions may not give the same result, owing to the stochastic nature of the imaging process, e.g., due to photon emission and absorption. This is what we generally call noise. Figure 2.17 illustrates how these three effects determine the quality of the images of a test object. In order to assess the performance of different imaging systems, precise definitions of contrast, resolution and noise are required.

2.6.1 Contrast

In the present context, contrast is used to describe the amplitude of an object or image component, compared to the average value. In other words, contrast describes how a structural feature stands out against the average. Note that contrast can be defined at several stages of the imaging process: In computed tomography, for example, we can quantify the input object contrast (differences in absorption coefficient μ), the output image contrast (differences in Hounsfield units after reconstruction) or the display

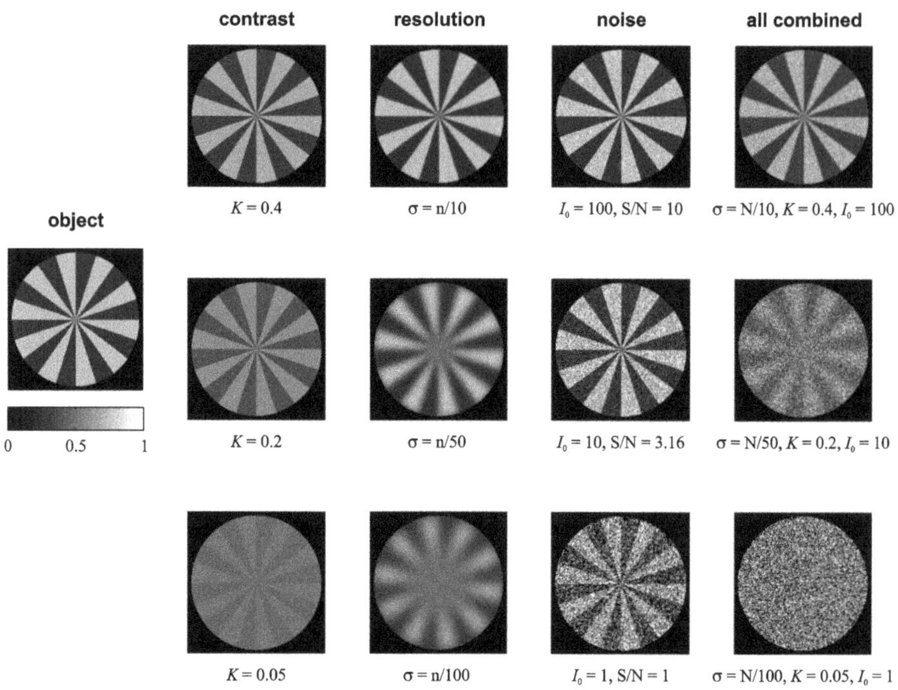

Fig. 2.17: Test object and its images for different parameters of contrast, resolution and noise, as well as the combination of all three effects. The color bar applies to all the images. Here, we use the definition of Michelson contrast (equation 2.81) as well as Poisson noise (equation 2.90). To simulate the different resolutions, the object was multiplied with Gaussians of different widths in Fourier space and back transformed into real space, i.e., $g(x,y) = \mathcal{F}^{-1}[\tilde{f}(k_x,k_y)\cdot \exp(-|k|^2/(2\sigma^2))]$. $n = 255$ denotes the number of pixels along each direction of the image, and $\mathrm{S/N} = \sqrt{I_0}$ the signal-to-noise ratio of Poisson noise.

contrast (differences in gray values on the monitor when displayed to the radiologist for final examination) [6].

The basic idea of contrast is that small absolute differences may lead to visible differences in an image, if the total signal is small, i.e., when the relative difference is still sufficiently strong. Contrarily, features are likely to escape detection if they are embedded in a high background signal level. Two different definitions of contrast are commonly used in medical imaging, see Figure 2.18. The local (or Weber) contrast is defined using the difference between the signal g_f of a certain image feature and the signal g_b of the background or the embedding environment. It is usually applied in images where small features are visible above a rather uniform background (Figure 2.18, left):

$$K_{\mathrm{W}} = \frac{g_f - g_b}{g_b} \,. \tag{2.80}$$

The modulation (or Michelson) contrast

$$K_{\mathrm{M}} = \frac{g_{\max} - g_{\min}}{g_{\max} + g_{\min}} \tag{2.81}$$

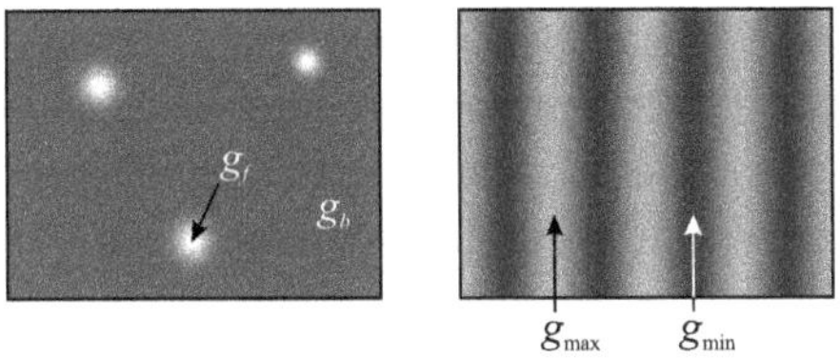

Fig. 2.18: Two different definitions of contrast: Weber contrast $K_W = \frac{g_f - g_b}{g_b}$ utilizes the typical brightness g_f of small, isolated features in front of a rather uniform background g_b, while Michelson contrast $K_M = \frac{g_{max} - g_{min}}{g_{max} + g_{min}}$ considers the ratio of the maximum variation, divided by its mean.

is used instead if high and low signal strength make up similar fractions of the image. As was already noted above, for a cosine signal with constant background (Figure 2.15), this definition yields the amplitude of the signal variation divided by the mean value (Figure 2.18, right)

$$g(x) = A + B\cos(2\pi x) \quad \Rightarrow \quad K_M = \frac{(A+B) - (A-B)}{(A+B) + (A-B)} = \frac{B}{A} . \tag{2.82}$$

2.6.2 Resolution

Resolution is a measure for the smallest structure that can be reliably detected under given imaging circumstances. Using the theory of LSIs, it can be defined as the width of the PSF or equivalently in Fourier space by the spatial frequency where the MFT falls below a certain threshold, see, e.g., the horizontal dashed lines in Figure 2.16. Either way, we have to define the threshold or width criterion, e.g., half width at half maximum (HWHM), full width at half maximum (FWHM), root mean square value σ or $1/e$ value. Here, some work is typically required to clarify prefactors when comparing different choices or conventions.

In some cases, the functional form of the PSF is known beforehand, and it may, therefore, lend itself for a certain definition, tailored to this function. This is the case for many optical systems, where the PSF takes the form of the so called Airy pattern.

In other instances, we may know very little about the imaging system beforehand. For example, we may neither know the PSF (or MTF), nor the type of noise nor whether our sampling was sufficient or whether systematic errors have spoiled the results. In this case, consistency can often be verified by splitting the dataset into two and comparing the images obtained from the two subsets. A typical result will show large correlation for small spatial frequencies and low correlation for high spatial frequencies. Based on interrogation of this Fourier shell correlation, a resolution can be defined such that it warrants a certain degree of consistency for structures larger than the resolution. Finally, we encounter situations where we suspect not so much image degradation from inconsistencies, poor sampling or a wide PSF, but simply from stochastic noise.

The resolution achievable based on the PSF can be far from being actually reached, if the signal-to-noise is very low. Now, if the noise is uncorrelated and hence flat in reciprocal space (such as Gaussian white noise, for example), the power spectral density (PSD) is well suited to evaluate the resolution. In the following, we briefly address each of these prototypical cases.

Airy pattern

Optical instruments like microscopes, telescopes, cameras, but also the human eye, can be treated by diffraction theory for given wavelength λ. If aberrations can be neglected, the resolution is limited by diffraction. The PSF of these linear and radially symmetric imaging systems is given by the Fourier transform (coherent illumination) or Fourier transform squared (incoherent illumination) of the pupil function, i.e., the disk function with radius a (radius of aperture or lens). The Fourier transform of the disk function leads to the Bessel function J_1 of first kind and first order, with a (Fraunhofer) intensity pattern

$$I_{\mathrm{Airy}}(k) = I_0\, A \left[\frac{J_1(ka)}{ka/2}\right]^2 \qquad (k = 2\pi\nu)\,, \tag{2.83}$$

where $A = \pi a^2$ is the area of the disk. The image of a point source, i.e., the PSF of this system, results in this circular symmetric intensity distribution I_{Airy} in the image plane at distance d behind the lens, evaluated for $k \to 2\pi\rho/\lambda d$ with $\rho = \sqrt{x^2+y^2}$ the lateral coordinate in the image plane

$$I_{\mathrm{Airy}}(\rho) = I_0\, A \left[\frac{J_1(2\pi\rho a/\lambda d)}{\pi\rho a/\lambda d}\right]^2 . \tag{2.84}$$

$J_1(z)$ has nonequidistant roots at $z \approx 3.83, 7.02, 10.17, \ldots$ and its central lobe ($z \leq 3.83$) is often called the Airy disk. The Airy pattern thus arises as the Fourier transform of a disk and quantifies the PSF for many radially symmetric imaging systems. To quantify the resolution, two incoherent point sources (of the same brightness) in the source plane located at distance f in front of the lens system are considered. According to the Rayleigh criterion, the minimum resolvable distance in the source plane $\rho_{\min}$, i.e., the resolution, is reached when the central maximum of one Airy pattern coincides with the first minimum of the other (Figure 2.19, center). Accounting for the magnification $M = d/f$, this occurs for $2\pi\rho_{\min} a/\lambda f = 3.832$, resulting in

$$\rho_{\min} = 0.61\,\frac{\lambda f}{a} = 0.61\,\frac{\lambda}{\theta} = 0.61\frac{\lambda}{\mathrm{NA}}\,, \tag{2.85}$$

where $\theta = a/f$ is the highest diffraction angle, written in small angle approximation. More exactly, without this approximation and taking the refraction in a medium n into account, the expression with the numerical aperture $\mathrm{NA} = n \sin\theta$ should be used.

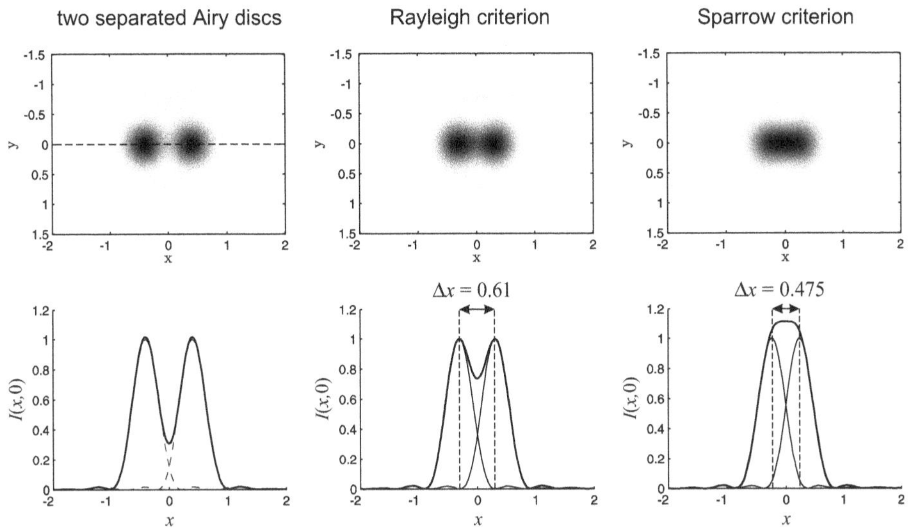

Fig. 2.19: Illustration of the Rayleigh and Sparrow criteria for the resolution in natural units. The Rayleigh criterion defines the resolution as the distance where the central maximum of one Airy disk coincides with the first minimum of the other (as indicated by the *dashed vertical lines*). The Sparrow criterion uses the distance where the local minimum at $x = 0$ in the total intensity (*thick line*) disappears and a plateau is formed.

Alternatively, the Sparrow criterion defines resolution when the dip between the two Airy disks vanishes, i.e., at a distance $\rho_{\min}$ where the two Airy disks merge to a single, broad maximum, resulting in $\rho_{\min} = 0.475\lambda/\theta$ (Figure 2.19, right).

Fourier shell correlation (FSC)

In 2d also denoted as Fourier ring correlation (FRC), this is a tool originally used to determine resolution in electron microscopy. However, it is a very general and useful concept, which should be more widely used. In very general terms, FRC evaluates the similarity of two independent reconstructions of the same object in frequency space to determine the resolution threshold (the spatial frequency) up to which both reconstructions are consistent. Up to this spatial frequency the object is considered to be resolved. To carry out this procedure, one splits the dataset into two, or compares two independent images of the same object or two independent reconstructions. The FRC is defined as [15, 16]

$$\mathrm{FRC}(k_i) = \frac{\sum_{k\in k_i} \tilde{g}_1(k)\cdot \tilde{g}_2^*(k)}{\sqrt{\sum_{k\in k_i} \tilde{g}_1^2(k)\cdot \sum_{k\in k_i} \tilde{g}_2^2(k)}}\,, \tag{2.86}$$

where $\tilde{g}_1$ and $\tilde{g}_2$ are the Fourier transforms of the two images (or tomographic reconstruction volumes) g_1, g_2. The variable k_i denotes the radial coordinate $\sqrt{k_x^2 + k_y^2}$ of spatial frequency and hence specifies the single ring/shell i in Fourier space. Summation is carried out over all elements k inside the ring/shell k_i. Note that the normalization is such that $\mathrm{FRC}(k_i) = 1$ if $\tilde{g}_1$ and $\tilde{g}_2$ are identical in amplitude and phase. The original images/volumes are real valued quantities and, thus, their Fourier transforms are symmetric $\tilde{g}_1(k_x, k_y) = \tilde{g}_1^*(-k_x, -k_y)$ (Friedel symmetry), so $\mathrm{FRC}(k_i)$ is also real valued. Since FRC measures the similarity between two images/volumes, a significance threshold has to be defined, which indicates the largest spatial frequency up to which two images/volumes can be considered as similar. This threshold must depend on the spatial frequency, since the number of independent entries in the shell $\tilde{g}_{1,k_i}(k)$, $N_{k_i}/2$ (factor 2 due to Friedel symmetry), increases with the radius of the shell. There exist a variety of different threshold criteria and curves, derived by statistical analysis under different assumptions [15, 16]. In essence, one compares the correlation of the two datasets to that expected for a random array of points and wants to identify the critical spatial frequency, below which the data is correlated to a statistically significant degree. A commonly used threshold curve is the so called σ factor curve, which in 2d is given by

$$\sigma_{\mathrm{th}}(k_i) = \frac{2}{\sqrt{N_{k_i}/2}} \,. \tag{2.87}$$

Figure 2.20 illustrates the FRC method for two images of different signal-to-noise ratio (SNR). Different rings in Fourier space are highlighted. The values of the corresponding rings in (b) and (d) are correlated following equation (2.86) and are plotted in (e).

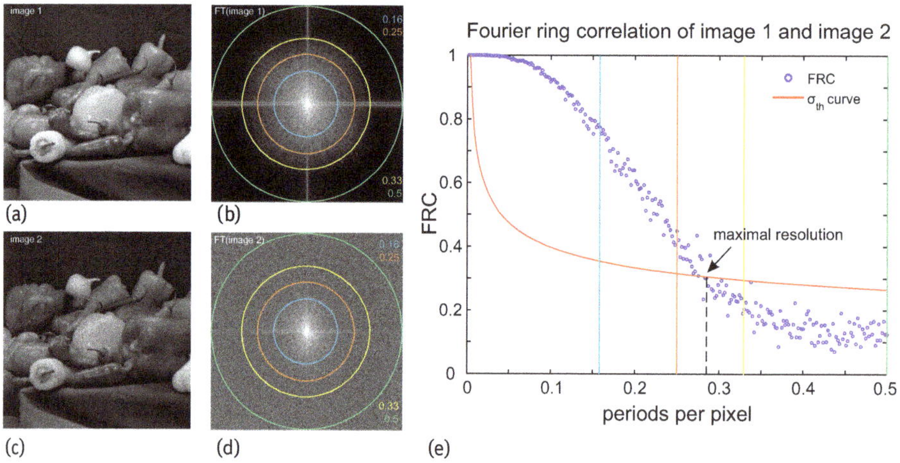

Fig. 2.20: (a) Noise-free image and (b) corresponding Fourier transform (logarithm of modulus). (c) Same image with noise (uncorrelated noise, SNR 6.2) and (d) corresponding Fourier transform (logarithm of modulus). (e) FRC calculated for image 1 and 2 (blue) and the threshold curve indicating a resolution of about 0.28 periods per pixel.

Power spectral density

If significant noise is present in an image, the resolution will be limited by the SNR, and the PSF alone is no longer a suitable criterion. The FRC will still work, if two datasets with statistically independent realizations of the noise are available. This is not always the case. Yet, even if only a single image is given, the statistical properties of the noise itself can be used to estimate the resolution. This is a particularly well suited approach, if the noise is uncorrelated, i.e., if there are no correlations between different pixels. Then, the noise will be distributed evenly over all spatial frequencies, i.e., there will be a flat noise floor in Fourier space. One can then interrogate the PSD, i.e., the modulus square of the Fourier transform, in view of the decrease of the signal as a function of spatial frequency. The spatial resolution can be defined from the crossover where the signal 'drowns' in the noise floor. To this end, the PSD is often inspected after azimuthal average. This procedure will be illustrated graphically in Section 2.6.4, after addressing in more detail some aspects of image noise as well as of the PSD.

2.6.3 Image noise

Even if two images are recorded under exactly the same conditions, they will never look exactly the same. This is due to random processes summarized as noise. Noise is a key issue of signal processing, and hence of image processing, in particular. Studying image processing without noise would be roughly equivalent to studying statistical mechanics without temperature. Consequently, there is a large amount of literature on the subject. In compiling the rather brief account of noise in this subsection, the work by [1, 6, 8] was found helpful and can also be recommended for further reading. Several types of noise can be distinguished:
- *photon noise*: random variation of a signal due to the quantization of electromagnetic radiation into statistically independent photons;
- *shot noise*: quantization of electric charge into multiples of an elementary charge e;
- *thermal noise* (or Johnson–Nyquist noise): random thermal motion of charge carriers (usually electrons) in absence of an applied voltage;
- *$1/f$ noise* (or pink noise): slow fluctuations of material properties, e.g., resistance.

In the last example, $1/f$ refers to the decay of the noise in the spectral domain, of the PSD to be more precise. This is linked to correlations in the noise, i.e., temporal correlations in 1d time series, or correlations between pixels in images. Contrarily, if the noise is uncorrelated, the spectrum will be flat, forming the noise plateau mentioned above. This is called white noise.

We will primarily consider photon noise here, for further types of noise see, e.g., [1]. Placing a detector in a temporally and spatially constant field of electromagnetic radiation, repeatedly recording photons over a fixed time interval Δt will not always give a constant number n, but rather a distribution of values, given by the Poisson distribution, described by a single parameter μ

$$P_\mu(n) = \frac{\mu^n}{n!} e^{-\mu} . \tag{2.88}$$

The parameter μ determines both the expectation value (or mean) $\langle n \rangle$ and the variance σ^2

$$\begin{aligned} \langle n \rangle &= \sum_{n=0}^{\infty} n \cdot P_\mu(n) = \sum_{n=0}^{\infty} n \cdot \frac{\mu^n}{n!} e^{-\mu} = \mu e^{-\mu} \sum_{n=1}^{\infty} \frac{\mu^{n-1}}{n-1} \overset{j=n-1}{=} \mu e^{-\mu} \underbrace{\sum_{j=0}^{\infty} \frac{\mu^j}{j!}}_{=e^\mu} = \mu \\ \sigma^2 &= \langle (n - \langle n \rangle)^2 \rangle = \cdots = \mu , \end{aligned} \tag{2.89}$$

where the calculation is carried out in Section 3.5. If we can expect a signal with a mean of $\mu = N$ photons, the standard deviation will be $\sigma = \sqrt{N}$, and hence the SNR is

$$\mathrm{SNR} := \frac{\text{mean number of photons}}{\text{standard deviation}} = \frac{\mu}{\sigma} = \frac{N}{\sqrt{N}} = \sqrt{N} . \tag{2.90}$$

This is the common definition of the SNR (or S/N) in medical image processing.[5] According to the Rose criterion, SNR ≥ 5 is required to distinguish image features with certainty. The SNR is obtained by repeated measurements of the photons N in a single detector pixel, for all pixels in parallel. If the noise is not spatially correlated, the SNR can be obtained from the fluctuations of the signal averaged over a homogeneously illuminated region. Strictly speaking, this does not hold for imaging systems that cause some blurring of the signal over adjacent pixels, i.e., with a PSF larger than a single pixel, since this introduces correlations in the noise [6]. If the signal to be measured is constant (apart from the intrinsic noise), one can simply increase the SNR by counting longer (equation 2.90). In Figure 2.21, this is illustrated for a circular object with a transmission of 0.97. If the SNR is too low (e.g., only 10 or 100 photons per pixel), the noise in the image will prevent detection of the object. A closely related quantity is the contrast-to-noise ratio (CNR)

$$\mathrm{CNR} = \frac{\Delta\mu}{\sigma} = \frac{|\mu_A - \mu_B|}{\sigma} = |\mathrm{SNR}_A - \mathrm{SNR}_B| , \tag{2.93}$$

5 Note that a different definition is used in other contexts such as electric engineering or ultrasound imaging. Here, the SNR is defined by

$$\mathrm{SNR} = \frac{\text{power of signal}}{\text{power of noise}} = \frac{P_{\text{signal}}}{P_{\text{noise}}} \tag{2.91}$$

Since both powers often differ by orders of magnitude, the SNR is sometimes given on a logarithmic scale in units of decibels [db]

$$\mathrm{SNR_{dB}} := 10 \cdot \log_{10} \left(\frac{P_{\text{signal}}}{P_{\text{noise}}} \right) = 20 \cdot \log_{10} \frac{A_{\text{signal}}}{A_{\text{noise}}} \tag{2.92}$$

where A_{signal} and A_{noise} denote the amplitudes of signal and noise, respectively.

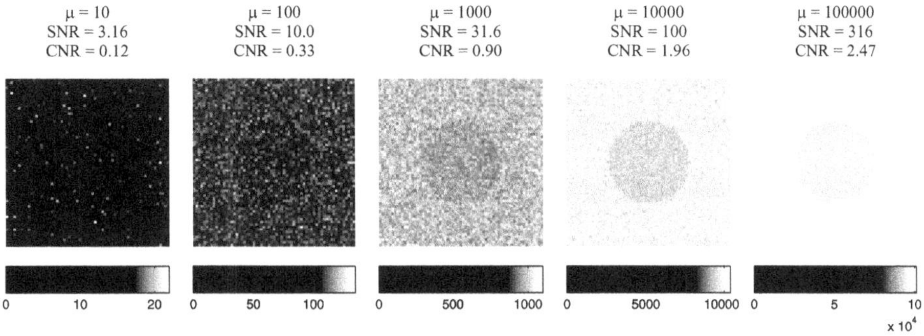

Fig. 2.21: Grayscale image of a circular object with a transmission of 0.97, resulting in a contrast of $\sim$ 3%. The mean number of photons of a pixel in the background is denoted by μ. With CNR < 1, the object is barely visible, while safe object detection is possible at CNR ≥ 2. Poisson noise has been added by the *imnoise* function of MATLAB.

where $\mu_{A,B}$ denotes the mean signal in two regions A and B to be distinguished and σ the image noise. As a rule of thumb, feature detection in an image is hardly possible for CNR < 1, while at CNR ≥ 2 features larger than a single pixel can be detected, see Figure 2.21.

In medical imaging, one always has to find a compromise between image quality and radiation dose to the patient, due to the harmful effects of ionizing radiation, as discussed in Chapter 5. Guidelines for x-ray diagnostics aim at a minimum dose needed for the recording of an image, which should be just sufficient for a certain diagnostic purpose. For preclinical biomedical research on biological tissues or biological cells, the dose can be increased by many orders of magnitude, but structural degradation due to radiation damage will also necessitate a careful assessment of SNR for the given dose budget. Since noise is an intrinsic property of each imaging process and the SNR can be made arbitrarily small by increasing the time of the measurement (and thus the dose), the total noise in an image cannot serve as a quality criterion for an imaging system [8]. Instead, only the additional noise added by the imaging system to the intrinsic noise determines its quality. To this end, one defines the detective quantum efficiency (DQE) as the ratio

$$\mathrm{DQE} = \frac{\mathrm{SNR}^2_{\mathrm{output}}}{\mathrm{SNR}^2_{\mathrm{input}}} = \frac{\langle n_{\mathrm{out}} \rangle}{\langle n_{\mathrm{in}} \rangle}, \tag{2.94}$$

where the last expression, i.e., the ratio between the expectation values of detected and incident quanta holds only for the Poisson distribution. An ideal imaging system would detect all incident quanta and not add any noise, hence DQE $= 1$. The more photons contribute to the detection of the image, the higher the DQE, and the lower the dose required to record an image with sufficient SNR or CNR.

2.6.4 Power spectral density

The PSD is defined as the absolute square of the Fourier transform of an image,

$$\mathrm{PSD}\left[f(x,y)\right] = \left|\mathcal{F}\left[f(x,y)\right](k_x,k_y)\right|^2 . \tag{2.95}$$

The PSD quantifies the intensities of the spatial frequencies which make up the image. High spatial frequencies correspond to small structures and sharp edges, while low spatial frequencies describe extended image components, see Figure 2.22. In practice, the PSD is computed by a discrete Fourier transform (DFT), giving rise to issues of sampling, periodization and truncation, as addressed in the next section. We can consider the DFT operations carried out on the discrete image as an approximation of the corresponding Fourier operations in the space of the object. In other words, the continuous physical variable of the imaged object is an element of a suitable functional space, and as such it makes sense to associate a spectrum of spatial frequencies and a PSD with the object itself. Contrarily, the PSD of the image bears information on the image object as well as the imaging system. In this respect, discretization can already be regarded as a prime distortion of the imaging process. The PSD provides an approach to determine the resolution of an image, since it monitors at which spatial frequency the signal has decayed to a critical value or below the noise floor. In practice, the resolution cutoff by the transition to a flat noise plateau is only a suitable criterion if the noise is sufficiently high and if the noise is white or spectrally broad, i.e., characterized by a flat background floor in the PSD, see Figure 2.24.

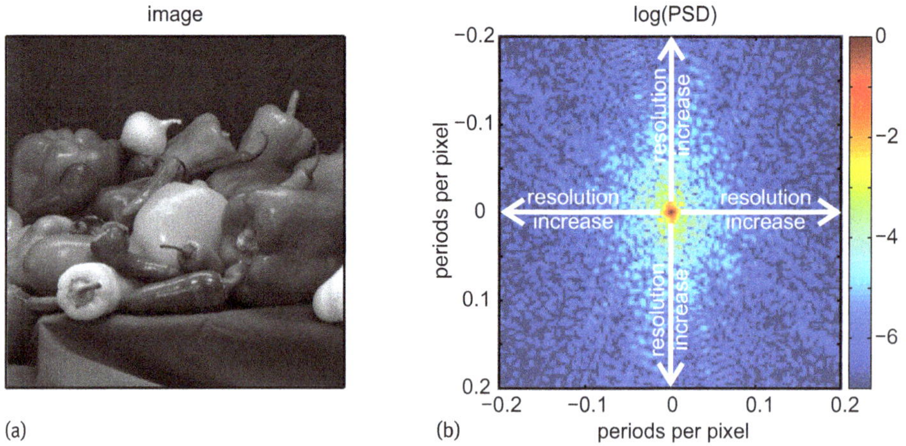

Fig. 2.22: (a) Image and (b) corresponding PSD. The PSD is plotted on logarithmic scale. The further the signal extends towards larger spatial frequencies, the higher the resolution of the image. The values at the boundary of the PSD plot represent spatial frequencies corresponding to a period of 2 pixels. The 2d PSD is often azimuthally averaged to obtain a 1d plot for better inspection of the signal decay.

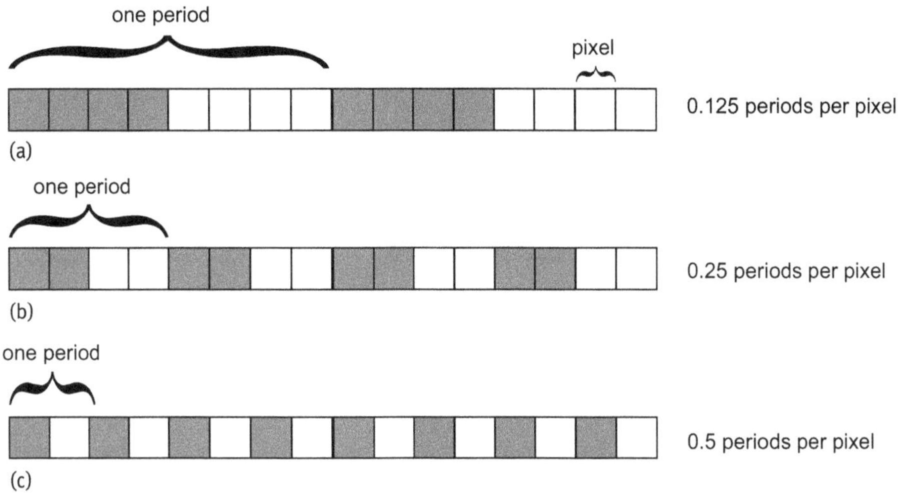

Fig. 2.23: (a) One full period is one black–white transition. (b) 0.25 periods per pixel. (c) Maximum resolution: 0.5 periods per pixel.

Spatial frequencies can be expressed in 'periods per pixel'. The unit 'periods per pixel' details how many pixels are needed to cover a full period of a given periodic structure in the image (sinusoidal variation, pattern of lines and spaces), see Figure 2.23 (a). For example, 0.25 periods per pixel states that 4 pixels are necessary to cover a full period, see Figure 2.23 (b). The maximum resolution is given at 0.5 periods per pixel, describing, for example, one black and one white pixel, see Figure 2.23 (c).

With the help of the PSD, it is possible to quantify the influence of noise. To this end, one can plot the angular averaged PSD as a function of spatial frequency $\sqrt{\nu_x^2 + \nu_y^2}$. For almost all relevant images this is a strongly decaying function, which has to be inspected on a semi or double logarithmic scale. When the decreasing PSD function reaches the noise floor, i.e., when the decay crosses over to a flat noisy plateau, the resolution limit is reached. Higher frequencies than this cross over or noise cut off correspond to noise in the image. To quantify the cut off frequency, one typically fits a power law to the decaying PSD

$$f(x) = a \cdot x^b \tag{2.96}$$

and a constant offset $g(x) = c$ to the data at high ν. The intersection of $f(x)$ and $g(x)$ determines the maximum detectable spatial frequency, e.g., in periods per pixel), and its inverse, the spatial resolution (full period resolution in number of pixels), see Figure 2.24.

Prior to calculation of the PSD, the image is often multiplied by a window function which is symmetric and rapidly decaying such that the windowed image is embedded in a bounded support [9]. The window has to be chosen such that its spectral range

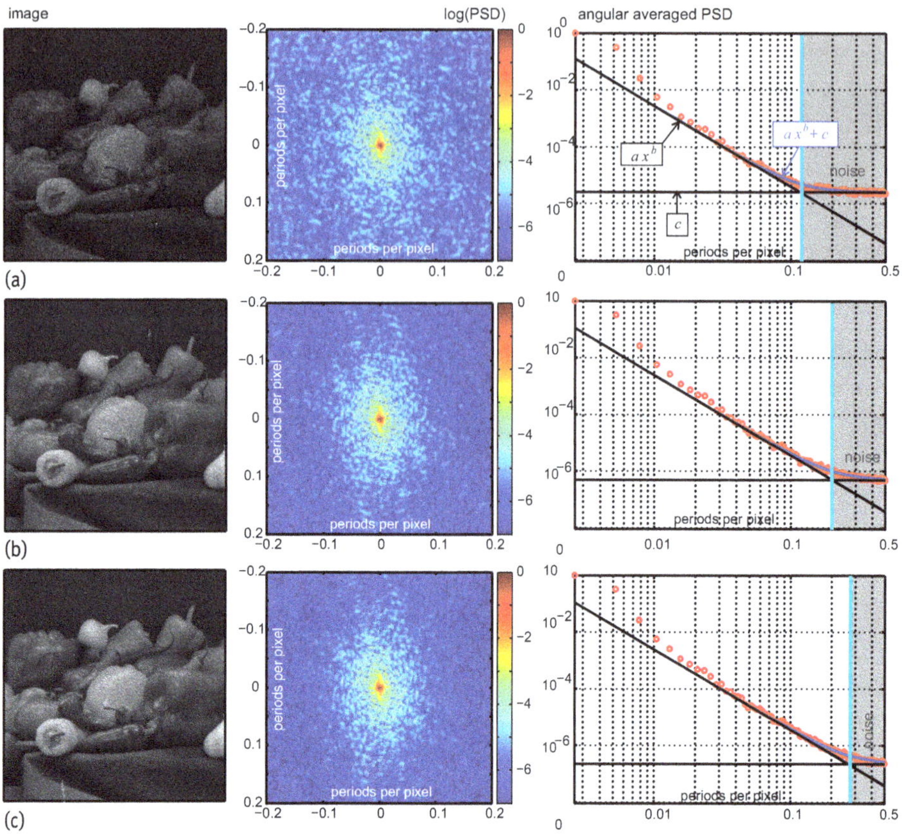

Fig. 2.24: Influence of noise visible in the PSD. (a) *Left*: Image corrupted with Poisson noise. *Center*: Corresponding PSD. *Right*: Angular averaged PSD. A power law $a \cdot x^b$ and an offset c were fitted to the angular averaged PSD indicating the smallest resolvable structure at the intersection of both functions. (b) and (c) Same analysis for less noise corrupting the image.

does not interfere with the spectrum of the image itself. Window functions reduce errors that result from truncation of the image at its edges and facilitate the detection of weak frequency components [9]. A possible choice of a window function is the Kaiser–Bessel window [4]

$$w(r) = \begin{cases} \frac{I_0\left(\beta\sqrt{1-(2r/N)^2}\right)}{I_0(\beta)} & \text{for } r \le N/2 \\ 0 & \text{else}\,. \end{cases} \tag{2.97}$$

Here, $r = \sqrt{x^2 + y^2}$ and N is the number of pixels of the image along one dimension. The function I_0 is the modified Bessel function and the parameter β is usually set to values $\in [2, 9]$.

2.7 Fourier transformation: fundamentals

The concept of Fourier transformation is an important analytical tool in biomedical imaging, indispensable for the treatment of propagation and diffraction of electromagnetic waves, reconstruction of CT data, signal acquisition in MRI, propagation and scattering of ultrasound waves or post-processing of images. Fourier theory allows us to decompose a physical signal into its spectral components and write it as a superposition of elementary waves, each of constant frequency. For the case of images, Fourier transformation decomposes the data into spatial frequencies, i.e., expands the image as a superposition of plane waves. Many other basis functions – different from the plane wave basis of Fourier theory – have also found important applications in image representation and processing, notably wavelet theory, but Fourier theory remains the basis of it all. The purpose of this section is hence to recall basic definitions and notations, for the sake of clarity when used throughout this book, and especially cover the relationship between continuous Fourier theory and discrete Fourier transformation, which is required for numerical implementation. For a proper mathematical introduction we refer to the exhaustive textbook literature on the subject, for example, [3, 4], as well as [9], which was a primary source of the summarizing purpose of the next two sections, regarding in particular the sampling issues of discrete image data.

2.7.1 Definitions and notations

Different notations are commonly found for the Fourier transform. For example, in the mathematical literature, the following definition of the two-dimensional Fourier transform prevails

$$\mathcal{F}[f(x,y)](\nu_x,\nu_y) = \tilde{f}(\nu_x,\nu_y) := \int_{-\infty}^{+\infty} dx \int_{-\infty}^{+\infty} dy\, f(x,y)\, e^{-2\pi i(\nu_x x+\nu_y y)} \,, \tag{2.98}$$

in terms of the spatial (image) frequencies ν_x and ν_y, along with the inverse Fourier transform of a function $\tilde{g}(\nu_x,\nu_y)$

$$\mathcal{F}^{-1}[\tilde{g}(\nu_x,\nu_y)](x,y) = g(x,y) := \int_{-\infty}^{+\infty} d\nu_x \int_{-\infty}^{+\infty} d\nu_y\, g(\nu_x,\nu_y)\, e^{+2\pi i(\nu_x x+\nu_y y)} \,. \tag{2.99}$$

Performing Fourier transform and inverse transform in sequence, one obtains

$$\mathcal{F}^{-1}[\mathcal{F}[f]] = f \,. \tag{2.100}$$

It can be shown that the Fourier transform defines an automorphism on the Schwartz space $\mathcal{S}(\mathbb{R}^2)$ of all infinitely differentiable functions on $\mathbb{R}^2$ that vanish faster than any polynomial as the arguments go to infinity, as do all of their derivatives, i.e., in this space there is a one-to-one correspondence between a function and its Fourier transform and vice versa:

$$f(x,y) \underset{\mathcal{F}^{-1}}{\overset{\mathcal{F}}{\rightleftharpoons}} \tilde{f}(\nu_x,\nu_y) \tag{2.101}$$

In other spaces, such as, for example, $\mathcal{L}^2(\mathbb{R}^2)$, the space of all square-integrable functions, the ideal bijectivity, is lost but there is still a close correspondence between a function and its Fourier transform. For example, a function g that differs from f for exactly one value of $\underline{x}$ will have the same Fourier transform as f.

In the physical literature, the Fourier transform is more often defined in terms of angular frequencies or momentum transfer (in diffraction theory) with coordinates $k_x = 2\pi\nu_x$ and $k_y = 2\pi\nu_y$ in Fourier space (reciprocal space), here for the 2d case. This eliminates the factor 2π in the exponential of the integrand, but gives rise to a prefactor in the inverse Fourier transform, according to

$$\mathcal{F}[f(x,y)](k_x,k_y) = \tilde{f}(k_x,k_y) := \int_{-\infty}^{+\infty} dx \int_{-\infty}^{+\infty} dy\, f(x,y)\, e^{-i(k_x x + k_y y)}\,, \tag{2.102}$$

with the inverse Fourier transform of a function $\tilde{g}(k_x,k_y)$

$$\mathcal{F}^{-1}[\tilde{g}(k_x,k_y)](x,y) = g(x,y) := \frac{1}{(2\pi)^2} \int_{-\infty}^{+\infty} dk_x \int_{-\infty}^{+\infty} dk_y\, g(k_x,k_y)\, e^{+i(k_x x + k_y y)}\,. \tag{2.103}$$

Alternatively, one can use a definition with symmetric prefactors for both directions of the transform, which is the choice that we will adopt in this book. Accordingly,

$$\mathcal{F}[f(x,y)](k_x,k_y) = \tilde{f}(k_x,k_y) := \frac{1}{(2\pi)} \int_{-\infty}^{+\infty} dx \int_{-\infty}^{+\infty} dy\, f(x,y)\, e^{-i(k_x x + k_y y)}\,, \tag{2.104}$$

with the inverse Fourier transform of a function $\tilde{g}(k_x,k_y)$

$$\mathcal{F}^{-1}[\tilde{g}(k_x,k_y)](x,y) = g(x,y) := \frac{1}{(2\pi)} \int_{-\infty}^{+\infty} dk_x \int_{-\infty}^{+\infty} dk_y\, g(k_x,k_y)\, e^{+i(k_x x + k_y y)}\,. \tag{2.105}$$

In medical imaging, Fourier transformation is not only used for two-dimensional (2d) data, but also in three dimensions (3d), for example, in CT, or in one dimension (1d), for example, regarding the temporal decay of an MRI signal. Accordingly, for the third choice of notation above we also formulate the definitions of the d-dimensional Fourier transform and its inverse of a (real or complex valued) function f on $\mathbb{R}^d$ as

$$\mathcal{F}[f(x_1,\dots,x_d)](k_1,\dots,k_d) = \frac{1}{\sqrt{2\pi}^d} \int_{\mathbb{R}^d} dx_1 \dots dx_d\, f(x_1,\dots,x_d)\, e^{-i[k_1x_1+\dots+k_dx_d]}\,, \tag{2.106}$$

$$\mathcal{F}^{-1}[\tilde{f}(k_1,\dots,k_d)](x_1,\dots,x_d) = \frac{1}{\sqrt{2\pi}^d} \int_{\mathbb{R}^d} dk_1 \dots dk_d\, \tilde{f}(k_1,\dots,k_d)\, e^{+i[k_1x_1+\dots+k_dx_d]}\,. \tag{2.107}$$

This can be written in more compact form by using vector notation

$$\mathcal{F}[f(\underline{x})](\underline{k}) = \frac{1}{\sqrt{2\pi}^d} \int_{\mathbb{R}^d} d^d x\, f(\underline{x})\, e^{-i\underline{k}\cdot\underline{x}} \tag{2.108}$$

$$\mathcal{F}^{-1}[\tilde{f}(\underline{k})](\underline{x}) = \frac{1}{\sqrt{2\pi}^d} \int_{\mathbb{R}^d} d^d k\, \tilde{f}(\underline{k})\, e^{+i\underline{k}\cdot\underline{x}} \tag{2.109}$$

All properties of the Fourier transform derived for $d = 1$ can be readily extended to $d > 1$.

Fourier theory defined for suitable vector spaces of (nonperiodic) square integrable functions can be linked to the analysis of Fourier series applicable to periodic functions. In fact, the definition of the Fourier transform is often motivated by the taking the limit $L \longrightarrow \infty$ of a periodic function, i.e., $f_L : x \in \mathbb{R} \longrightarrow f(x) \in \mathbb{C}$ with period L, i.e., $f_L(x + nL) = f_L(x)\, \forall\, n \in \mathbb{Z}$. This function can be represented by its Fourier series (provided that requirements such as square-integrability apply)

$$f(x) = \sum_{n=-\infty}^{+\infty} C_n\, e^{+ik_n x}\,, \tag{2.110}$$

the coefficients $C_n \in \mathbb{C}$ are given by the integrals

$$C_n = \frac{1}{L} \int_{-L/2}^{+L/2} dx\, f_L(x)\, e^{-ik_n x}\,. \tag{2.111}$$

Only frequency components that are integer multiples of a fundamental frequency $k_1 = \frac{2\pi}{L}$ can occur, since only these are periodic with L. Thus, a finite period length L implies a discrete frequency spectrum $k_n = n\Delta k = n\frac{2\pi}{L}$, $n \in \mathbb{Z}$. To extend this to nonperiodic functions, consider the limit $L \longrightarrow \infty$:

$$\begin{aligned} f(x) &= \lim_{L\to\infty} \sum_{n=-\infty}^{+\infty} \frac{1}{L} \left(\int_{-L/2}^{+L/2} dx\, f(x)\, e^{-ik_n x} \right) e^{+ik_n x} \\ &= \lim_{\Delta k\to 0} \sum_{n=-\infty}^{+\infty} \frac{\Delta k}{2\pi} \left(\int_{-L/2}^{+L/2} dx\, f(x)\, e^{-in\Delta k x} \right) e^{+in\Delta k x} \\ &= \frac{1}{\sqrt{2\pi}} \int dk \left(\frac{1}{\sqrt{2\pi}} \int_{-\infty}^{+\infty} dx\, f(x)\, e^{-ikx} \right) e^{+ikx} \end{aligned} \tag{2.112}$$

which brings us back to the previous definition of the Fourier transform of a function f

$$\mathcal{F}[f(x)](k) = \tilde{f}(k) := \frac{1}{\sqrt{2\pi}} \int_{-\infty}^{+\infty} dx\, f(x)\, e^{-ikx} \tag{2.113}$$

and the inverse Fourier transform of a function $\tilde{g}(k)$

$$\mathcal{F}^{-1}[\tilde{g}(k)](x) = g(x) := \frac{1}{\sqrt{2\pi}} \int_{-\infty}^{+\infty} dk\, \tilde{g}(k)\, e^{+ikx} . \tag{2.114}$$

Using the last three equations, it follows that the inverse Fourier transform is, indeed, the inverse of the Fourier transform

$$\mathcal{F}^{-1}[\mathcal{F}[f]] = f . \tag{2.115}$$

2.7.2 Properties and examples of Fourier pairs

The Fourier transform defines a bijection on the Schwartz space $\mathcal{S}(\mathbb{R})$, i.e., there is a one-to-one correspondence between a function and its Fourier transform and vice versa

$$f(x) \underset{\mathcal{F}^{-1}}{\overset{\mathcal{F}}{\rightleftharpoons}} \tilde{f}(k) . \tag{2.116}$$

Hence, we can represent the signal (or correspondingly the image) in either space, depending on what serves the purpose best. For 1d signals, the two spaces are often denoted as temporal and spectral domain, since most applications relate to time-dependent signals $f(t)$. For 2d and 3d signals, the prevailing examples of images and diffraction have coined the terms real and reciprocal space. Here, we present a selection of important functions in terms of the respective Fourier pairs, and some important theorems that are used throughout this book. We remain within the framework of continuous signals and their respective vector spaces. The discrete Fourier transform, which is needed for numerical treatment of measured data, is covered in Section 2.8.3.

Table 2.2 includes some important Fourier pairs. In the following, some useful properties of the Fourier transform are listed without proof:

- *linearity*:

$$\alpha f(x) + \beta g(x) \quad \underset{\mathcal{F}^{-1}}{\overset{\mathcal{F}}{\rightleftharpoons}} \quad \alpha \tilde{f}(k) + \beta \tilde{g}(k) \tag{2.117}$$

- *shifting*:

$$\begin{aligned} f(x - x_0) \quad &\underset{\mathcal{F}^{-1}}{\overset{\mathcal{F}}{\rightleftharpoons}} \quad \tilde{f}(k)\, e^{-ikx_0} \\ f(x) e^{ik_0 x} \quad &\underset{\mathcal{F}^{-1}}{\overset{\mathcal{F}}{\rightleftharpoons}} \quad \tilde{f}(k - k_0) \end{aligned} \tag{2.118}$$

- *scaling (similarity)*:

$$f(ax) \quad \underset{\mathcal{F}^{-1}}{\overset{\mathcal{F}}{\rightleftharpoons}} \quad \frac{1}{|a|} \tilde{f}\left(\frac{k}{a}\right) \tag{2.119}$$

- *derivatives*:

$$\frac{d^m}{dx^m} f(x) \quad \underset{\mathcal{F}^{-1}}{\overset{\mathcal{F}}{\rightleftharpoons}} \quad (ik)^m \tilde{f}(k) \tag{2.120}$$

Table 2.2: Some important Fourier pairs.

	$f(x)$	$\tilde{f}(k)$
Gaussian	$e^{-\alpha x^2}$	$\frac{1}{\sqrt{2\pi}} e^{-k^2/4\alpha}$
slit function	$\text{rect}_L(x) = \begin{cases} 1 & \lvert x\rvert \le \frac{L}{2} \\ 0 & \text{else} \end{cases}$	$\frac{1}{2\pi} \text{sinc}\left(\frac{kL}{2}\right)$
Dirac delta function	$\delta(x - x_0)$	$\frac{1}{\sqrt{2\pi}} e^{-ikx_0}$
Dirac comb	$\underbrace{\sum_{n=-\infty}^{+\infty} \delta(x - nL)}_{:=\text{III}_L(x)}$	$\frac{\sqrt{2\pi}}{L} \underbrace{\sum_{m=-\infty}^{+\infty} \delta(k - m\frac{2\pi}{L})}_{:=\text{III}_{\frac{2\pi}{L}}(k)}$

– *convolution theorem*: The convolution of two functions f and g is defined as

$$(f * g)(x) := \int d\tau\, f(\tau)\, g(x - \tau) = \int d\tau\, f(x - \tau)\, g(\tau) = (g * f)(x) \tag{2.121}$$

$$\begin{aligned} \Rightarrow \quad (f * g)(x) \quad &\underset{\mathcal{F}^{-1}}{\overset{\mathcal{F}}{\rightleftharpoons}} \quad \sqrt{2\pi}\, \tilde{f}(k) \cdot \tilde{g}(k) \\ f(x) \cdot g(x) \quad &\underset{\mathcal{F}^{-1}}{\overset{\mathcal{F}}{\rightleftharpoons}} \quad \frac{1}{\sqrt{2\pi}} \left(\tilde{f} * \tilde{g}\right)(k) \end{aligned} \tag{2.122}$$

Convolution of two functions in real space is equivalent to multiplication in Fourier space, and vice versa (Figure 2.25).

– *correlation theorem*: The cross correlation of two functions f and g is defined as

$$(f \otimes g)(x) := \int d\tau\, f(\tau)\, g^*(x + \tau) = \int d\tau\, f(\tau - x)\, g^*(\tau)$$

$$\Rightarrow \quad (f \otimes g)(x) \quad \underset{\mathcal{F}^{-1}}{\overset{\mathcal{F}}{\rightleftharpoons}} \quad \sqrt{2\pi}\, \tilde{f}(k) \cdot \tilde{g}^*(k) \tag{2.123}$$

For $f = g$, we have the autocorrelation of a function f. Fourier transformation yields the squared modulus of the Fourier transform of f, which is the PSD as introduced in Section 2.6.4:

$$\mathcal{F}[f \otimes f](k) = \sqrt{2\pi}\, \underbrace{\left|\tilde{f}(k)\right|^2}_{\text{PSD}} \tag{2.124}$$

– *Plancherel's theorem*: Inverse Fourier transform of equation (2.124) and setting $x = 0$ yields

$$\int dx\, |f(x)|^2 = \int dk\, |\tilde{f}(k)|^2 \,. \tag{2.125}$$

Fourier theorems are often simple to prove. For example, the proof of the derivative theorem, which is important, e.g., in filtered backprojection of CT data (Chapter 4), takes just one line. The Fourier pairs involving the Dirac δ function or the Dirac comb

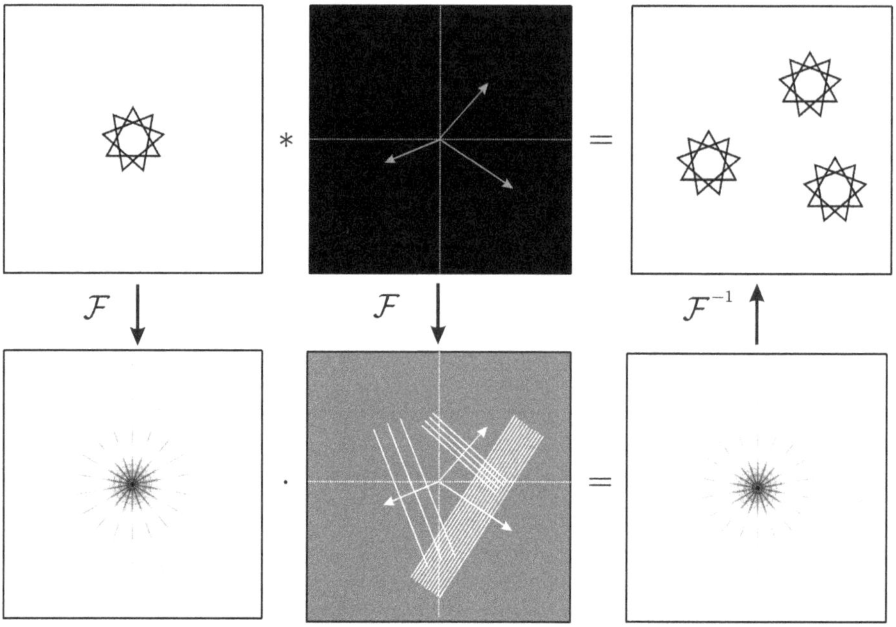

Fig. 2.25: Illustration of the convolution theorem. Convolution of an object (here: a star) with a delta function places the object at the position of the delta function (*top row*). The same result is obtained by multiplying the Fourier transforms (*bottom row*) and performing an inverse Fourier transform back to real space.

(as listed in Table 2.2) may require more work from a mathematical point of view, including in particular the theory of distributions, but all one would need here to establish the relationships with correct prefactors follows from the sifting property of the δ-function [9]

$$\int dx\, \delta(x - x_0) f(x) = f(x_0)\,. \tag{2.126}$$

Finally we should stress that the proper mathematical definition of Fourier transformation would require us to be more specific with respect to the functional spaces. The space of integrable functions $\mathcal{L}^1(\mathbb{R})$ is a natural candidate to declare the Fourier transform, but the Fourier transform does not form a bijection on this space, questioning the concept of Fourier pairs. For other properties stated above, such as the derivative theorem, the function obviously has to be differentiable. The so called Schwartz space $\mathcal{S}$ is a suitable space of well-behaved (in particular differentiable and rapidly decaying) functions that is preserved under the Fourier transform, and enables band limited signals in both source and target space.

2.8 Digital data and sampling

The purpose of this subsection is to give a short overview of discrete signal sampling and the discrete Fourier transformation, following [9] as a very useful reference, which presents the topic in more detail but is still quite concise and from a perspective of practical numerical work. For a more complete treatment, we refer to introductory mathematical textbooks, in particular [2, 4, 14].

2.8.1 Sampling continuous functions

The measurement of a continuous physical signal $f: x \longrightarrow f(x) \in \mathbb{C}$ brings about some difficulties (Figure 2.26):

1. Only a finite set of function values $\{f(x_n)\}$ at discrete (and usually equidistant) points $\{x_n\}$ is measured. This process is known as sampling. Both the fact that the total number of samples N is finite and, therefore, may not cover the entire domain of the function, as well as the fact that the sampling interval Δx is finite raise questions of how well the functions can be recovered.
2. Each sample $f(x_n)$ may either correspond to a single point of the signal, or as is mostly the case for imaging detectors to the mean over a finite interval $I = [x_n - \Delta/2, x_n + \Delta/2]$ in x, e.g., given by the dimensions of a pixel of a CCD. We denote this by

$$\langle f \rangle_I := \frac{1}{\Delta x} \int_{x_n-\Delta/2}^{x_n+\Delta/2} dx\, f(x) \tag{2.127}$$

3. The function values are discretized. Firstly, the analog signal $f(x)$ is converted, for example, into a (still analog) voltage by some sensor. Secondly, this voltage is converted into a digital signal described by a sequence of binary numbers by an ADC.

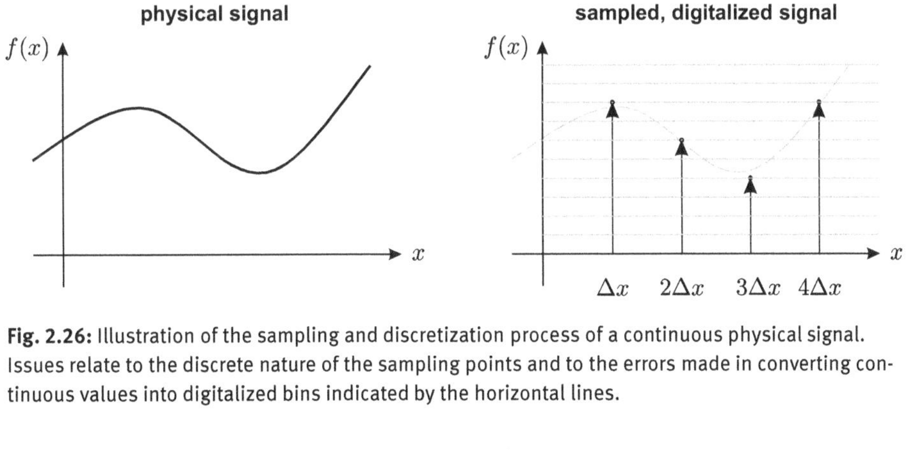

Fig. 2.26: Illustration of the sampling and discretization process of a continuous physical signal. Issues relate to the discrete nature of the sampling points and to the errors made in converting continuous values into digitalized bins indicated by the horizontal lines.

Thus, not only the sampling positions $\{x_n\}$, but also the corresponding (averaged) function values $\{f(x_n)\}$ can only adopt discrete values. One has to assure that this discretization is fine enough, so that the digitized values sufficiently approximate the continuous signal (note that electromagnetic signals are inherently discrete due to the quantum nature of photons.)

4. In addition, the 'true' signal is corrupted by noise, e.g., Poisson noise in the case of photons (cf., Section 2.6.3).

In view of sampled signals, the central question is the following: How fine do we have to sample to reconstruct f? As we will see, the answer depends on the 'smoothness' of the signal!

Next, we derive a quantitative criterion for sampling, assuming ideal sampling. We hence ignore the latter three of the above issues for a start, and assume that the values of a function f are known exactly at an infinite number of discrete and equidistant points $x_n = n\Delta x$, $n \in \mathbb{N}$. This ideal sampling process is described mathematically by multiplication with a Dirac comb $\text{III}_{\Delta x}(x)$ (Table 2.2):

$$S_{\Delta x}[f](x) := \text{III}_{\Delta x}(x) \cdot f(x) = \sum_n f(n\Delta x)\delta(x - n\Delta x) , \tag{2.128}$$

where $S_{\Delta x}[f]$ denotes the sampling operator. In addition, we introduce the periodization operator $P_L[f]$, defined by convolution of the function f with a Dirac comb

$$P_L[f](x) := \text{III}_L(x) * f(x) = \sum_n f(x - nL) , \tag{2.129}$$

and yielding an infinite sum of copies of f, each translated by nL. For arbitrary f and L, the translated copies of f will overlap. However, if f vanishes outside an interval $[-b, b]$ and if $L > 2b$, the copies do not overlap, and $P_L[f](x) = f(x)$ for $x \in [-L/2, +L/2]$. Hence, only for the latter case can we invert the operator P_L. With these definitions at hand, we now compute the Fourier transform of the sampled signal, which in short hand notation becomes

$$\mathcal{F}[S_{\Delta x}[f]] = \mathcal{F}[f \cdot \text{III}_{\Delta x}] = \frac{1}{\sqrt{2\pi}}\tilde{f} * \widetilde{\text{III}}_{\Delta x} = \frac{1}{\Delta x}\tilde{f} * \text{III}_{\frac{2\pi}{\Delta x}} = \frac{1}{\Delta x} P_{\frac{2\pi}{\Delta x}}[\tilde{f}] , \tag{2.130}$$

using first the definition of the sampling operator, then the convolution theorem, followed by the Fourier pair property of the Dirac comb, and finally the definition of P_L. The result shows that the sampling operation in real space corresponds to convolution in Fourier space with a Dirac comb, i.e., to a periodization of $\tilde{f}$.

Reciprocally, we can also carry out a Fourier transformation of the periodized signal. In summary, the Fourier transforms of the ideally sampled and the periodized function are

$$\mathcal{F}[S_{\Delta x}[f]] = \mathcal{F}[f \cdot \text{III}_{\Delta_x}] = \frac{1}{\sqrt{2\pi}} \cdot \tilde{f} * \widetilde{\text{III}}_{\Delta x} = \frac{1}{\Delta x} \cdot \tilde{f} * \text{III}_{\frac{2\pi}{\Delta x}} = \frac{1}{\Delta x} \cdot P_{\frac{2\pi}{\Delta x}}[\tilde{f}] , \tag{2.131}$$

$$\mathcal{F}[P_L[f]] = \mathcal{F}[f * \text{III}_L] = \sqrt{2\pi} \cdot \tilde{f} \cdot \widetilde{\text{III}}_L = \frac{2\pi}{L} \cdot \tilde{f} \cdot \text{III}_{\frac{2\pi}{L}} = \frac{2\pi}{L} \cdot S_{\frac{2\pi}{L}}[\tilde{f}] . \tag{2.132}$$

Hence, *(ideal) sampling in real space with sampling period Δx corresponds to periodization in reciprocal space with period $\frac{2\pi}{\Delta x}$, and vice versa.* The corresponding Fourier pairs are

$$S_{\Delta x}[f] \underset{\mathcal{F}^{-1}}{\overset{\mathcal{F}}{\rightleftharpoons}} \frac{1}{\Delta x} P_{\frac{2\pi}{\Delta x}}[\tilde{f}] \quad \text{and} \quad P_L[f] \underset{\mathcal{F}^{-1}}{\overset{\mathcal{F}}{\rightleftharpoons}} \frac{2\pi}{L} S_{\frac{2\pi}{L}}[\tilde{f}] \,. \tag{2.133}$$

2.8.2 The sampling theorem

We discuss the nature of the sampling operation from the following point of view: We can reconstruct f from the sampled data, if we can reconstruct its Fourier transform $\tilde{f}$ from the Fourier transform of the sampled data, i.e., from $P_{\frac{2\pi}{\Delta x}}[\tilde{f}]$. This is possible only if $\tilde{f}$ can be "cut out" from the periodized Fourier transform, i.e., if the translated copies do not overlap, see Figure 2.27. This can only be the case if firstly the signal f is band limited, i.e., $\exists\, b$ with $\tilde{f}(k) = 0 \;\forall\, |k| > b$, with $b \in \mathbb{R}$, and secondly, the sampling rate is high enough, i.e., when

$$\frac{2\pi}{\Delta x} > 2b \,. \tag{2.134}$$

In this case, the Fourier transforms of the original function $\tilde{f}$ and of the sampled function $P[\tilde{f}]$ are equal over the entire domain of $\tilde{f}$, i.e., the interval $[-b, +b]$. Contrarily, for $\frac{2\pi}{\Delta x} < 2b$, the translated copies of $\tilde{f}$ overlap, giving rise to 'flaws' in signal recovery, denoted by aliasing effects. Technically, cutting out the signal in Fourier space is achieved using an ideal low pass filter represented by a rectangular function $\mathrm{rect}_{\frac{2\pi}{\Delta x}}(k)$ to truncate the periodized Fourier transform,

$$\tilde{f}(k) = \mathrm{rect}_{\frac{2\pi}{\Delta x}}(k) \cdot P_{\frac{2\pi}{\Delta x}}[\tilde{f}](k) = \Delta x \cdot \mathrm{rect}_{\frac{2\pi}{\Delta x}}(k) \cdot \mathcal{F}\left[S_{\Delta x}[f]\right](k) \,. \tag{2.135}$$

The inverse Fourier transform of this equation yields an expression for the original function $f(x)$ (the full calculation is presented in, for example, [9])

$$f(x) = \sum_{n=-\infty}^{+\infty} f(n\Delta x)\, \mathrm{sinc}\left(\frac{\pi}{\Delta x}(x - n\Delta x)\right) \tag{2.136}$$

where $\mathrm{sinc}(x) = \sin(x)/x$. Most importantly, this procedure allows for an *exact* reconstruction of the signal $f(x)$, provided that it satisfies the criteria stated above. This result is also known as the

Nyquist–Shannon sampling theorem: *Let $f(x)$ be a band limited, possibly complex valued function, i.e., $\exists\; b \in \mathbb{R}^+ \,|\, \tilde{f}(k) = 0 \;\forall\; |k| > b$. Assume that f is only known at the discrete sampling positions $n\Delta x$, where $n \in \mathbb{Z}$. If the sampling period Δx is chosen in such a way that*

$$\frac{1}{\Delta x} \geq \frac{b}{\pi} \,,$$

the continuous function f is fully determined by its sampled values $f(n\Delta x)$.

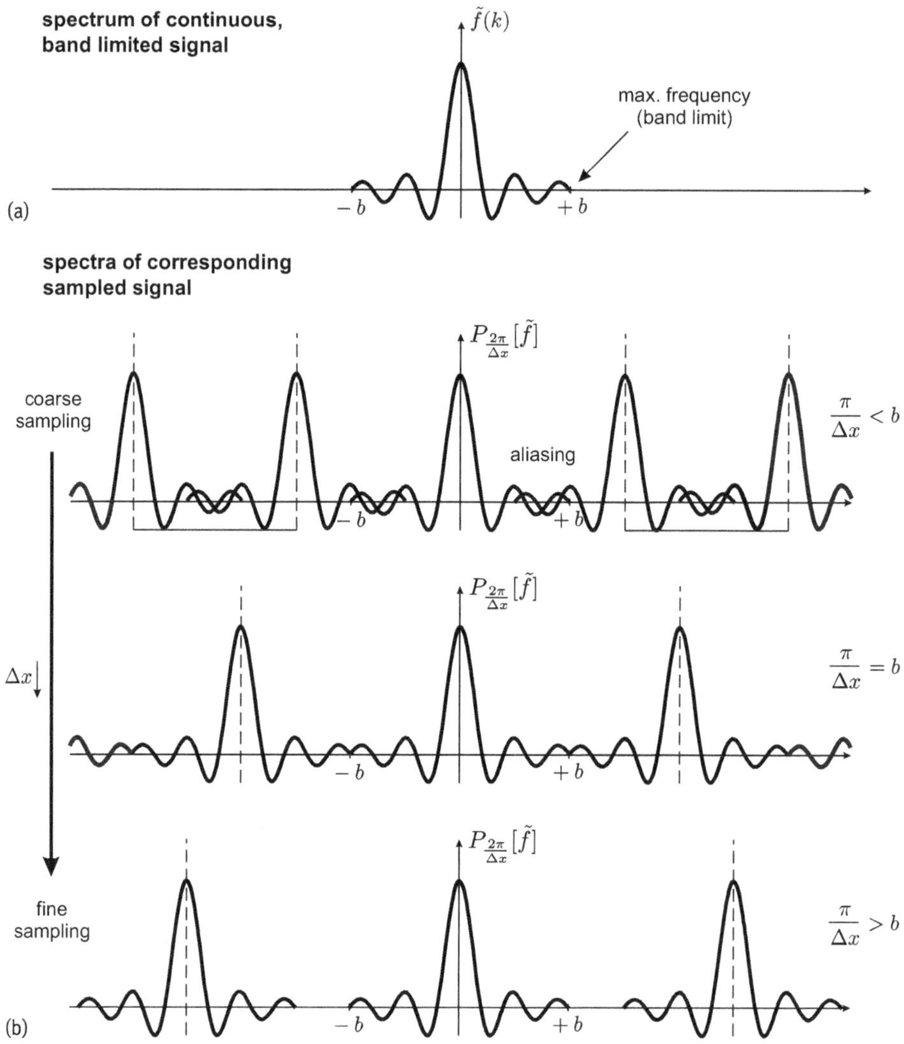

Fig. 2.27: (a) Fourier transform (or spectrum) $\tilde{f}(k)$ of a band limited signal. The spectrum is bounded by the band limit b (the signal $f(x)$ itself is not shown). (b) Sampling the signal $f(x)$ at positions $n\Delta x$, $n \in \mathbb{N}$ corresponds to periodization of $\tilde{f}$ in the spectral domain, and the distance of the shifted copies of $\tilde{f}$ depends on the sampling rate $1/\Delta x$. If $1/\Delta x < b/\pi$ (Nyquist rate), the copies of $\tilde{f}$ overlap and information is lost. In particular, Fourier components of high spatial frequencies are transported to smaller spatial frequencies, distorting the signal by aliasing.

For a given signal, the smallest sampling rate Δx^{-1} that suffices this condition is called the

$$\textit{Nyquist rate} \quad r_{\mathrm{Ny}} := \frac{b}{\pi} \,. \tag{2.137}$$

For a given sampling rate Δx^{-1}, the highest frequency a function f is allowed to have is the

$$\textit{Nyquist frequency} \quad k_{\mathrm{Ny}} = \frac{\pi}{\Delta x} \qquad \left(\text{or } \nu_{\mathrm{Ny}} = \frac{1}{2\Delta x}\right) \,. \tag{2.138}$$

The sampling theorem is quite remarkable from the point of view of information theory. If certain requirements are met, the *complete* determination of a continuous function is possible from its values at discrete sample positions. There is one concern left: An infinite number of samples $f(n\Delta x)$ is required for an exact reconstruction of the continuous function f, since a band limited function can never have a bounded support (or, vice versa, a function with bounded support can never be band limited). Mathematical resort can be provided by the concept of Schwartz space, i.e., function spaces with sufficiently rapid decay in real and reciprocal space, extending the concept of the band limited signal. In practice, some smooth truncation is always required.

2.8.3 Discrete Fourier transformation (DFT)

So far, we have considered the continuous Fourier transform. However, in most cases the signal is not known as a continuous, analytical function f, but only at a finite number N of discrete points (in time or space). This corresponds to the sampling process outlined above. The sampled values can be written as a signal sequence or signal vector

$$\underline{f}_N = \{f_0, f_1, f_2, \ldots, f_{N-1}\} \,. \tag{2.139}$$

If the sampling points x_n are equidistantly spaced, it suffices to store only the values f_N and note the sampling interval Δx. We saw that discrete sampling introduces periodization in Fourier space. In order to define a transform on a finite signal vector (in real and reciprocal space) we have to consider it as part of an infinite and periodic signal vector. Of course, it is sufficient to specify only one period ('unit cell'). This is achieved by the DFT. Excellent textbooks treat the subject of the DFT and its numerically efficient implementation in terms of the Fast Fourier Transform (FFT), see [2, 4, 14]. The present section largely follows the treatment of the subject by [4]. In essence, we need a numerical recipe to establish the "discrete Fourier pair"

$$\underline{f}_N = \{f_0, f_1, f_2, \ldots, f_{N-1}\} \quad \underset{\mathrm{iDFT}}{\overset{\mathrm{DFT}}{\rightleftharpoons}} \quad \underline{F}_N = \{F_0, F_1, F_2, \ldots, F_{N-1}\} \tag{2.140}$$

connected via DFT and inverse discrete Fourier transform (iDFT) as an approximation for the continuous case. As we will see, the following definitions meet our requirements.

DFT:

$$\underline{F}_N = \mathrm{DFT}[\underline{f}_N] := \left(\frac{1}{\sqrt{N}} \sum_{n=0}^{N-1} f_n \omega_N^{-nm} \right)_{m=0,1,\dots,N-1} \quad \text{where} \quad \omega_N := e^{\frac{2\pi i}{N}} \tag{2.141}$$

iDFT:

$$\underline{f}_N = \mathrm{iDFT}[\underline{F}_N] := \left(\frac{1}{\sqrt{N}} \sum_{m=0}^{N-1} F_m \omega_N^{+nm} \right)_{n=0,1,\dots,N-1} \quad \text{where} \quad \omega_N := e^{\frac{2\pi i}{N}} \tag{2.142}$$

A factor of the type $e^{\frac{2\pi i}{N}} := \omega_N$ is called the Nth *complex root of unity*, since $\omega_N^N = 1$. In the following, we give a brief summary of the most important properties of the DFT.

Reciprocity

At first, we show that subsequent application of DFT and iDFT yields the original signal, in analogy to the Fourier theorem $\mathcal{F}^{-1}[\mathcal{F}[f]] = f$ (2.100):

$$\begin{aligned}
\mathrm{iDFT}[\mathrm{DFT}[\underline{f}_N]] &= \left(\frac{1}{\sqrt{N}} \sum_{m=0}^{N-1} \left(\mathrm{DFT}[\underline{f}_N]\right)_m \omega_N^{+nm} \right)_{n=0,1,\dots,N-1} \\
&= \left(\frac{1}{\sqrt{N}} \sum_{m=0}^{N-1} \left(\frac{1}{\sqrt{N}} \sum_{n'=0}^{N-1} f_{n'} \omega_N^{-n'm} \right) \omega_N^{+nm} \right)_{n=0,1,\dots,N-1} \\
&= \left(\sum_{n'=0}^{N-1} f_{n'} \cdot \frac{1}{N} \sum_{m=0}^{N-1} \omega_N^{(n-n')m} \right)_{n=0,1,\dots,N-1} \\
&= \left(\sum_{n'=0}^{N-1} f_{n'} \cdot \delta_{n,n'} \right)_{n=0,1,\dots,N-1} \\
&= (f_n)_{n=0,1,\dots,N-1} \\
&= \underline{f}_N \,.
\end{aligned} \tag{2.143}$$

Here, we have used the following property of the ω_N:

$$\sum_{m=0}^{N-1} \omega_N^{(n-n')m} = N\delta_{n',n} \quad \text{where} \quad \delta_{n',n} = \begin{cases} 1 & \text{if } n' = n \\ 0 & \text{else} \end{cases} \tag{2.144}$$

This holds, since the terms in the sum are evenly distributed on the unit circle in the complex plane. Only for $n = n'$, they all point into the same direction and, thus, do not cancel out.

Sampling intervals Δx and Δk

Secondly, we note that the discrete Fourier pair $\underline{f}_n \rightleftharpoons \underline{F}_N$ defined by equation (2.141) and (2.142) comes without information about the intervals Δx and Δk in real and reciprocal space. If the vector $\underline{f}_N$ represents samples of a continuous function f at equidistant points $n\Delta x$, what is the corresponding interval Δk of the entries of its DFT $\underline{F}_N$? We have

$$f_n \mathrel{\widehat{=}} f(n\Delta x) \qquad F_m \mathrel{\widehat{=}} \tilde{f}(m\Delta k) \tag{2.145}$$

and thus (disregarding constant prefactors; for a more detailed derivation see [9])

$$\tilde{f}(m\Delta k) = \int dx\, f(x)\, e^{-im\Delta kx} \approx \sum_n f(n\Delta x)\, e^{-i(m\Delta k)(n\Delta x)} \stackrel{!}{=} \sum_n f_n\, e^{-2\pi inm/N} . \tag{2.146}$$

By comparison of the latter two expressions, we see that the intervals in real and reciprocal space are related via

$$\Delta x \Delta k = \frac{2\pi}{N} . \tag{2.147}$$

Matrix representation

The DFT can be considered as a linear operator and written in matrix form

$$\underbrace{\begin{pmatrix} F_0 \\ F_1 \\ F_2 \\ \vdots \\ F_{N-1} \end{pmatrix}}_{\underline{F}_N} = \frac{1}{\sqrt{N}} \underbrace{\begin{pmatrix} 1 & 1 & 1 & \cdots & 1 \\ 1 & \omega_N^{-1} & \omega_N^{-2} & \cdots & \omega_N^{-(N-1)} \\ 1 & \omega_N^{-2} & \omega_N^{-4} & \cdots & \omega_N^{-2(N-1)} \\ \vdots & \vdots & \vdots & \ddots & \vdots \\ 1 & \omega_N^{-(N-1)} & \omega_N^{-2(N-1)} & \cdots & \omega_N^{-(N-1)(N-1)} \end{pmatrix}}_{:=\mathbf{F}} \cdot \underbrace{\begin{pmatrix} f_0 \\ f_1 \\ f_2 \\ \vdots \\ f_{N-1} \end{pmatrix}}_{\underline{f}_N} . \tag{2.148}$$

The inverse DFT is given by

$$\underline{f}_N = \frac{1}{\sqrt{N}} \mathbf{F}^{-1} \cdot \underline{F}_N = \frac{1}{\sqrt{N}} \mathbf{F}^{*} \cdot \underline{F}_N . \tag{2.149}$$

Periodicity

The sampled function is only known on a finite sampled interval of length $L = N\Delta x$. The DFT automatically leads to a periodic spectrum. We illustrate this for the 1d case by extending the allowed values of m by an integer multiple $l \in \mathbb{Z}$ of N

$$\begin{aligned} F_{m+lN} = \left(\mathrm{DFT}[\underline{f}_N]\right)_{m+lN} &= \frac{1}{\sqrt{N}} \sum_{n=0}^{N-1} f_n\, \omega_N^{-n(m+lN)} \\ &= \frac{1}{\sqrt{N}} \sum_{n=0}^{N-1} f_n\, \omega_N^{-nm} \underbrace{e^{-2\pi inl}}_{=1\ \forall\, n,l \in \mathbb{Z}} \\ &= \left(\mathrm{DFT}[\underline{f}_N]\right)_m \\ &= F_m . \end{aligned} \tag{2.150}$$

Vice versa, its inverse leads to a periodic signal in real space:

$$\begin{aligned} f_{n+lN} = (\mathrm{iDFT}[\underline{F}_N])_{n+lN} &= \frac{1}{\sqrt{N}} \sum_{m=0}^{N-1} F_m\, \omega_N^{+(n+lN)m} \\ &= \frac{1}{\sqrt{N}} \sum_{m=0}^{N-1} F_n\, \omega_N^{+nm} \underbrace{e^{-2\pi i l m}}_{=1\ \forall\, l,m \in \mathbb{Z}} \\ &= (\mathrm{iDFT}[\underline{F}_N])_n \\ &= f_n \,. \end{aligned} \tag{2.151}$$

Therefore, the DFT does not simply relate the sampled version of a function and the sampled version of its Fourier transform, but their periodized versions. However, it is usually the case that the original signal $f(x)$ is *not* periodic. In addition, even if $f(x)$ is periodic, one would have to know the period in advance in order to sample an integer number of intervals. The periodicity of the DFT and its inverse thus requires some special attention. For example, to achieve sufficient sampling, it may be required to embed the series $\underline{f}_N$ in a large number of zeros (zero padding). The use of a window function (apodization) may also be important in order to evaluate the spectrum on a logarithmic scale and to avoid the otherwise dominating contributions from the boundaries (edges). If the sampling conditions are not fulfilled, artifacts are observed (leakage, aliasing), as explained in more detail in the literature.

Further properties

The DFT shares many properties with its continuous counterpart, such as linearity and shifting behavior. Importantly, operations like convolution and cross/autocorrelation can also be defined for the discrete case in a straightforward way. Likewise, derivatives and Plancherel's theorem have their counterparts.

Fast Fourier transform (FFT)

Considering the matrix formulation equation (2.148), one realizes that the DFT of a signal sequence of length N requires N^2 multiplications and $N(N-1)$ additions. The total number of elementary operations is thus of order $\mathcal{O}(N^2)$. However, this can be considerably reduced by the fast Fourier transform algorithm (FFT). In its original form proposed by Cooley and Tukey, the DFT of a signal of size $N = 2^n\ n \in \mathbb{N}$ is recursively split into DFTs of size $N/2$, ultimately resulting in $\mathcal{O}(N \log_2 N)$ operations. For example, for $N = 1024 = 2^{10}$, this reduces computation time by a factor of ≈ 100. Due to the enormous relevance of the Fourier transform in almost any field of signal processing, FFT is ranked among the most important algorithms of our time.

Discrete Fourier transform in multiple dimensions

Just like the continuous Fourier transform, also the discrete (and fast) Fourier transform can be extended to multiple dimensions d. For two-dimensional images of size $N \times M$, this reads

$$F_{kl} = \mathrm{DFT}[f_{mn}]_{kl} = \frac{1}{\sqrt{MN}} \sum_{m=0}^{M-1} \sum_{n=0}^{N-1} f_{mn}\, e^{-2\pi i\left(\frac{mk}{M}+\frac{nl}{N}\right)} \tag{2.152}$$

$$f_{mn} = \mathrm{iDFT}[F_{kl}]_{mn} = \frac{1}{\sqrt{MN}} \sum_{k=0}^{M-1} \sum_{l=0}^{N-1} F_{kl}\, e^{+2\pi i\left(\frac{mk}{M}+\frac{nl}{N}\right)} \tag{2.153}$$

References

[1] H. H. Barrett and K. J. Myers. *Foundations of Image Science.* Wiley, 2003.

[2] E. O. Brigham. *The Fast Fourier Transform and its Applications.* Prentice-Hall, 1974.

[3] R. Brigola. *Fourier-Analysis und Distributionen: Eine Einführung mit Anwendungen.* Edition SWK, 2013.

[4] T. Butz. *Fouriertransformation für Fußgänger.* Vieweg+Teubner Verlag / Springer Fachmedien Wiesbaden GmbH, Wiesbaden, 2012.

[5] J. Canny. A Computational Approach to Edge Detection. *IEEE Transactions on Pattern Analysis and Machine Intelligence*, 8(6):679–714, 1986.

[6] D. R. Dance, S. Christofides, A. D. A. Maidment, I. D. McLean, and K. H. Ng. *Diagnostic Radiology Physics – A Handbook for Teachers and Students.* International Atomic Energy Agency (IAEA), 2014.

[7] O. Dössel. *Bildgebende Verfahren in der Medizin – Von der Technik zur medizinischen Anwendung.* Springer Vieweg, 2nd edition, 2016.

[8] O. Dössel and T. M. Buzug. *Biomedizinische Technik – Biomedizinische Bildgebung.* DeGruyter, 2014.

[9] K. Giewekemeyer. *A study on new approaches in coherent x-ray microscopy of biological specimens.* PhD thesis, Universität Göttingen, 2011.

[10] P. V. C. Hough. Method and means for recognizing complex patterns. *US Patent 3069654*, 1962.

[11] B. Jähne. *Digitale Bildverarbeitung und Bildgewinnung.* Springer, 7th edition, 2012.

[12] R. A. Kirsch. Computer determination of the constituent structure of biological images. *Computers and Biomedical Research*, 4(3):315–328, 1971.

[13] T. Lehmann, W. Oberschelp, E. Pelikan, and R. Repges. *Bildverarbeitung für die Medizin.* Springer, 1997.

[14] A. Neubauer. *DFT – Diskrete Fourier-Transformation: Elementare Einführung.* Springer Vieweg, 2012.

[15] M. van Heel. Similarity measures between images. *Ultramicroscopy*, 21(1):95–100, 1987.

[16] M. van Heel and M. Schatz. Fourier shell correlation threshold criteria. *J. Struct. Biol.*, 151(3):250–262, 2005.

Symbols and abbreviations used in Chapter 2

B	black point of an image (lowest occurring gray value)
Br	brightness
CNR	contrast-to-noise ratio
D_x, D_y	differentiation filter kernels
DFT	discrete Fourier transform
DQE	detective quantum efficiency
$\delta(x)$	Dirac delta function (1d)
$\Delta = \frac{\partial^2}{\partial x^2} + \frac{\partial^2}{\partial y^2}$	2d Laplacian operator
FRC	Fourier ring correlation
FOV	field of view
FWHM	full width at half maximum
$\mathcal{F}, \mathcal{F}^{-1}$	continuous and inverse continuous Fourier transform
g	gray value
$\underline{G}$	vector of gray values
$\mathcal{H}$	operator representing an imaging system
$\text{III}_L(x)$	Dirac comb with period L
iDFT	inverse discrete Fourier transform
J_1	Bessel function of the first kind and first order
k	spatial frequency
K_M	Michelson contrast
K_W	Weber contrast
$K^{(i)}$	Kirsch filters
L	Laplacian filter kernel
LSI	linear, shift-invariant systems
MTF	modulation transfer function
NA	numerical aperture
OTF	object transfer function
P_L	periodization operator with period L
PSF	point spread function
PSD	power spectral density
rect(x)	rectangular function
$S_{\Delta x}$	sampling operator with period Δx
$S_x, S_y, S_/, S_\backslash$	Sobel filters
SNR	signal-to-noise ratio
$\mathcal{S}(\mathbb{R}^n)$	Schwartz space
$\Theta(x)$	Heaviside step function
W	white point of an image (highest occurring gray value)

3 Essentials of medical x-ray physics

Historically, x-rays provided the first noninvasive probe of the interior body. Soon after their discovery in 1895 by W. C. RÖNTGEN, x-ray imaging in form of radiography revolutionized medical diagnostics. Radiography is based on shadow imaging, or more precisely projection imaging. Image gray values result from the different attenuation of rays traversing tissues. In other words, in its conventional form, radiography is based on absorption contrast. This chapter presents fundamental properties of x-rays, addressing the generation of x-rays, the interaction processes of x-rays and matter, as well as the detection of x-rays. This provides a basis to understand more advanced techniques like tomography or phase contrast radiography, which will be presented in subsequent chapters.

3.1 Electromagnetic spectrum

The electromagnetic spectrum from radio waves to gamma rays is sketched in Figure 3.1. As we know, the human eye can perceive only a minute spectral range, approximately the wavelength interval between 400 and 800 nm. Visible light is bordered by infrared ($\lambda > 800$ nm) and ultraviolet ($\lambda < 400$ nm) light. Further towards smaller wavelengths or correspondingly higher photon energies E, we have the spectral ranges of extreme ultraviolet (EUV), soft x-rays and, finally, hard x-rays. While absorption in the EUV and soft x-ray range ($E < 1$ keV) is strong even in air, matter quickly becomes transparent in the multi-keV photon energy range. In x-ray microscopy, the so called water window in between the absorption edges of carbon at 284 eV and oxygen at

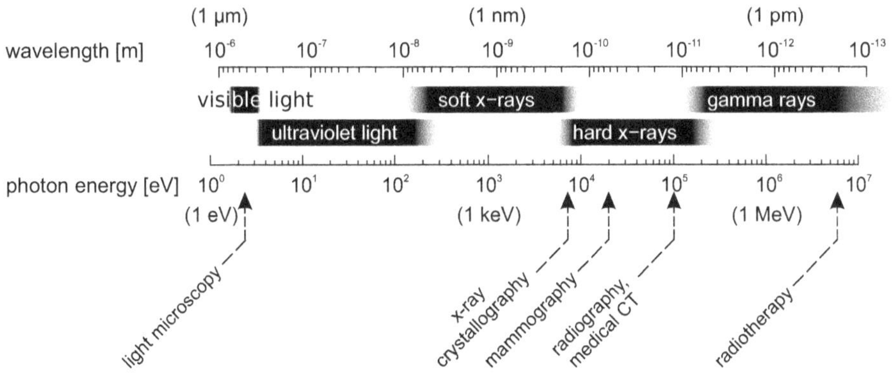

Fig. 3.1: Electromagnetic spectrum from visible light (narrow band between about 400 and 800 nm) to gamma rays. Energies around 100 keV are used in medical diagnostics, energies on the order of several MeV in radiotherapy.

DOI 10.1515/9783110426694-003

543 eV plays an important role. Radiation with wavelengths around $\lambda \simeq 0.1$ nm $= 1$ Å, or roughly $E \simeq 10$ keV is common in crystallography, x-ray diffraction and high resolution tomography of tissues, while medical diagnostic imaging requires higher photon energies, 20 to 120 keV. Photons with energies above 500 keV, as created by linear accelerators in radiotherapy, are sometimes referred to as ultra hard x-rays. However, the limits between soft, hard and ultra hard x-rays are not strictly defined. Since ultra hard x-rays are in the typical energy range of gamma radiation generated by transitions of radioactive nuclei, these two terms are often used interchangeably. Strictly speaking, photons generated by processes in the electron shell are called x-rays, while photons generated by nuclear transitions are called gamma rays.

For the sake of notational clarity, we list the terms for the description of photons according to quantum optics. Let ν denote the frequency, λ the wavelength, $\underline{k}$ the wave vector and $|\underline{k}| = k = \frac{2\pi}{\lambda}$ the wave number. With Planck's constant $h = 6.626 \cdot 10^{-34}$ Js (with the "reduced" form $\hbar = h/2\pi$) and the speed of light c, the following relations hold for photon energy E and momentum $\underline{p}$:

$$E = h\nu = \frac{hc}{\lambda} = \hbar c k \tag{3.1}$$

$$\underline{p} = \hbar \underline{k}\,, \quad |\underline{p}\,| = \hbar k = \frac{E}{c} \tag{3.2}$$

In units of 1 Å $= 10^{-10}$ m typically used in crystallography and 1 eV $= 1.602 \cdot 10^{-19}$ J, wavelength λ and photon energy E are related by

$$\lambda\,[\text{Å}] = \frac{12.398}{E\,[\text{keV}]}\,. \tag{3.3}$$

3.2 Interactions of x-rays and matter

In the following, we consider interaction processes that can take place if x-ray photons of a given energy E traverse a material of a given atomic composition. We start with the Beer–Lambert law describing the total attenuation of the primary beam intensity as a result of all possible interaction channels, before the different effects causing the attenuation are discussed in detail, in particular in view of their dependence on photon energy and material composition. In medical applications, the predominant interactions are photoelectric effect, Compton (or incoherent) scattering and pair production. Thomson (or elastic) scattering is important in view of its role in beam propagation, since this coherent interaction process also determines the x-ray index of refraction, and hence forms the microscopic basis of the novel phase contrast radiography techniques discussed in Chapter 6. An understanding of all relevant interaction probabilities is indispensable for the optimization of x-ray based medical imaging, radiation protection as well as dose calculation in radiotherapy. To this end, the respective interaction cross sections have to be known for given photon energy E and atomic number Z. Further interaction processes such as resonant scattering or multiple photon absorption play a minor role in biomedical imaging and certainly no role in current clinical applications. Along the same lines, we do not need to consider in great detail the many possible relaxation processes after x-ray photoionization, such as x-ray fluorescence and the Auger effect, which are more important for spectroscopy.

3.2.1 Beer–Lambert law

Consider the experimental situation sketched in Figure 3.2: A parallel and monochromatic photon beam with primary intensity I_0 (number of photons per second) propagates in z direction and impinges onto a thin layer of a given material with thickness dz. By measurement of the intensity I before and behind the sample, one observes that the beam is attenuated by an amount dI proportional to I and dz[1]

$$dI \propto -I\,dz \qquad \Rightarrow \qquad dI = -\mu I\,dz\,. \tag{3.4}$$

The proportionality constant μ is called the total linear attenuation coefficient. Integration yields the Beer–Lambert law

$$I(z) = I_0 \exp(-\mu z)\,. \tag{3.5}$$

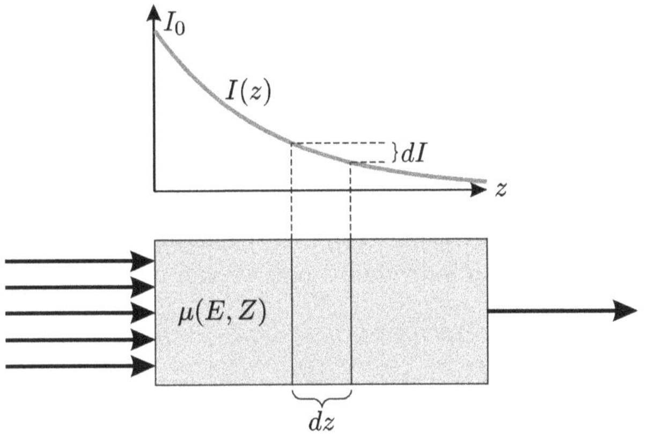

Fig. 3.2: Illustration of the attenuation of the primary beam intensity I_0 along the propagation direction while traversing a material with total linear attenuation coefficient μ. Since the reduction of the number of primary quanta in a thin layer is proportional to its thickness dz and to the number of quanta impinging on the layer, the intensity decays exponentially (Beer–Lambert law). Adapted from [1].

1 Importantly, I only includes the photons of the primary beam that have not interacted with the absorber, but no scattered photons generated in the sample. To assure this experimentally, both primary beam and detector slits must be chosen sufficiently narrow.

Table 3.1: Average elemental composition of several selected tissues [5]. The differences between adipose tissue (high content of hydrocarbon), muscle (high content of protein and thus nitrogen) and bone (high content of calcium and phosphorus) are obvious. The resulting differences in μ allow us to gain insight into the human body by absorption contrast radiography or tomography.

tissue	percentage by mass					$\rho\left[\frac{g}{cm^3}\right]$
	H	C	N	O	others	
adipose tissue	11.4	59.8	0.7	27.8	0.1 Na, 0.1 S, 0.1 Cl	0.95
brain	10.7	14.5	2.2	71.2	0.2 Na, 0.2 S, 0.3 Cl, 0.4 P, 0.3 K	1.04
muscle	10.2	14.3	8.4	71.0	0.1 Na, 0.3 S, 0.1 Cl, 0.2 P, 0.4 K	1.05
cartilage	9.6	9.9	2.2	74.4	0.5 Na, 0.9 S, 0.3 Cl, 2.2 P	1.10
cortical bone	3.4	15.5	4.2	43.5	0.1 Na, 0.3 S, 0.2 Mg, 10.3 P, 22.5 Ca	1.92

Varying the photon energy E and the material of the attenuating layer reveals a pronounced energy and material dependence of the linear attenuation coefficient μ,

$$\mu = \mu(E, Z) , \tag{3.6}$$

where Z denotes the atomic number of the absorbing element. In the case of hard x-rays up to $E \approx 100\,\text{keV}$ and biological tissue, one observes approximately $\mu \propto Z^4E^{-3}$. The variation with Z^4 leads to very different absorption of soft tissue (mostly made up of the light elements carbon, hydrogen, nitrogen and oxygen) and bone (containing considerable fractions of the "heavier" elements calcium and phosphorus, see also Table 3.1). This enabled Röntgen to record the shadow picture (or radiograph) of his wife's hand,[2] followed by the immediate clinical application of x-rays shortly after the discovery. By choice of the photon energy, the contrast can be adapted to the specific application, easily enabled by the power law dependence of μ with E^{-3}. The attenuation in a given material must be proportional to the number of possible interaction centers per unit volume, i.e., to the density of atoms. To separate this material property from the intrinsic properties of an individual atom, we can write the total linear attenuation coefficient μ as

$$\mu = \rho_a \sigma_a = \left(\rho \frac{N_A}{A}\right) \sigma_a , \tag{3.7}$$

where ρ_a denotes the density of atoms (number per volume). By definition, σ_a denotes the total attenuation cross section of an atom (index a). The number density is expressed by the mass density ρ (which is more directly measurable), using Avogadro's

2 The most famous of Röntgen's radiographies of human hands is actually not of his wife, but of the anatomist A. von Kölliker, recorded on the occasion of Röntgen's first (and only) oral presentation of his discovery [10].

constant $N_A = 6.022 \cdot 10^{23}\ \text{mol}^{-1}$ and the atomic mass number A. Databases[3] typically contain the mass attenuation coefficient μ/ρ.

In biomedical applications, one usually has to consider materials or tissues that are composed of different elements (an exception is, e.g., radiation protection calculations with pure lead, which is typically used as a shielding material). Therefore, one has to work with mean values of different elements weighted according to the given elemental composition. In the case of molecules composed of several elements i (or other materials of fixed stoichiometry), μ is given by

$$\mu = \rho N_A \frac{\sum_i \sigma_{a,i} x_i}{\sum_i A_i x_i} \tag{3.8}$$

where x_i denotes the number of atoms of element i and $\sum_i A_i x_i$ the molecular mass [9]. For tissues, it is more practical to compute (μ/ρ) using the mass percentages w_i (values for some typical human tissues are given in Table 3.1) and mass attenuation coefficients $(\mu/\rho)_i$ of element i [23]:

$$\left(\frac{\mu}{\rho}\right)_{\text{mix}} = \sum_i w_i \left(\frac{\mu}{\rho}\right)_i \tag{3.9}$$

In the energy range up to about 100 keV, which is relevant for diagnostic imaging, the attenuation of the primary intensity is due to three effects: Thomson (or coherent) scattering, Compton (or incoherent) scattering and the photoelectric effect. In addition, at photon energies $E \geq 2m_e c^2 = 2 \cdot 511$ keV (m_e: rest mass of an electron), pair production (i.e., generation of electron–positron pairs) is enabled. Such high photon energies are relevant for radiotherapy since they allow for sufficient dose deposition in regions deep within the body, while keeping skin doses at a tolerable level.

In order to account for these different interaction channels between x-ray photons and matter, the absorption cross section per atom (index a) is written as the sum of partial cross sections

$$\sigma_a = \sigma_a^{\text{p.e.}} + \sigma_a^{\text{Th}} + \sigma_a^{\text{C}} + \sigma_a^{\text{pair}} . \tag{3.10}$$

Each of the partial cross sections σ_a^i is a measure for the probability with which a certain interaction takes place and thus contributes to the attenuation of the primary beam. Figure 3.3 shows a plot of Equation (3.10) for the cases of water and lead for a wide range of photon energies relevant for medical applications. In the following, the physics of each interaction process and the corresponding cross sections $\sigma_a^i(E, Z)$ are discussed in view of their functional dependence on photon energy E and atomic number Z.

To this end, we first need to widen the definition of cross section, so that it allows us not only to describe the probability (or rate) at which photons are lost in the primary

---

3 Several very useful databases that provide interaction cross sections, attenuation coefficients, atomic form factors and so forth are provided by the National Institute of Standards and Technology (NIST) at http://www.nist.gov/pml/data/xray_gammaray.cfm. These were used to create several figures in this book.

beam (absorption), but also to describe the rate at which the photons and the subsequent interaction particles are scattered in a given direction, leading to the concept of a differential cross section. In this concept, the emission direction describing the final state of the interaction can specify either the scattered photons or – if applicable – the secondary particle, such as the photo or Compton electron.

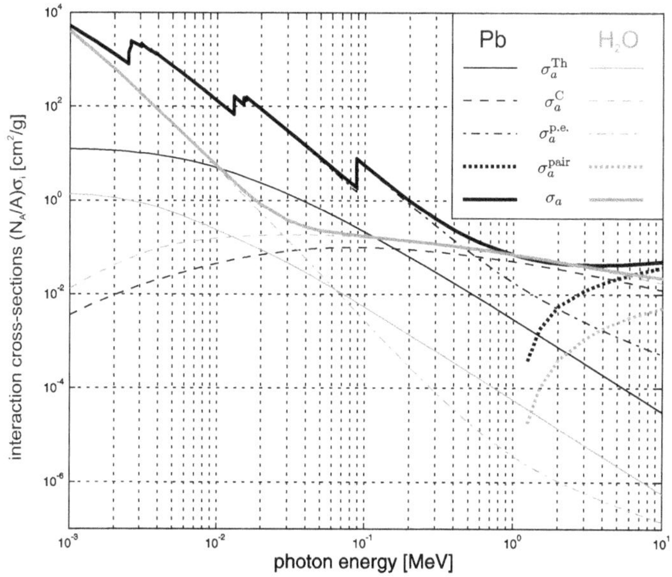

Fig. 3.3: Photon cross sections (divided by mass density ρ) for coherent (Thomson) scattering, incoherent (Compton) scattering, photoelectric effect and pair production in lead (Pb, *black*) and water (H_2O, *gray*), respectively, for energies ranging from 1 keV to 10 MeV. The resulting total interaction cross sections σ_a are indicated by thick lines. In case of lead, several absorption edges can clearly be seen. Pair production becomes possible at $E \geq 2m_e c^2 = 1022$ keV and is the dominant interaction process in lead above 5 MeV.

3.2.2 Differential and integral cross sections

Consider the situation sketched in Figure 3.4. A parallel, monochromatic x-ray beam impinges on a layer containing scattering centers with uniform density ρ_a. The layer is large enough to intercept the entire beam, has a constant thickness Δz, resulting in a surface density of scatterers $\rho_a \Delta z$, which we assume low enough to neglect multiple scattering events or extinction of the beam within the layer Δz. We consider a measurement of the photon fraction scattered into a detector that is placed at the scattering angle θ (with respect to the direction of the incident beam) and azimuthal angle ϕ, subtending a solid angle $\mathrm{d}\Omega = \sin\theta\,\mathrm{d}\theta\,\mathrm{d}\phi$. In the limit of $\mathrm{d}\Omega \longrightarrow 0$, and dividing out the extrinsic parameters of the experiment, we gain information on the microscopic nature of the scattering process.

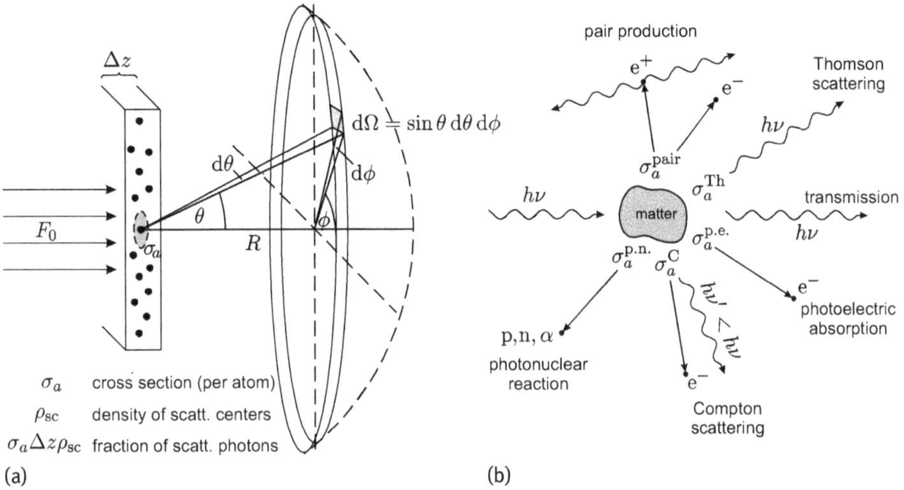

Fig. 3.4: (a) Schematic scattering experiment for the definition of the differential cross section. A thin layer with random distribution of point like scattering centers is exposed to the monochromatic photon beam. Each center "captures" photons in proportion to its cross section characteristic for the process. (b) Schematic illustration of the different interaction channels of a photon beam with matter. The relative strength of each channel is given by its respective cross section.

We start with defining the quantities proportional to the incident number of photons. These are

photon flux $\frac{N_0}{t}$ (number of photons per unit time in the beam)

fluence $\Phi = \frac{N_0}{A}$ (number of photons per unit area perpendicular to the beam)

flux density $F_0 = \frac{N_0}{At}$ (number of photons per unit time and area)

intensity $I_0 = \frac{N_0 h\nu}{At}$ (energy per unit time and area)

The total number of quanta (photons, particles) scattered per unit time into the solid angle $d\Omega$ for a sufficiently thin layer[4] irradiated by the flux density F_0 is proportional to F_0, to the number of scatterers N, and to the solid angle $d\Omega$ (e.g., subtended by a detector pixel). Normalizing by these extrinsic parameters, we obtain an intrinsic quantity, which is characteristic for the interaction on a microscopic level, the differential cross section and its angular integral, the cross section. Its geometric meaning in terms of an area is illustrated in Figure 3.4a. Accordingly, the differential cross sec-

4 In order to measure the cross section, the layer must be thin enough so that F_0 is constant over Δz.

tion per atom (index a) for interaction of type i is defined by

$$\left(\frac{\mathrm{d}\sigma}{\mathrm{d}\Omega}\right)_\mathrm{a}^i = \frac{\text{number of photons scattered per second into } \mathrm{d}\Omega}{F_0\, N\, \mathrm{d}\Omega}\,. \tag{3.11}$$

and is hence a measure of how efficiently an atom in the sample scatters into the direction (θ, ϕ) defined by $\mathrm{d}\Omega$. In most, but not all cases (Thomson scattering!), the cross section depends on the energy E and on the atomic number Z. Integration over the unit sphere yields the integral or total cross section of the type of interaction considered

$$\sigma_\mathrm{a}^i = \int \left(\frac{\mathrm{d}\sigma}{\mathrm{d}\Omega}\right)_\mathrm{a}^i \mathrm{d}\Omega = \int_0^{2\pi} \mathrm{d}\phi \int_0^{\pi} \left(\frac{\mathrm{d}\sigma}{\mathrm{d}\Omega}\right)_\mathrm{a}^i \sin\theta\, \mathrm{d}\theta\,. \tag{3.12}$$

In the case of symmetry with respect to the azimuthal direction, the integration over ϕ is trivial and yields the distribution of scattering angles θ

$$\left(\frac{\mathrm{d}\sigma}{\mathrm{d}\theta}\right)_\mathrm{a}^i = 2\pi \left(\frac{\mathrm{d}\sigma}{\mathrm{d}\Omega}\right)_\mathrm{a}^i \sin\theta\,. \tag{3.13}$$

After these definitions, the next step is to consider how the cross sections can be calculated from first principles. While this can be carried out within the framework of classical electrodynamics for Thomson scattering (as demonstrated in Section 3.2.4), Compton scattering and pair production require advanced quantum mechanics (e.g., [14, 15]).

In first-order perturbation theory, the transition rate (number of transitions per second) between the initial eigenstate $|i\rangle$ and a continuum of final eigenstates $|f\rangle$ of a quantum-mechanical system is given by Fermi's Golden Rule:

$$\lambda_{i\to f} = \frac{2\pi}{\hbar} \left|\langle f|\hat{\mathcal{H}}'|i\rangle\right|^2 \rho(E_f) \tag{3.14}$$

Here, $\hat{\mathcal{H}}'$ denotes the Hamiltonian of the perturbation, $\langle f|\hat{\mathcal{H}}'|i\rangle$ the corresponding matrix element and $\rho(E_f)$ the density of final states. It can be shown that the Hamiltonian $\hat{\mathcal{H}}'$ describing the interaction between photon and electron (neglecting the spin) is given by

$$\hat{\mathcal{H}}' = \frac{e}{m}\left[\underline{\hat{A}}\cdot\underline{\hat{p}} + \frac{e}{2}\underline{\hat{A}}^2\right], \tag{3.15}$$

where $\underline{\hat{A}}$ and $\underline{\hat{p}}$ denote vector potential and momentum in operator form. The first term in the square bracket is linear in the vector potential and describes photoelectric absorption (single photon process), the quadratic term describes elastic scattering (two photon process). Details of the calculations in this framework exceed the scope of this book. For a complete treatment both of the semiclassical model where the particles (electrons) are treated quantum mechanically, interacting with classical fields, as well as the full quantum electrodynamics (QED) where also the fields are quantized, we refer to the respective textbook literature, in particular to HAU-RIEGE and HEITLER [14, 15]. In the following, we focus on the relevant results and their implications for biomedical imaging and medical applications, and provide additional literature references where needed.

3.2.3 Photoelectric absorption

For free electron states, the inner product over the first term in (3.15) vanishes. In a transition between free electron states, the entire momentum and energy of the photon cannot be simultaneously transferred to the electron[5], requiring instead the presence of an atom to take up momentum. Photoabsorption, therefore, requires transitions from the discrete set of bound electron states in the atom (i.e., the occupied orbitals) to an unbound (continuum) state, resulting in photoionization of the atom.[6] The orbitals are referenced in terms of their quantum numbers, with the principal (radial) quantum number n denoting the respective shell. Each shell is subdivided into $(2l + 1)$ slightly different energy levels with the orbital quantum number $l = 0, \ldots, n - 1$, i.e., 1 for the K-shell, 3 for the L-shell, 5 for the M-shell and so forth. Binding energies E_b of the lowest orbitals ($1s \rightarrow$ K-shell) range between a few tens of eV for low Z elements (starting from hydrogen with 13.6 eV) up to $\sim$ 100 keV for heavy (high Z) elements. If the energy of an incident photon exceeds the binding energy, $E_\gamma = \hbar\omega \geq E_\mathrm{b}$, the photon can be absorbed. Its energy is transferred to the electron, which is emitted from the atom with kinetic energy

$$E_\mathrm{e} = \hbar\omega - E_\mathrm{b} \,, \tag{3.16}$$

leaving behind a positively charged ion with a vacancy in an inner shell. This explanation of the photoelectric effect was given by A. Einstein in 1905. It is the dominant interaction between photons and atoms at photon energies up to about 100 keV, with some variation between different materials (Figure 3.3). If $E_\gamma = \hbar\omega \leq E_\mathrm{b}$ for a certain energy level, photoelectric absorption is not possible for these electrons. Successively increasing photon energy, one therefore observes characteristic absorption edges at energies $\hbar\omega = E_\mathrm{b}$, when emission of additional electrons becomes possible. In the case of a K shell edge, the linear attenuation coefficient $\mu(E, Z)$ suddenly increases by approximately one order of magnitude. Binding energies E_b and energy differences between two energy levels are tabulated in databases.[7] As an approximation, one can use the following formulas for the three lowest shells (K,L,M)

$$E_\mathrm{b}^\mathrm{K}(Z) \approx \mathrm{Ry}(Z - 1)^2 \tag{3.17}$$

5 Writing down the momentum and energy balance one obtains the formula for Compton scattering, which requires an outgoing photon to take away some of the incident momentum and energy. Hence an incident photon cannot transfer all of its energy and momentum onto a free electron.

6 Resonant transitions between two bound states are possible for photon energies coinciding with the respective energy differences.

7 For example, http://xdb.lbl.gov/ (x-ray data booklet, Lawrence Berkeley Laboratory) or http://www.nist.gov/pml/data/xraytrans/index.cfm (x-ray transition energies, National Institute of Standards and Technology).

$$E_b^L(Z) \approx \frac{Ry}{4}(Z-5)^2 \quad (3.18)$$

$$E_b^M(Z) \approx \frac{Ry}{9}(Z-13)^2 \quad (3.19)$$

where the effective atomic charge results from a screened nuclear charge Z, and

$$Ry = \frac{m_e e^4}{8c\epsilon_0^2 h^3} = 13.61\,\text{eV} \quad (3.20)$$

denotes the Rydberg energy [23]. Reflecting the fine structure of energy levels, each shell with principal quantum number $n > 1$ results in $2l+1$ closely adjacent absorption edges. For example, in the plot of the total cross section σ_a(Pb) in Figure 3.3, one can distinguish one edge at 88 keV (K-edge), three edges around 15 keV (L-edges) and five around 3 keV (M-edges).[8]

The ionized atom is in an excited state. Within a short lifetime, typically on the order of $\sim 10^{-14}$ s, the inner shell vacancy is filled by a transition of an electron from a less tightly bound outer shell. Importantly, not all transitions between two arbitrary energy levels are possible. The quantum numbers of initial and final state must fulfill certain selection rules to conserve total angular momentum and spin, as treated in atomic physics. As an example, an energy diagram and allowed transitions of copper are shown in Figure 3.6. For the transition from the excited to the ground state, two competing mechanisms exist. Firstly, the binding energy difference (Figure 3.5: difference between K and L or K and M shells, respectively) can be emitted as a photon, which is known as fluorescence. In the case of heavier elements, the fluorescence photons themselves exhibit energies in the x-ray range. They originate from inner shells and not from molecular orbitals as in visible light fluorescence. Accordingly, their presence does not depend on a chemical binding state such as in a chromophore, and they are characteristic for each element, providing information on the chemical stoichiometry of a sample. In spectra of x-ray tubes, they are superimposed as characteristic x-rays on the continuous bremsstrahlung spectrum (Section 3.3).

Alternatively, the electron which fills the vacancy can transfer the binding energy difference to yet another electron that is expelled from the atom (Figure 3.5, right). This radiation-less auto-ionization effect by emission of secondary electrons following the photoelectric effect is known as inner conversion or Auger effect (P. V. AUGER, 1925), but was first discovered by L. MEITNER in 1922. Let N_K denote the probability for emitting a photon per K shell vacancy (fluorescence yield) and $1 - N_K$ the corresponding

8 Near the absorption edge, additional peaks and modulations of σ_a are observed related to the resonant absorption (transitions to unoccupied bound states) forming the near edge x-ray absorption fine structure (NEXAFS). Since the unoccupied states are highly sensitive to molecular coordination and binding states, elemental specification (oxidation states) and chemical binding can be deduced from the absorption spectra. This near edge regime is followed by an oscillatory pattern above the edge, the so called extended x-ray absorption fine structure (EXAFS) resulting from interactions of the photoelectron with neighboring atoms. This can be exploited for structural analysis [1].

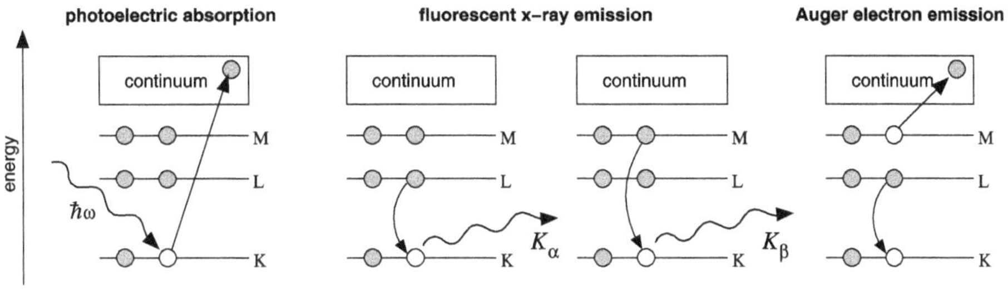

Fig. 3.5: Schematic energy level diagram showing the three lowest shells (K,L,M) of an atom and the continuum of unbound states. For simplicity, the subshells ($L_{I\text{-}III}$, $M_{I\text{-}V}$) are not shown. (*left*) Photoelectric absorption of a photon with sufficiently high energy $\hbar\omega$ leads to ejection of an electron from the atom. The vacancy is filled by transition of a less tightly bound electron to the K shell. The energy difference is either emitted as a photon (fluorescence, *center*) or alternatively transferred to another electron that is expelled from the atom (Auger electron emission, *right*). Adapted from [1].

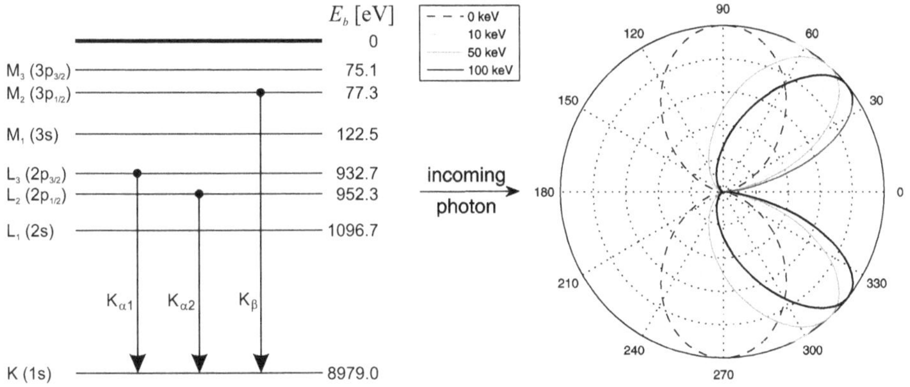

Fig. 3.6: *Left*: Atomic energy levels of copper ($Z = 29$) and allowed transitions leading to emission of characteristic K_α and K_β radiation. *Right*: Angular distribution (equation 3.23) of K shell photoelectrons at different photon energies. Each curve has been normalized to its maximum.

probability for ejecting an Auger electron (Auger yield). An empirical formula for the ratio of both probabilities is

$$\frac{N_\text{K}}{1 - N_\text{K}} = (-6.4 + 3.4Z - 0.000103Z^3)^4 \cdot 10^{-8} \,, \tag{3.21}$$

as plotted in Figure 3.7 (left). For light elements, Auger electron emission is the dominant effect and considerably reduces the fluorescence intensity, posing a challenge for fluorescence based methods. At $Z \approx 30$, fluorescence and Auger effect occur with approximately equal probabilities [23].

Next we give quantitative expressions for the differential cross section describing photon absorption for given photon energy and polarization, and subsequent photoelectron emission. To this end, we differentiate between directions with respect to the

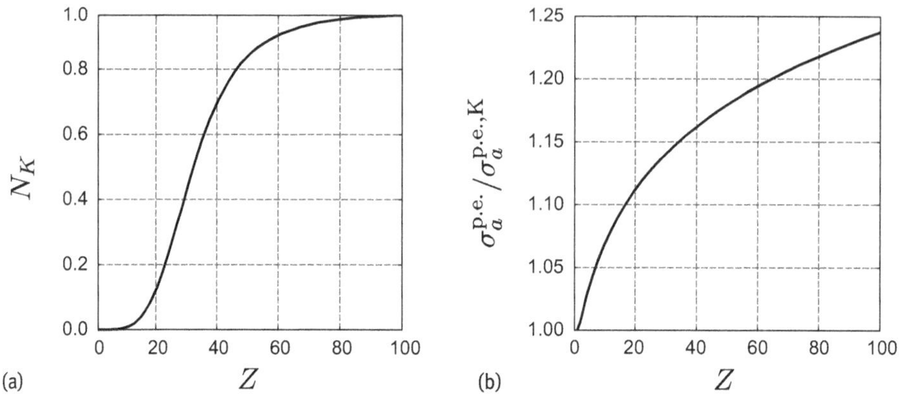

Fig. 3.7: (a) Plot of the fluorescence yield N_K of the K shell as a function of atomic number Z computed from equation (3.21). At low Z, Auger electron emission (with Auger yield $1 - N_K$) is the dominating process and considerably reduces the fluorescence intensity. (b) Plot of equation (3.27). For all Z, photoelectric absorption is dominated by K shell ionization.

incident photon. By integration over the solid angle, the corresponding total absorption cross section can then be computed. The derivations are based on perturbation theory and are quite involved [1, 15, 23]; only the most important results are summarized here. For the nonrelativistic case

$$E_\gamma = \hbar\omega \ll m_e c^2 \quad \text{or} \quad \mathcal{E} := \frac{\hbar\omega}{m_e c^2} \ll 1 , \tag{3.22}$$

but sufficiently far above all absorption edges, the differential cross section for photoelectric absorption and emission of a photoelectron from the K shell is

$$\left(\frac{\mathrm{d}\sigma}{\mathrm{d}\Omega}\right)_{\mathrm{a}}^{\mathrm{p.e.,K}} = 4\sqrt{2} r_0^2 \alpha^4 Z^5 \mathcal{E}^{-7/2} \frac{\sin^2\theta \cos^2\phi}{(1-\beta\cos\theta)^4} , \tag{3.23}$$

where θ denotes the angle between the incoming photon and the emitted photoelectron, ϕ the angle between the scattering plane (spanned by the direction of the incoming photon and the emitted electron) and the polarization of the incident radiation, and $\beta = v/c$ the velocity of the emitted electron divided by the speed of light. The prefactor

$$r_0 = \frac{e^2}{4\pi\epsilon_0 m_e c^2} \tag{3.24}$$

collects the relevant natural constants of the interaction, and is known as the Thomson scattering length or classical electron radius.[9] α denotes the fine structure constant

$$\alpha = \frac{e^2}{4\pi\epsilon_0 \hbar c} \approx \frac{1}{137} . \tag{3.25}$$

9 This name originates from the historical idea that the relativistic rest energy $m_e c^2$ of the electron could be accounted for by the electrostatic self energy of a (hypothetical) spherical charge distribu-

For low energies (UV and x-rays), the term $\beta \cos\theta$ in the denominator of equation (3.23) goes to zero, the emission of photoelectrons follows a dipole pattern with maximum probability along the polarization of the incoming photons. No photoelectrons are emitted in forward direction. As the photon energy increases, electrons are emitted more and more into the forward hemisphere (Figure 3.6, right). Integration over the full solid angle $d\Omega$ and multiplication by 2 (two K-shell electrons of opposite spin) yields the integral cross section for the photo effect with emission of K shell electrons ($\epsilon \ll 1$)

$$\sigma_a^{\text{p.e.,K}} = \sigma_e^{\text{Th}} 4\sqrt{2}\alpha^4 Z^5 \varepsilon^{-7/2} , \tag{3.26}$$

which is written here in terms of $\sigma_e^{\text{Th}} = \frac{8}{3}\pi r_0^2$, known as the classical Thomson scattering cross section, as it also appears in Thomson scattering, see the following section. The calculation for higher shells (L,M,...) is even more complex. However, it turns out that K shell photoelectric absorption dominates the photoelectric effect. For E_γ above K shell binding energies, the ratio of the total photoelectric cross section and K shell photoelectric cross section is given (with a maximum error of ±3%) by [23]

$$\frac{\sigma_a^{\text{p.e.}}}{\sigma_a^{\text{p.e.,K}}} = 1 + 0.01481 \ln^2 Z - 0.000788 \ln^3 Z , \tag{3.27}$$

which is plotted in Figure 3.7 (right). In the case of light elements ($Z < 10$), keeping only the K shell contribution to the absorption cross section while neglecting all higher shells, results in a total error of only a few percent, while this approximation would result in a 25% error for very heavy elements ($Z \approx 100$). We will encounter this predominance of K shell absorption again in x-ray spectra, where the K_α line is by far the most intense characteristic line (Section 3.3). Experimental data deviate slightly from the theoretical result equation (3.23). One reason is that the Coulomb interaction between the photoelectron and the positively charged ion has been neglected. For practical purposes, one often uses the approximation

$$\sigma_a^{\text{p.e.}} \propto \frac{Z^n}{E^3} \tag{3.28}$$

with $n \approx 4.5$ for light elements and $n \approx 4$ for heavy elements. At photon energies E where photoelectric absorption is the dominant interaction process (Figure 3.3), the linear attenuation coefficient, therefore, varies approximately as ($Z \approx \frac{A}{2}$)

$$\mu \approx \left(\rho \frac{N_A}{A}\right) \sigma_a^{\text{p.e.}} \propto \frac{Z^{n-1}}{E^3} . \tag{3.29}$$

This strong variation of x-ray attenuation with atomic number Z is the basis of conventional radiography and computed tomography based on absorption contrast.

---

tion of the electron, $e^2/(4\pi\epsilon_0 R)$, resulting in the expression for r_0 in equation (3.24). However, as we know today, the electron must be considered as a point like particle without internal structure (with a confidence interval of 10^{-19} m!).

3.2.4 Thomson scattering

In this section, we treat x-ray scattering within the framework of classical electrodynamics. We start with a single, free point charge, which oscillates in the electric field of an electromagnetic wave and radiates an electromagnetic field of the same frequency. In the corresponding photon picture, the photon energy does not change upon scattering, i.e., the scattering process is elastic. This requires that the oscillating charge, on average, does not pick up any drift velocity or momentum and is thus spatially fixed. In addition, atomic electrons can be approximated as free only if resonances can be neglected. The limitations of these assumptions will be discussed further below, and the results obtained for a single, free electron will be extended to bound electrons and continuous charge distributions such as atoms. In the derivations, we closely follow [1].

In honor of J. J. THOMSON (who also discovered the electron), elastic scattering by point charges is called Thomson scattering. It accounts only for a small contribution to the overall absorption damping of an x-ray beam (e.g., Figure 3.3), but is important as the fundamental process underlying diffraction and hence structure analysis by interference of elastically scattered radiation. It also plays an important role as the microscopic source of the index of refraction, which is briefly outlined at the end of this section, providing the basis for the concept of x-ray phase contrast.

(i) A single, free electron

Consider the situation sketched in Figure 3.8. A plane incident electromagnetic wave with frequency ω drives a free electron, oscillating around a mean position chosen as the coordinate origin. Electric and magnetic field of the plane wave are given by

$$\underline{E}_{\text{in}}(\underline{r}, t) = \underline{E}_0 e^{i(\underline{k}\cdot\underline{r}-\omega t)} \qquad \text{and} \qquad \underline{B}_{\text{in}}(\underline{r}, t) = \underline{B}_0 e^{i(\underline{k}\cdot\underline{r}-\omega t)} , \tag{3.30}$$

where $\underline{k}$ denotes the wavevector pointing along the propagation direction with wavenumber $k = \omega/c$. From Maxwell's equations, it follows that the wave is transverse, i.e., $\underline{k}, \underline{E}_{\text{in}}, \underline{B}_{\text{in}}$ form a right-handed system of orthogonal vectors and fulfill (e.g., [18])

$$\underline{B}_{\text{in}} = \frac{1}{c}\underline{e}_k \times \underline{E}_{\text{in}} \qquad \Rightarrow \qquad |\underline{E}_{\text{in}}| = c|\underline{B}_{\text{in}}| . \tag{3.31}$$

Due to the Lorentz force, the wave causes an oscillation of the electron described by the following equation of motion:

$$m_{\text{e}}\underline{a} = -e\left(\underline{E}_{\text{in}} + \underline{v} \times \underline{B}_{\text{in}}\right) \tag{3.32}$$

For nonrelativistic oscillation velocities $v \ll c$, the magnetic term can be neglected (equation 3.31). In addition, we choose the coordinate system so that $\underline{E}_0 = E_0\underline{e}_p \parallel \underline{e}_z$, i.e., the incoming wave is polarized in the z direction. The acceleration of the electron

$$\underline{a}(t) = -\frac{e}{m_{\text{e}}}\underline{E}_{\text{in}}(t) = \left(0, 0, -\frac{eE_0}{m_{\text{e}}}\right)^{\text{T}} e^{-i\omega t} \tag{3.33}$$

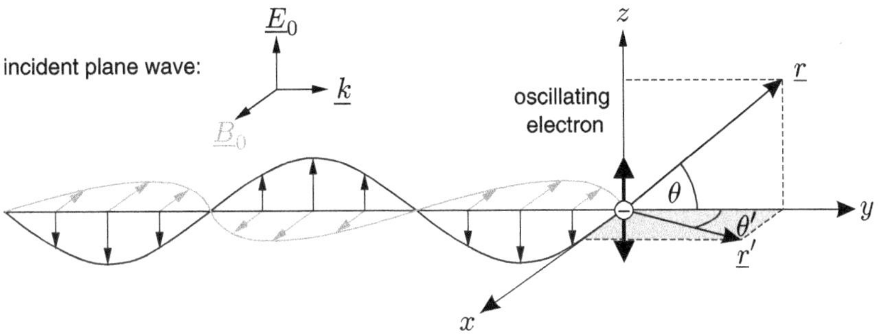

Fig. 3.8: Thomson scattering of a plane electromagnetic wave by a free electron located at the coordinate origin. The polarization of the incoming electric field is chosen in z direction. Driven by the electric field, the electron undergoes oscillatory motion, leading to the emission of dipole radiation. At the point $\underline{r}$ in the plane spanned by $\underline{k}$ and $\underline{E}_0$, the observable acceleration of the electron is reduced by a factor $\cos\theta$. At $\underline{r}'$ in the plane perpendicular to the electron's acceleration (indicated in *gray*), the full acceleration is observed for all angles θ'.

then is solely due to the electric field of the incident wave polarized in the z direction, and the electron oscillates with the frequency of the incident wave. We are interested in deriving the amplitude of the electromagnetic radiation that is emitted by the oscillating electron. The following derivation of the Thomson scattering cross section is based on the book of Als-Nielsen [1], Appendix B:[10]

At first, we have to gather some results from classical electrodynamics (e.g., [12, 18]): The scalar potential $\phi(\underline{r}, t)$ and the vector potential $\underline{A}(\underline{r}, t)$ are defined in such a way that electric and magnetic fields follow by the differential operations

$$\underline{E} = -\nabla\phi - \frac{\partial \underline{A}}{\partial t} \quad \text{and} \quad \underline{B} = \nabla \times \underline{A}\,. \tag{3.34}$$

The vector potential of a charge distribution that varies in time is given by

$$\underline{A}(\underline{r}, t) = \frac{\mu_0}{4\pi} \int \frac{\underline{J}\left(\underline{r}', t - \frac{|\underline{r}-\underline{r}'|}{c}\right)}{|\underline{r} - \underline{r}'|} d\underline{r}' \overset{r' \ll r}{\approx} \frac{\mu_0}{4\pi}\frac{1}{r} \int \underline{J}\left(\underline{r}', t - \frac{r}{c}\right) d\underline{r}'\,. \tag{3.35}$$

The current density $\underline{J}$ is equal to the product of charge density ρ and velocity $\underline{v}$, and retardation (changes in the sources propagate with the speed of light) is taken into account by using the retarded time $t' = t - |\underline{r}-\underline{r}'|/c$ in the integrand. In addition, the dipole approximation $r' \ll r$ is used. For a single, point like charge such as the electron, the integral over the current density simplifies to

$$\int \underline{J}\, d\underline{r}' = \int \rho \underline{v}\, d\underline{r}' = -e\underline{v}\,. \tag{3.36}$$

10 Alternatively, one can start from the equations for the electric and magnetic fields of a Hertzian dipole as given in any electrodynamics textbook.

Since we assumed that the magnetic field is negligible and, thus, $\underline{v} \parallel \underline{e}_z \, \forall t$, we obtain

$$\underline{A}(\underline{r}, t) = -\frac{\mu_0 e}{4\pi}\left(0, 0, \frac{v(t')}{r}\right)^{\mathrm{T}} . \tag{3.37}$$

Using the definition of the vector potential (equation 3.34), the magnetic field resulting from the oscillating electron is

$$\underline{B}_{\mathrm{sc}}(\underline{r}, t) = -\frac{\mu_0 e}{4\pi}\left(+\frac{\partial}{\partial y}\left[\frac{v(t')}{r}\right], -\frac{\partial}{\partial x}\left[\frac{v(t')}{r}\right], 0\right)^{\mathrm{T}} . \tag{3.38}$$

For the x component, we obtain

$$\frac{\partial}{\partial y}\left[\frac{v(t')}{r}\right] = \frac{1}{r}\frac{\partial v(t')}{\partial y} - \frac{v(t')}{r^2}\frac{\partial r}{\partial y} \tag{3.39}$$

We are interested in the far field limit and thus neglect the second term $\propto r^{-2}$. The first term is evaluated by noting that

$$\frac{\partial}{\partial y} = \frac{\partial}{\partial t'}\frac{\partial t'}{\partial y} = \frac{\partial}{\partial t'}\frac{\partial}{\partial y}\left(t - \frac{\sqrt{x^2+y^2+z^2}}{c}\right) = -\frac{1}{c}\left(\frac{y}{r}\right)\frac{\partial}{\partial t'}$$

$$\Rightarrow \quad \frac{1}{r}\frac{\partial v(t')}{\partial y} = -\frac{1}{c}\frac{y}{r^2}a(t') . \tag{3.40}$$

The y component follows by interchanging x and y, which leads to

$$\underline{B}_{\mathrm{sc}}(\underline{r}, t) = -\frac{\mu_0 e a(t')}{4\pi c r^2}(-y, x, 0)^{\mathrm{T}} \qquad \text{(far field limit for } \underline{a} \parallel \underline{e}_z) . \tag{3.41}$$

Using the definition of the vector product and the unit vector $\underline{e}_r = \underline{r}/r$ in the direction of $\underline{r}$, this can be generalized to the arbitrary direction of $\underline{a}$:

$$\underline{B}_{\mathrm{sc}}(\underline{r}, t) = -\frac{\mu_0 e}{4\pi c r}\left[\underline{a}(t') \times \underline{e}_r\right] \tag{3.42}$$

With equation (3.33) for the acceleration of the electron, $\epsilon_0\mu_0 = c^{-2}$ and $e^{-i\omega t'} = e^{ikr-i\omega t}$, this yields

$$\underline{B}_{\mathrm{sc}}(\underline{r}, t) = \frac{e^2}{4\pi\epsilon_0 m_e c^3}\left[\underline{E}_{\mathrm{in}}(0, t) \times \underline{e}_r\right]\frac{e^{ikr}}{r} . \tag{3.43}$$

The magnetic field caused by the oscillating electron in the far field is perpendicular to the distance vector $\underline{r}$ and decays with r^{-1}; retardation is taken into account by the phase factor e^{ikr}. For a propagating EM field in free space, the electric and magnetic fields are always perpendicular, and the same orthogonality relation $\underline{E}_{\mathrm{sc}} = c\underline{B}_{\mathrm{sc}} \times \underline{e}_r$ holds as for a plane wave (equation 3.31). Applying the cross product on equation (3.43) allows us to compute the electric field vector of the scattered radiation; the fundamental constants appearing in the prefactor can be regrouped into the Thomson scattering length r_0 defined by equation (3.24)

$$\underline{E}_{\mathrm{sc}}(\underline{r}, t) = \frac{e^2}{4\pi\epsilon_0 m_e c^2}\left[\left[\underline{E}_{\mathrm{in}}(0, t) \times \underline{e}_r\right] \times \underline{e}_r\right]\frac{e^{ikr}}{r} = r_0 E_{\mathrm{in}}(0, t)\left[\left[\underline{e}_p \times \underline{e}_r\right] \times \underline{e}_r\right]\frac{e^{ikr}}{r} . \tag{3.44}$$

Now, two particular cases must be distinguished (Figure 3.8). In the first case, $\underline{r}$ is located within the plane of electric polarization (spanned by $\underline{k}$ and $\underline{E}_0$), and the ratio of the radiated to incident field amplitude becomes

$$\frac{E_{sc}(\underline{r}, t)}{E_{in}(0, t)} = -r_0 \frac{e^{ikr}}{r} \cos\theta \, . \tag{3.45}$$

An observer at $\theta = 0$ would "see" the full acceleration of the electron, while the apparent acceleration is zero for $\theta = 90°$. The leading minus sign indicates that there is a phase shift of π between incident and scattered fields (this becomes most obvious by considering the double cross product in (3.44) for the case $\theta = 0$). Following Section 3.2.2, the differential cross section for Thomson scattering by a free electron (index e) is defined as

$$\left(\frac{d\sigma}{d\Omega}\right)_e^{\mathrm{Th}} = \frac{\text{radiated energy into } d\Omega \text{ per unit time}}{\text{incident energy per area and unit time}} \, . \tag{3.46}$$

This is now readily evaluated by using the Poynting vector $\underline{S} = \frac{1}{\mu_0}\underline{E} \times \underline{B}$, which represents the direction and rate of energy transport per unit area (perpendicular to the flow) and the unit time of an electromagnetic field. The denominator is given by the temporal average of the Poynting vector of the incident plane wave,

$$\langle|\underline{S}_{in}|\rangle = \left\langle \frac{1}{\mu_0}|\underline{E}_{in} \times \underline{B}_{in}| \right\rangle = \frac{E_0^2}{\mu_0 c} \cdot \frac{1}{T}\int_0^T \cos^2(\omega t)\, dt = \frac{E_0^2}{2\mu_0 c} \, , \tag{3.47}$$

where equation (3.31) and the harmonic time-dependence of the incident field have been used. In the same way, we obtain

$$\langle|\underline{S}_{sc}|\rangle = \frac{E_0^2}{2\mu_0 c} \frac{r_0^2 \cos^2\theta}{r^2} \tag{3.48}$$

for the field radiated by the oscillating charge. As required, the radiated energy per area decays $\propto r^{-2}$. Hence, the differential cross section for Thomson scattering is given by

$$\left(\frac{d\sigma}{d\Omega}\right)_e^{\mathrm{Th}} = \frac{\langle|\underline{S}_{sc}|\rangle \cdot r^2}{\langle|\underline{S}_{in}|\rangle} = r_0^2 \cos^2\theta \tag{3.49}$$

in the case of a linearly polarized incident EM wave and $\underline{r}$ in the plane spanned by $\underline{k}$ and $\underline{E}_0$. The factor r^2 appears in converting the radiated energy density from surface area to solid angle and cancels with the r^{-2} dependence of the scattered radiation.

In the second case to be considered, an observer at $\underline{r}'$ in the plane spanned by $\underline{k}$ and $\underline{B}_0$ (i.e., perpendicular to the plane of electric polarization, see Figure 3.8) would "see" the full acceleration of the electron for any angle θ', and hence

$$\frac{E_{sc}(\underline{r}, t)}{E_{in}(0, t)} = -r_0 \frac{e^{ikr}}{r} \, . \tag{3.50}$$

For an unpolarized source of x-rays (such as the x-ray tube), one can average over both directions. Therefore, depending on the polarization of the x-ray source and the experimental geometry, the differential cross section for Thomson scattering can be written as

$$\left(\frac{d\sigma}{d\Omega}\right)_{\mathrm{e}}^{\mathrm{Th}} = r_0^2 P \quad \text{with} \quad P = \begin{cases} \cos^2\theta & \underline{r} \text{ in plane spanned by } \underline{k}, \underline{E}_0 \,, \\ 1 & \underline{r} \text{ in plane spanned by } \underline{k}, \underline{B}_0 \,, \\ \frac{1}{2}(1+\cos^2\theta) & \text{unpolarized source} \,. \end{cases} \tag{3.51}$$

The total cross section for Thomson scattering by a single, free electron is obtained by integration over all directions (ϕ, θ)

$$\sigma_{\mathrm{e}}^{\mathrm{Th}} = r_0^2 \int_0^{\pi} \cos^2\theta \sin\theta \, d\theta \int_0^{2\pi} d\phi = 2\pi r_0^2 \int_0^{\pi} \sin^3\theta \, d\theta = \frac{8\pi}{3} r_0^2 = 0.665 \cdot 10^{-28}\,\mathrm{m}^2 \,. \tag{3.52}$$

We can see that the cross section for Thomson scattering is independent of the photon energy. We also note that the proportionality to $(e/m_{\mathrm{e}})^2$ included in r_0^2 implies that scattering from protons or atomic nuclei can safely be neglected. Scattering of electromagnetic waves is entirely dominated by contributions from oscillating electrons.

It is important to keep in mind that the above results are obtained in a framework of classical electrodynamics for a free (unbound) electron that was fixed in space. This is a problematic assumption and an oversimplification in many respects. Firstly, it can be only valid in a nonrelativistic regime of electron motion. Secondly, classical electrodynamics should be applied not to a point charge, but to a (radiating) charge distribution as given by the electron's wave function of the respective atomic orbital. Thirdly, due to the fact that atomic electrons are bound and exhibit resonances, one would expect that the scattering amplitude shows some variation with the frequency of the incident plane wave. These points are addressed in the following.

(ii) Extended electron distribution

Consider two point charges, one located at the coordinate origin, the other at position $\underline{r}$ (Figure 3.9). Both charges oscillate in the electric field of an incoming plane wave characterized by the wavevector $\underline{k}$. In a very distant observation point in direction $\underline{k}'$, the elastically ($|\underline{k}| = |\underline{k}'|$) scattered waves from both charges have a phase difference

$$\Delta\phi(\underline{r}) = 2\pi \frac{(\underline{e}_k - \underline{e}_{k'}) \cdot \underline{r}}{\lambda} = (\underline{k} - \underline{k}') \cdot \underline{r} = \underline{q} \cdot \underline{r} \tag{3.53}$$

where the quantity

$$\underline{q} = \underline{k} - \underline{k}' \qquad \text{with} \qquad q = \frac{4\pi}{\lambda} \sin\left(\frac{\theta}{2}\right) \tag{3.54}$$

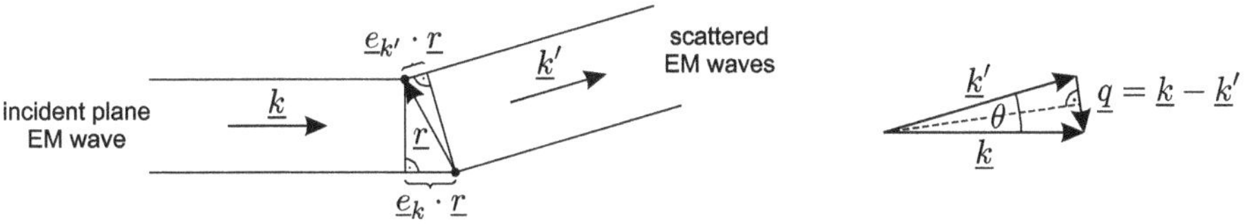

Fig. 3.9: Thomson scattering by a system of two point charges. The incident x-ray beam is represented by a plane monochromatic wave with wavevector $\underline{k}$. Both charges exert forced oscillations and emit EM radiation with the same wavelength ($|\underline{k}| = |\underline{k}'|$). In the far field (distance $\gg r$), the path length difference of the scattered waves in direction $\underline{k}'$ is $(\underline{e}_k - \underline{e}_{k'}) \cdot \underline{r}$, causing a phase difference $\Delta\phi = (\underline{k} - \underline{k}') \cdot \underline{r} = \underline{q} \cdot \underline{r}$.

denotes the wavevector transfer.[11] The sum of the scattered amplitudes of two electrons is hence proportional to $-r_0(1 + e^{i\underline{q}\cdot\underline{r}})$. This can be generalized for an arbitrary number n of electrons or an extended electron density distribution ρ_e

$$- r_0 \sum_n e^{i\underline{q}\cdot\underline{r}_n} \longrightarrow -r_0 \int_V \rho_e(\underline{r})\, e^{i\underline{q}\cdot\underline{r}}\, d^3r\,. \tag{3.55}$$

The integral can be identified as the Fourier transform of $\rho_e(\underline{r})$ with respect to $\underline{q}$. If $\rho_e(\underline{r})$ represents the electron density distribution of an atom (which is given by the sum of absolute squares of the wavefunctions of the occupied atomic orbitals), it is called the atomic form factor

$$f_0(\underline{q}) = \int \rho_e(\underline{r}) e^{i\underline{q}\cdot\underline{r}}\, d^3r = \begin{cases} Z & |\underline{q}| \to 0\,, \\ 0 & |\underline{q}| \to \infty\,. \end{cases} \tag{3.56}$$

The differential cross section for Thomson scattering by an atom is obtained by multiplying equation (3.51) with the absolute square of the atomic form factor

$$\left(\frac{d\sigma}{d\Omega}\right)^{\mathrm{Th}}_{\mathrm{atom}} = r_0^2 P\, |f_0(\underline{q})|^2\,. \tag{3.57}$$

The step from an isolated point charge to a continuous charge distribution introduces both energy and directional dependence of the scattered intensity. For $\underline{q} \to 0$, i.e., scattering in forward direction or large x-ray wavelength, the scattering contributions of all Z electrons in an atom interfere constructively, while $f_0(\underline{q})$ decreases for large scattering angles or small x-ray wavelengths, since the phase factors $e^{i\underline{q}\cdot\underline{r}}$ of all volume elements contributing to ρ_e tend to cancel each other. Integrated over all directions, this accounts for a decrease of the total Thomson scattering cross section σ^{Th}, which is no longer constant with the photon energy. In Figure 3.3, the cross sections σ^{Th} for

11 Note that, except for a factor $\hbar$, this is equivalent to the change in momentum of a photon upon scattering.

H_2O and for Pb are found to be approximately constant at photon energies of few keV, but decrease for higher photon energies.

Beyond isolated atoms, this formalism can be extended to molecules and especially to crystals. The exploitation of Thomson scattering through interference of scattered radiation from an ensemble of scattering centers is the basis of structure analysis by diffraction and crystallography, see, for example, [1].

(iii) Bound electrons

It is instructive to consider a model where the binding of electrons is taken into account in terms of a simple oscillator model. To this end, the equation of motion (3.33) needs to be modified. In a classical picture, an oscillating electron driven by an external electric field can be described by

$$\ddot{z} + \Gamma \dot{z} + \omega_r^2 z = -\left(\frac{e}{m_e} E_0\right) e^{-i\omega t} , \tag{3.58}$$

where $\Gamma \dot{z}$ denotes a velocity-dependent damping term and $-m_e \omega_r^2 z$ a restoring force, with resonance frequency ω_r. The usual harmonic ansatz $z(t) = z_0 e^{-i\omega t}$ yields the solution

$$z(t) = \frac{eE_0}{m_e} \frac{1}{\omega^2 - \omega_r^2 + i\omega\Gamma} e^{-i\omega t} , \tag{3.59}$$

which leads to the following equivalent of equation (3.45)

$$\frac{E_{sc}(\underline{r}, t)}{E_{in}(0, t)} = -r_0 \underbrace{\frac{\omega^2}{\omega^2 - \omega_r^2 + i\omega\Gamma}}_{f_s} \frac{e^{ikr}}{r} \cos\theta . \tag{3.60}$$

The modification of the scattered amplitude with respect to the case of a free electron is given by the factor f_s (scattering length in units of r_0). For the total cross section, one obtains

$$\sigma_e^R = \frac{8\pi}{3} \frac{\omega^4}{(\omega^2 - \omega_r^2)^2 + (\omega\Gamma)^2} r_0^2 = \begin{cases} \sigma_e^{Th} \left(\frac{\omega}{\omega_r}\right)^4 , & \omega \ll \omega_r \\ \sigma_e^{Th} , & \omega \gg \omega_r \end{cases} \tag{3.61}$$

which is plotted in Figure 3.10. The former case applies to the visible part of the electromagnetic spectrum (cf., Figure 3.1) and is known as Rayleigh scattering.[12] For the case $\omega \gg \omega_r$, equation (3.52) is retrieved. With $\Gamma \ll \omega_r$, the atomic scattering length f_s, which describes the amplitude of the scattered wave $\exp(ikr)/r$, can be approximated as

$$f_s = \frac{\omega^2}{\omega^2 - \omega_r^2 + i\omega\Gamma} \overset{\Gamma \ll \omega_r}{=} 1 + \frac{\omega_r^2}{\omega^2 - \omega_r^2 + i\omega\Gamma} = 1 + \chi(\omega) = 1 + f_s' + if_s'' \tag{3.62}$$

12 The standard example for the manifestation of Rayleigh scattering found in many textbooks on experimental physics is the blue color of the sky.

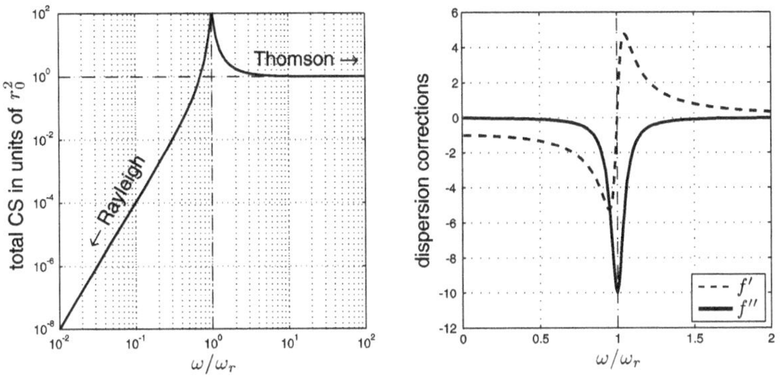

Fig. 3.10: (*left*) Plot of equation (3.61) illustrating the two limiting cases of Rayleigh and Thomson scattering far below and above the resonance frequency ω_r of a bound electron, respectively. (*right*) Plot of the real and imaginary part f' and f'' of the dispersion correction (equation 3.63) close to ω_r ($\Gamma = 0.1\omega_r$).

where

$$f_s' = \frac{\omega_r^2(\omega^2 - \omega_r^2)}{(\omega^2 - \omega_r^2)^2 + (\omega\Gamma)^2} \quad \text{and} \quad f_s'' = -\frac{\omega_r^2 \omega\Gamma}{(\omega^2 - \omega_r^2)^2 + (\omega\Gamma)^2} \tag{3.63}$$

denote the real and imaginary part of the dispersion correction $\chi(\omega)$ (Figure 3.10, right). This principle can be extended to atoms. The atomic form factor $f_0(\underline{q})$ is modified by the corresponding dispersion correction terms

$$f(\underline{q}, \omega) = f_0(\underline{q}) + f'(\omega) + if''(\omega) \tag{3.64}$$

with contributions from all orbitals. The real valued term f' takes into account the reduction of the scattering amplitude of an atom due to the fact that the electrons are bound. The imaginary term f'' is related to absorption, as we will see below. Atomic form factors are tabulated, e.g., in the International Tables for Crystallography.

(iv) Refractive Index of x-rays

Finally, Thomson scattering provides the physical basis of the index of refraction [1, 7], defined as the ratio of the vacuum speed of light $c = 1/\sqrt{\epsilon_0 \mu_0}$ and the phase velocity of an electromagnetic wave in a given material, $v = 1/\sqrt{\epsilon_0 \epsilon_r \mu_0 \mu_r}$,

$$n = \frac{c}{v} = \sqrt{\epsilon_r \mu_r} \approx \sqrt{\epsilon_r}\,, \tag{3.65}$$

which follows from Maxwell's equations (see Section 6.2.1). Here, ϵ_r denotes the relative permittivity and μ_r the relative permeability of the chosen material. The latter is very close to unity for nonferromagnetic materials. The relative permittivity $\epsilon_r = 1 + \chi$ also determines the proportionality coefficient between the macroscopic polarization P (not to be confused with the polarization factor of Thomson scattering used above) and the driving external electric field[13]

$$P = \epsilon_0(\epsilon_r - 1)E\,, \tag{3.66}$$

where the polarization is defined by the number of electric dipoles per volume element. If the displacement of all electrons in an infinitesimal volume element is given by equation (3.59) and the electron density (number per volume) is denoted by ρ_e,

$$P(t) = -e\rho_e z(t) = -\frac{e^2\rho_e}{m_e}\frac{1}{\omega^2 - \omega_r^2 + i\omega\Gamma}E(t)\,. \tag{3.67}$$

Combining these three equations, the index of refraction in the oscillator model is given by

$$n^2 = 1 + \frac{1}{\epsilon_0}\frac{P}{E} = 1 - \frac{e^2\rho_e}{\epsilon_0 m_e}\frac{1}{\omega^2 - \omega_r^2 + i\omega\Gamma}\,, \tag{3.68}$$

and expressing the numerical constants by r_0 and using $\omega = ck = 2\pi c/\lambda$ yields

$$n^2 = 1 - \frac{r_0\lambda^2\rho_e}{\pi}\frac{\omega^2}{\omega^2 - \omega_r^2 + i\omega\Gamma}\,. \tag{3.69}$$

For typical x-ray wavelengths ($\lambda \approx 1$ Å) and electron densities of soft matter ($\rho_e(H_2O) = 0.334\,$Å^{-3}) as well as $\omega \gg \omega_r$, the second term is much smaller than unity. Using the Taylor expansion $\sqrt{1-x} \simeq 1 - \frac{x}{2}$ $(x \ll 1)$ yields the x-ray index of refraction

$$n \simeq 1 - \frac{r_0\lambda^2\rho_e}{2\pi}\underbrace{\frac{\omega^2}{\omega^2 - \omega_r^2 + i\omega\Gamma}}_{f_s} = 1 - \frac{r_0\lambda^2\rho_e}{2\pi}\left[1 + f_s' + if_s''\right]\,. \tag{3.70}$$

Note that by the above derivation, the index of refraction as a macroscopic quantity has been linked to the atomic phenomenon of scattering. The inherent averaging as well as fluctuations and correlation effects in matter are, of course, nontrivial issues. An alternative derivation, which is not based on polarization, is also given in [1]. Replacing electron density ρ_e by atom density ρ_a and the scattering length of a single electron f_s by the atomic form factor (3.64) with $\underline{q} = 0$ (forward direction) yields

$$n = 1 - \frac{r_0\lambda^2\rho_a}{2\pi}\left[Z + f'(\omega) + if''(\omega)\right]\,, \tag{3.71}$$

which is usually abbreviated as

$$n = 1 - \delta + i\beta \tag{3.72}$$

with

$$\delta = \frac{r_0\lambda^2\rho_a}{2\pi}\left[Z + f'(\omega)\right] \quad \text{and} \quad \beta = -\frac{r_0\lambda^2\rho_a}{2\pi}f''(\omega)\,. \tag{3.73}$$

This notation is instructive in view of judging contrast of elements arising either from absorption or phase contrast, as is discussed further in Chapter 6.

---

13 For simplicity, we assume that the chosen material is isotropic and, hence, ϵ_r is a scalar quantity.

3.2.5 Compton scattering

In 1922, A. H. COMPTON studied the scattering of x-rays by graphite, using Bragg diffraction by a crystal for spectral analysis of the scattered radiation [6] (Figure 3.11a). He observed that the spectra measured at different angles θ_γ contained two peaks, one corresponding to the wavelength of the incident x-rays (K_α line of molybdenum, $\lambda = 0.709$ Å), which can be explained by elastic Thomson scattering, and a second one at a slightly larger wavelength $\lambda' > \lambda$, with a separation $\Delta\lambda = \lambda' - \lambda$ increasing with the scattering angle θ_γ (Figure 3.11b). The scattering process resulting in this second peak hence needed to be inelastic. In addition, the angular distribution of the scattered intensity was not symmetric in forward and backward direction. Both observations could not be reconciled with classical Thomson scattering.

Compton was able to explain his observation by treating photons as quantum-mechanical particles, which undergo collisions with the electrons and lose some part of their energy. The collision process is sketched in Figure 3.11c. The kinematics, i.e., the relations between energy and momentum of photon and electron before and after the

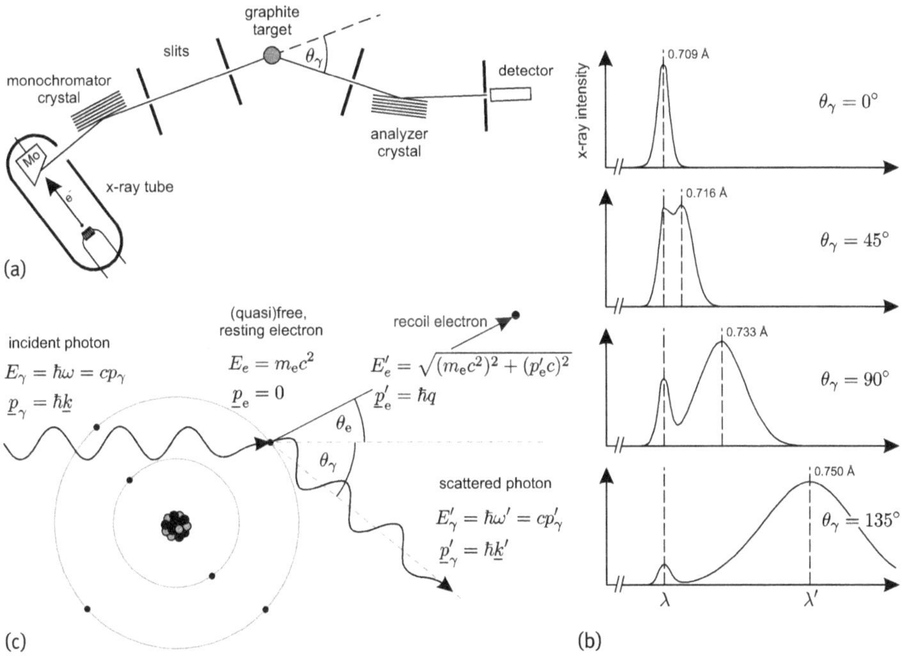

Fig. 3.11: The Compton effect. (a) Experimental setup, (b) measured x-ray spectra at different scattering angles θ_γ (following [13]) and (c) sketch for derivation of the wavelength shift $\Delta\lambda = \lambda' - \lambda$. The electron is assumed to be free (binding energy $\ll \hbar\omega$) and initially at rest. It takes up some momentum from the photon and is ejected from the atom. Both energy and momentum conservation must apply.

scattering event, including the scattering angles, can be derived using equation (3.1) and (3.2) for photon energy and momentum and the relativistic mass–energy relationship

$$E = \sqrt{m_e^2 c^4 + p^2 c^2} \tag{3.74}$$

for the electron (m_e: electron rest mass). For simplicity, it is assumed that the electron is free (since the binding energies of light elements are negligible compared to typical x-ray photon energies) and initially at rest. In the collision, some energy and momentum are transferred from the photon to the electron, resulting in a scattered photon with lower energy and momentum and a recoil electron. Equations for all quantities are given in Figure 3.11c. Imposing conservation of energy and momentum requires that

$$E_e + E_\gamma = E'_e + E'_\gamma \,, \tag{3.75}$$

$$\underline{p}_e + \underline{p}_\gamma = \underline{p}'_e + \underline{p}'_\gamma \,. \tag{3.76}$$

Plugging in the relations given in Figure 3.11c yields

$$m_e c^2 + c\hbar k = \sqrt{(m_e c^2)^2 + (c\hbar q)^2} + c\hbar k' \,, \tag{3.77}$$

$$\hbar \underline{k} = \hbar \underline{q} + \hbar \underline{k}' \,. \tag{3.78}$$

Dividing both sides of equation (3.77) by $m_e c^2$ leads to

$$1 + \frac{\hbar}{m_e c}(k - k') = \sqrt{1 + \left(\frac{\hbar}{m_e c} q\right)^2} \,. \tag{3.79}$$

By squaring both sides and collecting terms, this can be recast into an expression for q^2,

$$q^2 = (k - k')^2 + 2\frac{m_e c}{\hbar}(k - k') \,. \tag{3.80}$$

A second expression for q^2 is obtained from condition (3.78) for momentum conservation by taking the scalar product of $\underline{q} = \underline{k} - \underline{k}'$ with itself and using the law of cosines,

$$q^2 = \underline{q} \cdot \underline{q} = (\underline{k} - \underline{k}') \cdot (\underline{k} - \underline{k}') = k^2 + k'^2 - 2kk' \cos\theta_\gamma \,. \tag{3.81}$$

Combining equation (3.80) and (3.81) yields

$$\frac{k - k'}{kk'} = \frac{1}{k'} - \frac{1}{k} = \frac{\hbar}{m_e c}(1 - \cos\theta_\gamma) \,. \tag{3.82}$$

Finally, using $k = \frac{2\pi}{\lambda}$, this yields the angle dependent wavelength shift

$$\Delta\lambda = \lambda' - \lambda = \lambda_C(1 - \cos\theta_\gamma) \tag{3.83}$$

where

$$\lambda_C = \frac{h}{m_e c} = 2.42631 \cdot 10^{-12}\,\text{m} \tag{3.84}$$

denotes the Compton wavelength.[14] This equation was in excellent agreement with Compton's experimental results, and the inelastic scattering of photons by quasi free electrons has hence been termed the Compton effect or Compton scattering. Importantly, for a given scattering angle θ_γ, the wavelength shift $\Delta\lambda$ does *not* depend on the initial photon energy. The relative energy loss, therefore, increases, i.e., Compton scattering becomes more 'inelastic' for higher photon energies. The ratio of the energies of incident and Compton-scattered photons is

$$\frac{E'_\gamma}{E_\gamma} = \frac{\lambda}{\lambda'} = \frac{1}{1 + \mathcal{E}(1 - \cos\theta_\gamma)} \tag{3.85}$$

where $\mathcal{E} = E_\gamma/(m_e c^2)$. Neglecting binding energies (which are on the order of few eV in outer shells of light elements), the kinetic energy of the recoiling electron is

$$E_{\text{kin,e}} = E_\gamma - E'_\gamma = E_\gamma \frac{\mathcal{E}(1 - \cos\theta_\gamma)}{1 + \mathcal{E}(1 - \cos\theta_\gamma)}\,, \tag{3.86}$$

and approaches its maximum value $E_\gamma \frac{2\mathcal{E}}{1+2\mathcal{E}}$, the so-called Compton edge, for $\theta_\gamma = 180°$ (backscattering of photons). By enforcing conservation of momentum, one can show that the scattering angles of photon and electron are related by

$$(1 + \mathcal{E}) \tan\theta_e = \cot\left(\frac{\theta_\gamma}{2}\right) \quad \Rightarrow \quad 0 \le \theta_e \le 90°\,, \tag{3.87}$$

which yields a maximum electron recoil angle of $\theta_e = 90°$ for $\theta_\gamma = 0°$ and $\theta_e = 0°$ for $\theta_\gamma = 180°$, i.e., Compton electrons are emitted in the forward hemisphere.

For given photon energy E_γ, the equations derived so far allow us to compute the energy of Compton scattered photons observable at a given angle θ_γ. However, to determine the probability by which an incident photon is scattered by an angle θ_γ, the differential cross section (per electron, index e) for Compton scattering is required. Its derivation was given by O. Klein and Y. Nishina a few years after Compton's experiments and requires full quantum electrodynamic theory. It is given (for the case of unpolarized photons) by

$$\left(\frac{d\sigma}{d\Omega}\right)_e^C = \frac{r_0^2}{2}\left(\frac{\omega'}{\omega}\right)^2 \left(\frac{\omega'}{\omega} + \frac{\omega}{\omega'} - \sin^2\theta_\gamma\right)\,, \tag{3.88}$$

---

14 Note that the photon energy of electromagnetic radiation with wavelength λ_C equals the electron rest energy $m_e c^2 = 0.511\,\text{MeV}$.

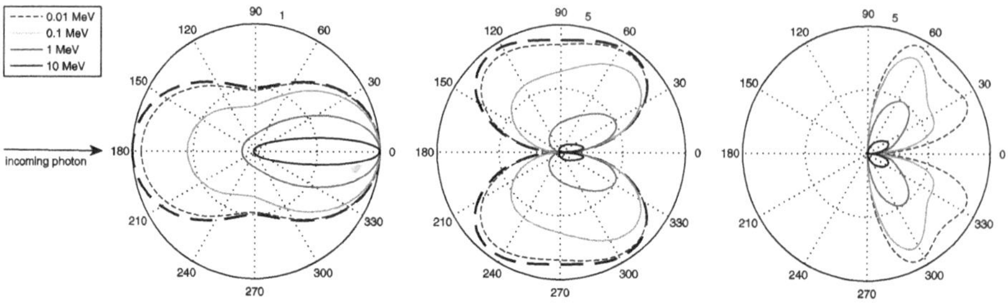

Fig. 3.12: *Left:* Differential cross section for Compton scattering at different photon energies E_γ calculated by equation (3.89). It approaches the classical Thomson scattering differential cross section (*thick dashed line*) for $E_\gamma \ll m_e c^2$. *Right*: Corresponding distributions of scattering angles θ_γ (photons) and θ_e (electrons) calculated by equation (3.13) and (3.87). Scattering becomes more forward directed with increasing photon energy E_γ.

which is known as the Klein–Nishina equation, one of the first results of quantum electrodynamics [21, 23]. Using equation (3.85), it can be rewritten as

$$\left(\frac{d\sigma}{d\Omega}\right)_e^C = \underbrace{\frac{r_0^2}{2}\left(1+\cos^2\theta_\gamma\right)}_{=\left(\frac{d\sigma}{d\Omega}\right)_e^{Th}} \times \underbrace{\left(\frac{1}{1+\mathcal{E}(1-\cos\theta_\gamma)}\right)^2\left(1+\frac{\mathcal{E}^2(1-\cos\theta_\gamma)^2}{(1+\cos^2\theta_\gamma)(1+\mathcal{E}(1-\cos\theta_\gamma))}\right)}_{:=F_{KN}(\mathcal{E},\theta_\gamma)}, \tag{3.89}$$

or

$$\left(\frac{d\sigma}{d\Omega}\right)_e^C = F_{KN}(\mathcal{E},\theta_\gamma)\cdot\left(\frac{d\sigma}{d\Omega}\right)_e^{Th}. \tag{3.90}$$

The differential cross section for Compton scattering is equal to the differential cross section for Thomson scattering multiplied by the "Klein–Nishina factor" F_{KN}. In Figure 3.12, the angular distribution of scattered photons and electrons is plotted for several energies. For $\mathcal{E} \ll 1$, $F_{KN} \longrightarrow 1$, hence, Thomson scattering can be considered as the nonrelativistic limit of Compton scattering [23].

The total cross section is obtained by integration over the solid angle $d\Omega$ and expresses the overall probability for a Compton scattering event, including all scattering angles θ_γ. The resulting analytical expression is rather lengthy (see, e.g., [23]); in the nonrelativistic and extreme relativistic limits one obtains

$$\sigma_e^C = \begin{cases} \sigma_e^{Th}\cdot\left[1-2\mathcal{E}+\frac{26}{5}\mathcal{E}^2\dots\right], & \mathcal{E}\ll 1, \\ \sigma_e^{Th}\cdot\frac{3}{8\mathcal{E}}\left[2\ln(2\mathcal{E})+1\right], & \mathcal{E}\gg 1, \end{cases} \tag{3.91}$$

where $\sigma_e^{Th} = \frac{8}{3}\pi r_0^2$ denotes the total Thomson cross section for elastic scattering (Section 3.2.4). As can be seen in Figure 3.3 for the case of lead and water, $\sigma_a^C = Z\sigma_e^C$ assumes its maximum value for intermediate energies where $\mathcal{E} \approx \mathcal{O}(1)$ and decreases both towards lower and higher energies. Interestingly, the mass attenuation coefficient due to the Compton effect is approximately constant for different elements [25]

$$\left(\frac{\mu}{\rho}\right)^{C} = \frac{N_A}{A}\sigma_a^C = \frac{N_A}{A}Z\sigma_e^C \propto \frac{Z}{A} \approx 0.5\,, \tag{3.92}$$

where the last approximation holds for all biologically relevant elements except for hydrogen. Therefore, in the photon energy range where the Compton effect is the dominant interaction, the total mass attenuation coefficients (μ/ρ) of different materials are approximately the same. In Figure 3.3, one can see that this is actually the case for water and lead at photon energies of few MeV.

So far, electrons have been treated as free and initially at rest. However, they are actually bound in an atom and have a statistical momentum distribution (which can hence be inferred by Compton scattering). The calculations can be generalized to this case, requiring many body quantum mechanics. Such a treatment results in two cross sections, an elastic one (Thomson) and an inelastic (Compton) one. Thus, these opposing views in the treatment of free electrons are reconciled in the treatment of the atom (see, e.g., [23] and references therein). In particular, the Compton cross section dominates if the photon energy is high and the electronic wave functions are extended (low Z). Since there is no fixed phase relation between the Compton scattered photons and energy is transferred to an electron, Compton scattering is incoherent and inelastic and gives rise to a smoothly varying background. No sharp interference effects are observed, even if the sample is crystalline. Consequences of Compton scattering in the clinical setting, where it is the dominant interaction process of x-ray photons and matter at photon energies around 100 keV, are discussed in Section 3.6. Exploitation of the Compton effect in so called Compton cameras to record both energy and momentum (and, hence, direction) of incoming photons is in development for medical imaging [24].

3.2.6 Pair production and annihilation

Another important interaction process that comes into effect in the case of ultrahard x-rays is pair production (Figure 3.13). When the photon energy reaches a certain threshold value $E_{\gamma,\mathrm{th}}$ it can generate an electron/positron pair

$$\gamma \longrightarrow e^+ + e^- \tag{3.93}$$

in the presence of the Coulomb field of a charged particle, such as an atomic nucleus. The photon energy in excess of $E_{\gamma,\mathrm{th}}$ is converted into kinetic energy of the generated electron and positron. Subsequently, these lose kinetic energy by collisions with electrons of the surrounding medium and slow down (the slowing down process is not

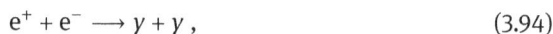

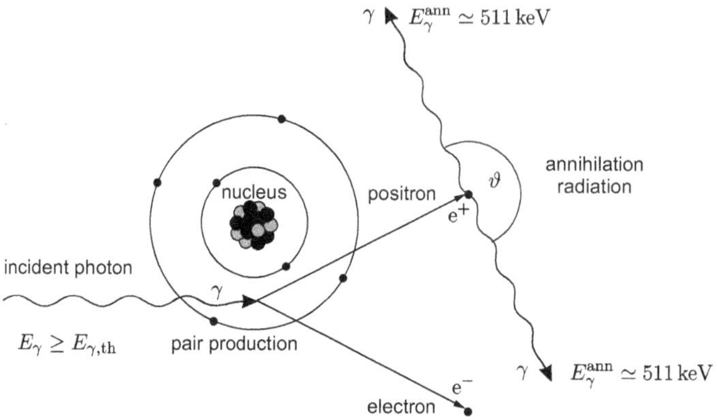

Fig. 3.13: Above a threshold energy $E_{\gamma,\mathrm{th}} \approx 2m_\mathrm{e}c^2$, a photon can be converted into an electron–positron pair in the presence of the Coulomb field of an atomic nucleus. Electron and positron interact with the surrounding medium (see Chapter 5). The positron finally annihilates with an electron of the medium, emitting two photons with $E_\gamma \simeq m_\mathrm{e}c^2 = 511$ keV in diametrically opposed directions ($\theta \simeq 180°$).

illustrated in Figure 3.13). The positron finally annihilates with an electron from the medium,

$$\mathrm{e}^+ + \mathrm{e}^- \longrightarrow \gamma + \gamma\,, \tag{3.94}$$

emitting two photons ($E \simeq 0.511$ MeV) in diametrically opposed directions. This annihilation radiation is exploited in positron emission tomography (PET), where the positron is generated by β^+ decay of selected radioisotopes (e.g., ${}^{15}_{8}\mathrm{O}$, ${}^{18}_{9}\mathrm{F}$ or ${}^{11}_{6}\mathrm{C}$). The exact angle ϑ between the two annihilation photons, as well as their energy, depend on the remaining kinetic energy and momentum of the positron and the electron when annihilation takes place. If these are negligible, $\vartheta = 180°$ and $E_\gamma^{\mathrm{ann}} = m_\mathrm{e}c^2 = 0.511$ MeV.

The threshold energy $E_{\gamma,\mathrm{th}}$ for pair production follows from conservation of energy and momentum. If the energy of the incident photon is exactly $E_{\gamma,\mathrm{th}}$, the created electron and positron are at rest. Therefore, the momentum $p = \hbar\omega/c$ of the photon must be completely transferred to a recoil particle. If this is an atomic nucleus with mass m_N, the threshold energy is

$$E_{\gamma,\mathrm{th}} = 2m_\mathrm{e}c^2 + \frac{p^2}{2m_\mathrm{N}}\,. \tag{3.95}$$

Using the concept of relativistic invariant mass [23], one obtains the simple expression

$$E_{\gamma,\mathrm{th}} = 2m_\mathrm{e}c^2\left(1 + \frac{m_\mathrm{e}}{m_\mathrm{N}}\right) \quad \text{(threshold energy for pair production)}\,, \tag{3.96}$$

i.e., the energy threshold for pair production is slightly larger than twice the rest energy of an electron, $2m_\mathrm{e}c^2 = 1.022$ MeV. Since $m_\mathrm{e}/m_\mathrm{N} \ll 1$, the factor $(1 + m_\mathrm{e}/m_\mathrm{N})$ is often neglected. Detailed calculations also allow us to determine the angular distribution of the produced electrons and positrons, as well as the distribution of the photon energy exceeding $2m_ec^2$ between them [25]. It turns out that higher photon energy leads to more forward directed electrons and positrons. Interestingly, the theoretical description of pair production in quantum electrodynamics is closely linked to that of bremsstrahlung production [15].

Due to the threshold energy, pair production is not relevant at typical radiography or CT energies. However, it may become very important in radiotherapy where photons with energies up to about 20 MeV are routinely generated by linear accelerators. The atomic cross section σ_a^{pair} for pair production can be derived from quantum electrodynamics; an approximation formula is [23]

$$\sigma_a^{\text{pair}} = \alpha r_0^2 Z^2 f(\mathcal{E}, Z) \quad \text{where} \quad f(\mathcal{E}, Z) \approx \begin{cases} \frac{28}{9}\ln[2\mathcal{E}] - \frac{218}{27} & 1 \ll \mathcal{E} \ll \alpha^{-1}Z^{-1/3}\,, \\ \frac{28}{9}\ln\left[\frac{183}{Z^{1/3}}\right] - \frac{2}{27} & \mathcal{E} \gg \alpha^{-1}Z^{-1/3}\,. \end{cases} \tag{3.97}$$

In contrast to the other interactions of photons with matter considered so far, the atomic cross section for pair production increases with photon energy. If pair production is the dominant interaction, this leads to the counter intuitive effect that photons of higher energy are *less* penetrating. For the case of lead, it can be seen in Figure 3.3 that this is the case for photon energies above 4 MeV.

Instead of transferring momentum to the nucleus, the recoil particle can also be a bound electron, which then leaves the atom. Since one incident photon then leads to emission of three particles from the atom, this process is called triplet production. The energy threshold is obtained by replacing m_N with m_e in equation (3.96), yielding

$$E_{\gamma,\text{th}} = 4m_ec^2 \qquad \text{(threshold energy for triplet production)}\,. \tag{3.98}$$

However, compared to pair production, the probability of triplet production is negligible for photon energies $\ll$ 100 MeV.

3.2.7 Photonuclear reactions

For the sake of completeness, we briefly mention another interaction channel [21, 22, 27]. If the photon energy is sufficiently high, absorption of a photon (γ) by an atomic nucleus can result in the emission of a proton (p) or neutron (n):

$$(\gamma, \text{n}) \text{ reaction:} \qquad {}^A_Z\text{X} + \gamma \rightarrow {}^{A-1}_{Z}\text{X} + \text{n} \tag{3.99}$$

$$(\gamma, \text{p}) \text{ reaction:} \qquad {}^A_Z\text{X} + \gamma \rightarrow {}^{A-1}_{Z-1}\text{Y} + \text{p} \tag{3.100}$$

The energy threshold is given by the binding energy of the most weakly bound nucleon. This is on the order of 10 MeV or higher for most nuclei; only in the case of ${}^2_1\text{H}$ and ${}^9_4\text{Be}$ can it be as low as 2 MeV. The interaction cross section is small compared to the other effects and contributes only few percent to the total linear attenuation coefficient even far above the energy threshold. Its energy dependence is determined by the energy levels of the nucleus and can, hence, be quite complex. In addition, the reaction product can be a radioactive nucleus (denoted by an asterisk *, see the reactions below); the irradiated material is activated.

In clinical practice, photonuclear reactions are relevant from the point of view of radiation protection. In radiotherapy, linear accelerators can generate maximum photon energies that exceed the threshold for the (γ, n) reactions in the air of the treatment room,

$$ {}^{14}_{7}\mathrm{N} + \gamma \rightarrow {}^{13}_{7}\mathrm{N}^* + \mathrm{n}\,, \qquad E_\gamma \geq 10.5\,\mathrm{MeV}\,, \tag{3.101} $$

$$ {}^{16}_{8}\mathrm{O} + \gamma \rightarrow {}^{15}_{8}\mathrm{O}^* + \mathrm{n}\,, \qquad E_\gamma \geq 15.7\,\mathrm{MeV}\,. \tag{3.102} $$

Subsequently, the free neutrons decay into a proton and an electron by β^- decay

$$ \mathrm{n} \rightarrow \mathrm{p} + \mathrm{e}^- + \overline{\nu}_\mathrm{e} \qquad (T_{1/2} = 611\,\mathrm{s})\,, \tag{3.103} $$

while the activated nuclei ${}^{13}_{7}\mathrm{N}^*$ ($T_{1/2}$ = 10 min) and ${}^{15}_{8}\mathrm{O}^*$ ($T_{1/2}$ = 2 min) undergo a β^+ decay with subsequent annihilation radiation (equation 3.94). This cascade of particles following a photonuclear reaction poses a risk for the treatment personnel. This can be lowered by strong air ventilation of the treatment room and by equipping the treatment room door and walls with an additional layer (e.g., paraffin and boron) that moderates and finally absorbs the free neutrons. In addition, photonuclear reactions and activation can also take place within the accelerator components. This is taken into account by using materials with low interaction cross section and short half life of the radioactive reaction products.

3.3 Generation of x-rays

This section addresses the generation of x-rays. A very recent book covering this topic in much more technical detail than we are able to present here is [3]. The classical x-ray sealed vacuum tube or the so called rotating anode are by far the most common sources for medical applications. The design principles have essentially remained unchanged for about 100 years. An x-ray tube consists of a source of electrons, the cathode, which is usually a heated filament, and a thermally conducting anode, for example made of copper, molybdenum or tungsten, enclosed in an evacuated glass tube (Figure 3.14). Free electrons are generated by thermal emission from the cathode and accelerated towards the anode by a voltage U on the order of about 10 to 100 kV, depending on the application. In the anode, the free electrons interact with bound electrons and atomic nuclei of the anode material.

The x-ray flux produced by a conventional x-ray tube is generally limited by the amount of heat that is transferred into the anode material. In fact, about 99% of the

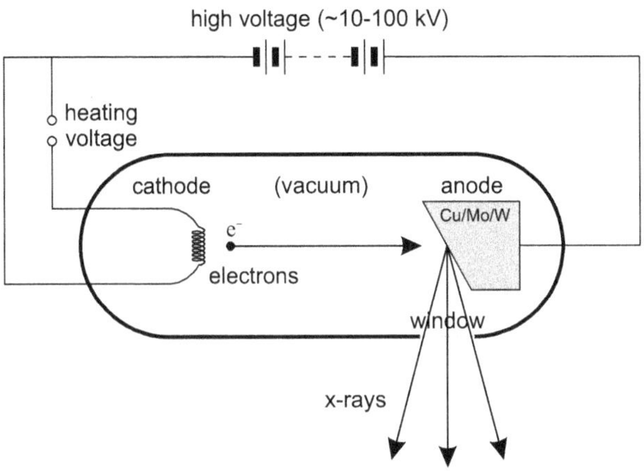

Fig. 3.14: Schematic of an x-ray tube. Electrons are emitted from the heated cathode, and accelerated by a high voltage U to the anode, made of a metal with high heat conductivity. The power of the tube is given by $P = I\,U$, where I is the electron current. Typical values are in the range of a few kW power of the electron beam, out of which only a tiny fraction is converted to x-rays by deacceleration (bremstrahlung) in the anode or by x-ray fluorescence (characteristic radiation) after K shell ionization of the anode material. Depending on the power per spot size on the anode, it may become necessary to distribute the head load by rotation of the anode. A more recent innovation uses a liquid metal jet injected into vacuum as the anode, which increases the allowable power density.

kinetic energy of the electron beam is directly converted into heat (Section 3.3.1). A typical limit for micro-CT sources in continuous operation, for example, is only $\approx 1\,\frac{\mathrm{W}}{\mu\mathrm{m}}$, before local melting of the anode. Different techniques have been developed to manage the heat, increasing either the effective area of the electron spot on the anode,[15] or to improve the conduction of heat away from the anode material. A rotating anode, for example, is constantly rotated in the electron beam (at the order of 5000 rpm) to distribute the heat load over a larger area and thus prevent the anode from melting. A more recent approach is the liquid metal jet anode [26], where a gallium (Ga) based alloy, which is liquid at $T \approx 30\,^\circ\mathrm{C}$, is injected through a thin nozzle with high pressure ($p \approx 200$ bar) into vacuum. In this way, a Ga jet with a diameter of $\approx 100\,\mu\mathrm{m}$ is produced which stays laminar for several millimeters. The speed of the jet is typically $50\,\frac{\mathrm{m}}{\mathrm{s}}$. In this way, the maximum power load is increased by a factor of ten up to $P_{\mathrm{max}} \approx 10\,\frac{\mathrm{W}}{\mu\mathrm{m}}$, which is advantageous for phase contrast micro-CT applications, see Chapter 6. However, liquid anode sources are not yet very suitable for diagnostic x-ray imaging, for example, since the photon spectrum is not hard enough.

15 This is typically unfavorable in terms of resolution, focusability or – if applicable – coherence.

Two distinct spectral components of x-rays are generated in a tube. The continuous bremsstrahlung spectrum reflects the negative acceleration of the free electrons in the Coulomb fields of bound electrons and nuclei, while the discrete characteristic spectrum results from ionization of the anode atoms and subsequent electronic transitions, thus reflecting the energy levels of the anode material.

3.3.1 Bremsstrahlung spectrum of an x-ray tube

The continuous bremsstrahlung spectrum results from the deflection of a large number of electrons in the Coulomb field of the electrons and nuclei of the anode material. Historically, this process has been described at several levels of accuracy. The first calculations were based on classical electrodynamics, later works included the emerging quantum theory. Summaries of this development are given, e.g., in [4, 8], on which this section is based. All theories for the continuous spectrum of x-ray tubes require a considerable amount of approximations.[16] More recently, also Monte Carlo simulations have been employed in order to correctly take into account all interaction processes that occur in the anode [28]. Below, we sketch some key elements of the derivation of the bremsstrahlung spectrum by classical electrodynamics.

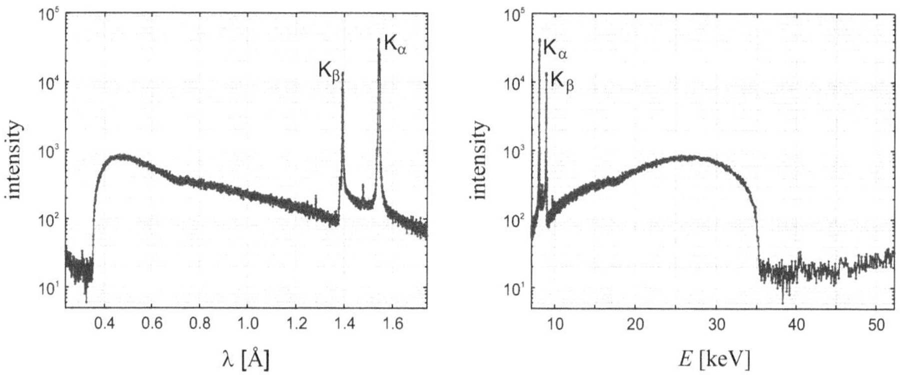

Fig. 3.15: Measured spectrum of an x-ray tube in terms of wavelength (*left*) and photon energy (*right*). The anode consists of copper (Cu), yielding the characteristic lines $\lambda_{\text{Cu,K}_\alpha} = 1.54$ Å and $\lambda_{\text{Cu,K}_\beta} = 1.39$ Å. The accelerating voltage $U = 35$ kV determines the maximum photon energy or minimum wavelength, respectively.

16 Remarkably, the calculation of synchrotron radiation spectra requires much less approximations than bremsstrahlung. Since electrons are forced to oscillations by a magnetic field in vacuum, the complex interactions with matter need not be considered.

Energy distribution

H. A. KRAMERS considered the deflection of electrons in the field of atomic nuclei [20]. The Coulomb force $\propto \frac{Ze^2}{r^2}$ leads to hyperbolic electron trajectories and causes the electron to irradiate bremsstrahlung. The upper frequency limit ν_0 is given by the accelerating voltage according to $h\nu_0 = eU$. Using a Fourier representation of the components of the electron's acceleration, Kramers obtained an expression for the spectral intensity of bremsstrahlung emitted by electrons traversing a thin anode layer of thickness dx (e.g., a metal film generated by vapor deposition), in which the electron velocity v remains approximately constant:

$$\frac{dI}{d\nu dx}\, d\nu\, dx = \begin{cases} \text{const.} \times \rho_a Z^2 \frac{1}{v^2}\, d\nu\, dx & \nu \leq \nu_0 \\ 0 & \nu > \nu_0 \end{cases} \tag{3.104}$$

Essentially, the spectral intensity from such a thin anode is constant up to ν_0 (corresponding to the conversion of the entire kinetic energy of an electron into a single photon) and proportional to number density ρ_a and squared atomic number Z^2 of the anode material. For nonrelativistic electrons with $\beta = \frac{v}{c} \ll 1$, we have $\frac{1}{2} m_e v^2 = eU$ and obtain the spectral intensity of a fixed frequency as a function of the accelerating voltage

$$\frac{dI}{d\nu dx}\, d\nu\, dx = \begin{cases} \text{const.} \times \rho_a Z^2 \frac{1}{U}\, d\nu\, dx & U \geq U_0 \\ 0 & U < U_0 \end{cases} \tag{3.105}$$

where $U_0 = \frac{h\nu}{e}$. The intensity radiated from such a thin anode decreases with higher voltage, since faster electrons change less in direction. For a massive anode, one has to take into account the slowing down of electrons. Using the empirical energy range relation, the Thomson–Whiddington law $R \propto (\rho_a Z)^{-1} U^2$ (which is a good approximation for energies from 8 to 30 keV), traversed distance in the anode and velocity loss are related by

$$dx \propto -\frac{1}{\rho_a Z} v^3\, dv\,. \tag{3.106}$$

Plugging this relation into equation (3.104) yields

$$\frac{dI}{d\nu dx}\, d\nu\, dx = -\text{const.} \times Zv\, dv\, d\nu \tag{3.107}$$

which has to be integrated from the initial electron velocity $\sqrt{2h\nu_0/m_e}$ to the minimum velocity $\sqrt{2h\nu/m_e}$, where the generation of a photon with frequency ν is still possible. The resulting spectral intensity of bremstrahlung from a massive anode is

$$\frac{dI}{d\nu}\, d\nu = \begin{cases} \text{const.} \times Z(\nu_0 - \nu)\, d\nu & \nu \leq \nu_0 \\ 0 & \nu > \nu_0 \end{cases}. \tag{3.108}$$

Instead of using the frequency as the independent variable, we can also consider the spectral intensity as a function of wavelength, where it turns into a nonmonotonous

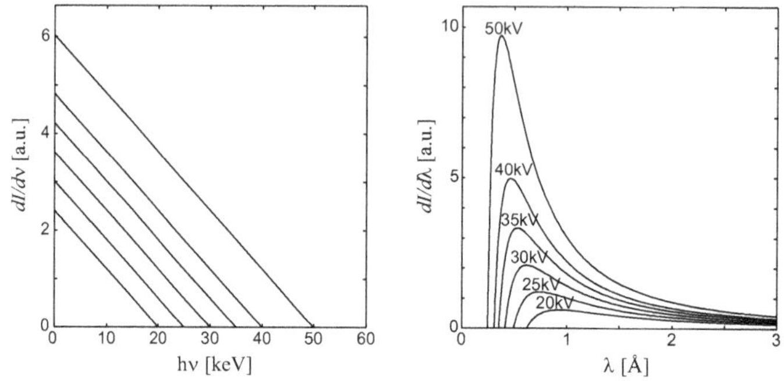

Fig. 3.16: Calculated spectral intensity (without absorption) of x-rays emitted from a massive anode as a function of frequency (*left*) and wavelength (*right*) for several voltages U from 20 to 50 kV. By variable transformation the linear spectral distribution $I(\nu)$ becomes a nonmonotonous distribution $I(\lambda)$. No absorption effects are included.

distribution due to the transformation of differentials. With $\nu = c/\lambda \Rightarrow d\nu = -(c/\lambda^2)d\lambda$ we obtain

$$\frac{dI}{d\lambda} = \frac{dI}{d\nu}\frac{d\nu}{d\lambda} = \frac{c}{\lambda^2}\frac{dI}{d\nu} = \begin{cases} \text{const.} \times Z\left(\frac{\lambda-\lambda_0}{\lambda_0\lambda^3}\right) & \lambda \geq \lambda_0 \\ 0 & \lambda < \lambda_0 \end{cases} \tag{3.109}$$

Resulting spectra are shown in Figure 3.16. From the last equation, one finds the wavelength of maximum spectral intensity

$$\lambda_{\text{max}} = \frac{3}{2}\lambda_0 \,. \tag{3.110}$$

By integration over all frequencies up to ν_0, one finds the total intensity

$$I = \int \frac{dI}{d\nu}\,d\nu = \text{const.} \times Z \int_0^{\nu_0} (\nu_0 - \nu)\,d\nu \propto ZU^2 \,. \tag{3.111}$$

Using all the constants that we have omitted here for simplicity and dividing this result by the kinetic energy of the incident electrons yields the efficiency η of bremsstrahlung production by an x-ray tube

$$\eta \approx 10^{-6} \times Z \times U\,[\text{kV}] \,. \tag{3.112}$$

For a typical voltage of 100 kV and anode materials like tungsten ($Z = 74$), copper ($Z = 29$) or molybdenum ($Z = 42$), this yields $\eta < 1\%$. The rest is ultimately converted into heat.

This theory involves some rather crude approximations. Interactions of the accelerated electron with shell electrons of the anode atoms, as well as their deflection from the initial direction (as visible in the Monte Carlo generated electron tracks in Chapter 5) are completely neglected. In addition, describing the slowing down of electrons in the anode material by the Thomson–Whiddington law is only an approximation. However, it yields remarkably good agreement with experimental results. More advanced theories are discussed, e.g., in [4, 8].

Angular distribution

The angular intensity distribution of bremsstrahlung radiation is described by a theory of A. SOMMERFELD [31]. According to special relativity, the instantaneous intensity emitted by a single electron with speed $\beta = \frac{v}{c}$ and acceleration a (in the plane of the anode) is

$$I(r, \theta) \propto \frac{a^2}{r^2} \frac{\sin^2 \theta}{(1 - \beta \cos \theta)^6} \tag{3.113}$$

where θ is the angle to the direction of the electron. The maximum of the emitted intensity shifts more and more from 90° at $\beta = 0$ (dipole pattern) into forward direction $\theta = 0$ for $\beta \longrightarrow 1$. This is plotted in Figure 3.17 (left). However, equation (3.113) only holds for very thin anodes. When entering a massive anode, electrons are scattered from their original direction and slow down. Therefore, the angular intensity distribution becomes smeared out, and some intensity is radiated into forward and backward directions, e.g., like in the spectrum sketched in Figure 3.17 (right). Note that via the factor β in equation (3.113), the spectral distribution varies with the angle θ.

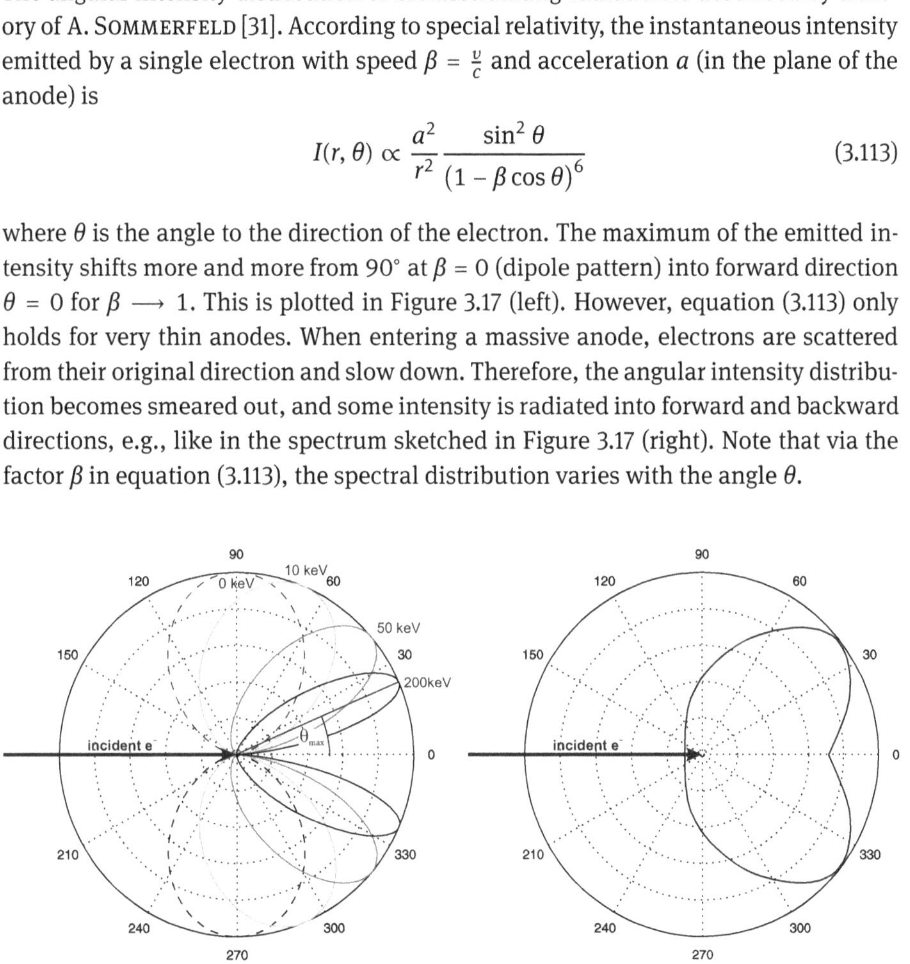

Fig. 3.17: *Left*: Polar diagram illustrating x-ray emission of bremstrahlung in a thin anode according to equation (3.113). The emission cone becomes more and more tilted into the forward direction for higher electron energies, as a result of relativistic transformation. *Right*: For a massive anode, successive events of deacceleration and directional changes of the electrons lead to a smearing of the emission pattern and nonzero intensity in forward and backward directions. The curve is artistically drawn, no absorption is included.

3.3.2 Characteristic spectrum of an x-ray tube

A set of discrete (sharp) emission lines, which are characteristic for the anode material (Figure 3.15) are superimposed on the continuous bremsstrahlung spectrum. In a collision with an atom, the incident electron can ionize the inner shells, creating a vacancy. The subsequent relaxation of an electron from an outer shell to the vacancy leads to transition radiation in the x-ray range with characteristic energies $h\nu = E_i - E_f$ corresponding to the energy difference between the two shells (c.f., Section 3.2.3). For experiments requiring monochromatic radiation, one often uses the K_α line, which is several orders of magnitude more intense than the bremsstrahlung spectrum. However, only a very small fraction of the photons emitted into the solid angle of 2π can be utilized in a beam requiring an angular divergence of a few squared milliradian. For the hydrogen atom, the well known energy transitions are given by

$$\hbar\omega = E_i - E_f = \frac{m_e e^4}{32\pi^2 \epsilon_0^2 \hbar^2}\left(\frac{1}{n_f^2} - \frac{1}{n_i^2}\right), \tag{3.114}$$

where n_i and n_f are the principal quantum numbers of the incident and final stationary electron states, E_i and E_f the energies corresponding to these states, m_e the electron mass and Z the nuclear charge. The constant $\frac{m_e e^4}{32\pi^2 \epsilon_0^2 \hbar^2} = hcR_\infty = 13.606\,\text{eV}$, known historically as the Rydberg constant from earlier studies of hydrogen spectra, gives the ionization potential (the energy required to remove an electron from an atom) of the ground state ($n_i = 1$, $n_f = \infty$) of the hydrogen atom ($Z = 1$). In terms of the Rydberg constant, the characteristic emission lines of a single electron atom of nuclear charge Z are

$$h\nu = (13.606\,\text{eV})Z^2\left(\frac{1}{n_f^2} - \frac{1}{n_i^2}\right). \tag{3.115}$$

According to Moseley, one additionally has to consider the screening of the inner electrons, so that

$$h\nu = (13.606\,\text{eV})(Z - \sigma)^2\left(\frac{1}{n_f^2} - \frac{1}{n_i^2}\right), \tag{3.116}$$

with σ the screening constant (c.f., Section 3.2.3). Importantly, not every electronic transition is possible; transitions are subject to the selection rules reflecting the corresponding conservation laws and symmetries, as presented in textbooks on atomic physics.

3.3.3 Synchrotron radiation

The acceleration of a relativistic charged particle in an external magnetic field leads to emission of EM radiation known as synchrotron radiation (SR). First discovered in particle accelerators (where the magnetic field has to be increased synchronously with the particle energy to keep it on the same circular track), synchrotron radiation is nowadays generated in electron storage rings, into which highly relativistic electrons of fixed energy

$$E = E_{\text{kin}} + m_e c^2 = \frac{m_e c^2}{\sqrt{1 - \left(\frac{v}{c}\right)^2}} = \gamma m_e c^2 \tag{3.117}$$

are injected and kept on periodic orbits. Electrons (or positrons) are first accelerated by radio-frequency (RF) cavities of a linear accelerator and booster synchrotron to the desired energy E, typically around a few GeV for hard x-rays, and then injected into a storage ring equipped with different magnets for generation of x-ray light by circular or sinusoidal acceleration. In the storage ring the energy loss from radiation is constantly re-fed into the electron beam by radio frequency cavities. For high electron energy in units of the rest mass energy, i.e., $\frac{E}{m_e c^2} = \gamma \gg 1$, we are dealing with the super relativistic limit, so that the relativistic energy–momentum relation

$$E = \sqrt{m_e^2 c^4 + p^2 c^2} \tag{3.118}$$

becomes asymptotically $p = \frac{E}{c}$, and the velocity in units of the speed of light becomes

$$\beta = \frac{v}{c} = \sqrt{1 - \frac{1}{\gamma^2}} \simeq 1 - \frac{1}{2\gamma^2} \,. \tag{3.119}$$

For concreteness, let us assume $E = 5$ GeV. With a rest energy $m_e c^2 \simeq 0.511$ MeV, this results in $\gamma \simeq 10^4$, and $\beta = 0.999999995$!

Synchrotrons are large instruments, as one can easily understand by considering a circular electron orbit in the super relativistic limit. The radius ρ of the orbit is found by setting the centripetal force equal the Lorentz force evB. For the nonrelativistic case, the centripetal force is mv^2/ρ, and expressed in terms of electron momentum $mv = p$, the relation $p = \rho eB$ follows. By substituting the relativistic momentum $p = \gamma mv$, and with $v \simeq c$, we have

$$\gamma mc = \rho eB \,, \tag{3.120}$$

and in practical units

$$\rho\,[\text{m}] = 3.3\,\frac{E\,[\text{GeV}]}{B\,[\text{T}]} \tag{3.121}$$

resulting in typical values of a few tens of meters for typical B fields in the range 0.1 to 1 T. In practice, such fields can only be confined to small regions, and many different magnet structures are required on a polygon for a stable orbit. The characteristic features of the radiation emitted depend on two key parameters: the cyclic frequency ω_0 of the orbiting electron and $\gamma = E/m_e c^2$. The instantaneous direction of the radiation cone is that of the instantaneous velocity of the electron, and the opening angle (in radians) of the cone is given by relativistic compression $\gamma^{-1} = m_e c^2/E$, i.e., 0.1 milli-rad for $\gamma = 10^{-4}$. Thus, the radiation from an electron orbiting at relativistic speed in

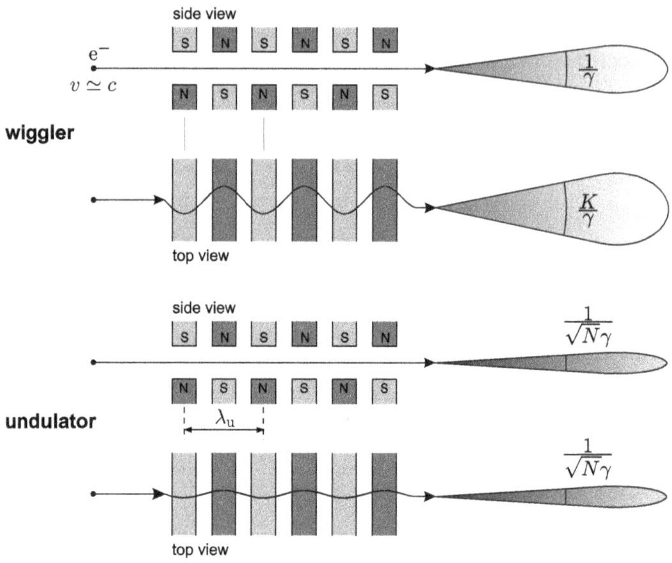

Fig. 3.18: Schematic setup of a wiggler/undulator. Adapted from [1].

a circle is similar to a sweeping search light. The emitted spectrum can be calculated from relativistic electrodynamics and is spectrally broad, ranging from the far infrared to the hard x-ray region, up to photon frequencies of $\gamma^3\omega_0$. The angular frequency of an electron in the storage ring ω_0 is typically on the order of 10^6 cycles per second, resulting in x-ray frequencies up to 10^{18} Hz.

The electron beam traverses numerous magnetic fields. While magnetic dipoles or bending magnets are needed to change the beam direction (and ultimately reach a closed circle), magnetic quadrupoles and sextupoles are arranged on a lattice to tailor the beam properties (such as controlling dispersion and beam size, re-focusing, etc.). Synchrotron radiation is 'harvested' at bending magnets, or even more efficiently at specific magnetic chicanes called wigglers and undulators (Figure 3.18), which induce oscillatory trajectories with periods on the cm scale. For a textbook treatment of SR including bending magnets, wigglers, undulators, as well as free electron laser (FEL) radiation, we refer to Als-Nielsen and McMorrow [1].

In a wiggler, magnets are placed in a row with alternating directions. In this way, the radiated intensity is increased by the number of turns N. For small deflections of the electron trajectory and for geometrically very precise arrangements of the magnets, the radiation emitted by a given electron at one oscillation is in phase with the radiation from the following oscillations. This is the case in a so called undulator. As a result, the amplitudes of the radiated waves add up coherently, and the gain factor is increased to N^2. This coherent addition of amplitudes, however, is only valid at a particular fundamental wavelength λ_1 and the corresponding higher harmonics, so that undulator radiation is quasi monochromatic. For bending magnet and wiggler radiation, the basic parameters are γ and the cyclic frequency ω_0. For undulator radiation, the basic parameters are γ and the undulator spatial period λ_u. To obtain the fundamental wavelength λ_1, consider an electron traversing an undulator as shown in Figure 3.18. The electron emits radiation of a wavelength $\lambda' = \lambda_u/\gamma$, corresponding to the period of forced oscillation that it experiences. Note that in the co-moving frame of the electron, the magnetic periodicity is relativistically contracted (Lorentz contraction). To the observer on the optical axis in the laboratory frame, this radiation is again subject to a relativistic Doppler compression $\nu = 2\gamma\nu'$, for $v \simeq c$. These two factors together result in the generation of x-rays with $\lambda = 2\gamma^2\lambda_u$. In a more complete treatment, which also takes the sinusoidal electron trajectory into account, this simple factor is modified to the so called undulator equation, which has an additional unitless parameter K taking into account the electron deflection [1]

$$\lambda_1 = \frac{\lambda_u}{2\gamma^2}\left(1 + \frac{K^2}{2}\right), \tag{3.122}$$

with $\beta = \frac{v}{c}$ and $K = \frac{eB_0}{mck_u}$. With γ^{-2} on the order of 10^{-8}, and λ_u on the order of 1 cm, λ_1 is on the order of 1 Å. By changing the magnetic field by varying the gap between the poles, the wavelength λ_1 is tunable.

The most important properties of SR can be summarized as follows:

- SR is intense at the moment when the instantaneous electron velocity points directly towards the observer, since at that instant the Doppler effect is maximal.
- The angular size of the radiation cone is in the order of γ^{-1} for wigglers and $\simeq \gamma^{-1}/\sqrt{N}$ for undulators (Figure 3.18).
- A typical frequency in the spectrum of a bending magnet is γ^3 times the cyclic frequency ω_0 of the orbiting electron, and $\simeq c/(2\gamma\lambda_u)$ of an undulator.
- SR is linearly polarized in the horizontal plane.
- SR is pulsed, the pulse duration being the electron bunch length divided by c.

Several aspects of an x-ray source determine the quality of the x-ray beam it produces. These aspects can be combined into a single quantity, called the brilliance, which allows us to compare the quality of x-ray beams from different sources. First, we consider the total number of photons emitted per second. Next, we consider the collimation of the beam both for the horizontal and for the vertical direction, quantifying how much the beam diverges, or spreads out, as it propagates. Typically, the collimation of the beam is given in milliradian. Third, the source area is important; it is usually given in mm^2. If it is small, the x-ray beam can be focused to a correspondingly small image size. Finally, we have to address the spectral distribution, i.e., which photon energies contribute to the measured intensity. The convention is, therefore, to define the photon energy range as a fixed relative energy bandwidth, which has been chosen to be 0.1%. Altogether then, one defines the brilliance of a source as

$$\text{brilliance} = \frac{\text{photons/second}}{(\text{mrad})^2\,(\text{mm}^2\text{ source area})\,(0.1\%\text{ bandwidth})}. \tag{3.123}$$

The maximum brilliance of third generation synchrotron radiation is about 10 orders of magnitude higher than that from a rotating anode at the K_α line! For this reason, SR has become a powerful analytical tool for biomedical research in general, including protein crystallography, pharmaceutical and biophysical research. However, it currently plays only a limited role in biomedical imaging, restricted to high profile research, where high resolution, sensitivity, and elemental specificity is required. Finally, apart from some exploratory studies at medical beamlines, it certainly plays no role in medical diagnostics. This may change in future, depending upon the availability of more compact synchrotron sources, which are amenable to the scale of university hospitals, such as offered by the technology of so called compact light sources (CLS), [11, 17, 30] where the magnetic field of an ultrashort infrared laser pulse serves as the 'undulator' for an MeV electron beam. This combination is well suited to produce hard x-rays, as can easily be seen from plugging in realistic values in equation (3.122). Equivalently, one can describe this x-ray emission as an inverse Compton effect. In contrast to GeV electron beams, MeV electrons are rather cheap and can be stored in much more compact rings. Finally, other types of sources such as laser-generated plasma sources or higher harmonic generation may develop some potential for biomedical imaging, even if they currently cannot reach the required photon energies [2].

3.4 Detection of x-rays

X-ray detectors can serve their purposes in many different ways and operation modes, see, e.g., the textbook by Knoll [19] for an in depth treatment. In the primary detection mechanism of converting x-ray radiation into electric, chemical, thermal or optical signals, we again encounter the physical principles of x-ray interaction with matter. We first have to specify what quantity is actually measured. Is the detector output proportional to the instantaneous radiation intensity such as a photocurrent or to the integrated flux density (fluence) such as in photographic film? Can the detector discriminate photons one-by-one (single photon counting) and can the photon energy be measured (energy sensitive detection)? Next, we have to specify the spatial properties of the detector. Can the detector record radiation only spatially integrated over the detection volume (point detector), or with spatial resolution along one (line detector) or two directions (area detector)? Finally, what are the temporal properties of detection? Over what time does the detector signal relax back to the resting state after input of a radiation quantum (dead time)? For biomedical imaging we almost always need area detectors, starting from Röntgen's use of photographic film to modern semiconductor detectors such as image plates and pixel detectors. However, for high end specifications of quantum efficiency (ratio of recorded photons over incoming photons), dynamic range (the ratio of strongest allowed signal to lowest detectable signal) or the dark signal (often called dark current), one must sometimes turn to point detectors. Recording images by point detectors would, in principle, be possible by scanning the point detectors and using detector slits. However, this would require prohibitively long exposures and high doses. Thus, point detectors are only used in dosimetry and radiation protection.

The detector types that are briefly addressed below are all based on the photoelectric effect, i.e., absorption of an x-ray photon, which then leads to an electric signal such as in an ionization chamber or a semiconductor detector, to optical signals such as in a scintillation camera or a chemical turnover such as in photographic film. In exploiting the photoelectric effect for detection, the photon is lost, so that its direction or momentum cannot be inferred from two subsequent measurements. This is different for the Compton effect, where measurement of the Compton scattering (by way of the recoil electron) can be combined with the subsequent detection of the Compton scattered radiation (by photo-absorption), cf. Figure 3.11. This gives access to the photon's direction (momentum), which would be extremely useful, in particular for SPECT [24]. Such detectors are not yet readily available, but under current development. Contrarily, for other purposes one may be interested in detecting radiation without any regard for the detailed interaction processes, summing up over all conversion routes. This is achieved by a calorimeter, where the temperature increase in response to irradiation is recorded.

Photographic film

Photographic film is based on the turnover of a redox reaction and its associated color changes (darkening) of the film. The most prominent example is the reaction

$$2\,\mathrm{Ag}^+ + 2\,\mathrm{Br}^- \xrightarrow{h\nu} 2\,\mathrm{Ag} + \mathrm{Br}_2\,. \tag{3.124}$$

Since silver bromide (AgBr) can be dispersed as very small crystallites, for example, in a gelatine matrix, the spatial resolution can be quite high. A concern in biomedical imaging is the proper tuning of parameters so that the response (in darkening) is linear to the photon fluence. Interesting microscopic structural changes and dynamics of silver bromide crystals during x-ray illumination have been revealed in a recent study [16]. Over the last decade, x-ray films have gradually been replaced in most medical applications by semiconductor detectors, where the digital readout makes the chemical processing (film development) and additional digitalization efforts obsolete.

Scintillation detectors

Scintillation converts x-ray photons to visible light. The classical scintillation detector is the scintillation counter (Figure 3.19), a point detector in which a bulk scintillation crystal (often NaI or CsI) generates light that ionizes electrons in a photo cathode at one side of the crystal. The electron signal is then amplified on a photomultiplier tube. The electric output signal is further amplified as a voltage signal, which is treated by a pulse height analyzer for digital conversion, the so called single channel analyzer

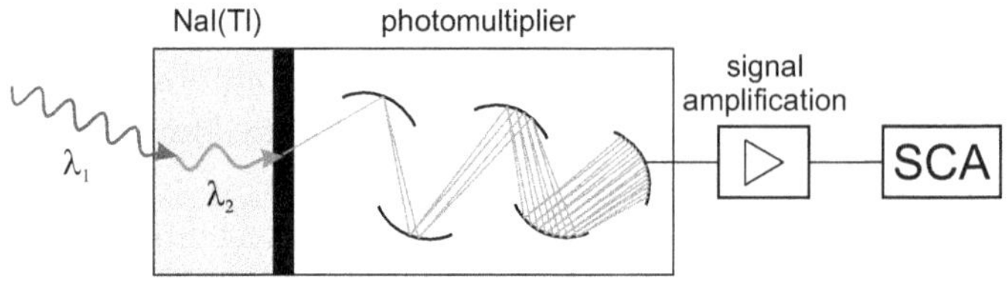

Fig. 3.19: Schematic of a scintillation counter. Conversion from x-ray wavelength (λ_1) to visible light (λ_2) occurs in the scintillation crystal. High amplification for a single event makes the scintillation counter a single photon counting detector.

(SCA). The voltage signal relaxes over a dead time of few ms, so that the maximum count rate to be detected is on the order of a few 10^5 photons per second. Ionization counters typically have a high detection volume, high quantum efficiency (stopping power of a bulk crystal!), short dead time, low background and, hence, a high dynamic range, in combination with medium energy resolution.

The scintillation effect can also be used for area detectors, but in this case, the crystal has to be very thin to achieve lateral spatial resolution. In this case, there is not enough space to convert the generated light to an electron signal that can easily be amplified. Instead, an imaging system for visible light (microscope objective or fiber optics) is used to form an image of the scintillation area (the plane of x-ray input) to a CCD or sCMOS chip. Such detectors exhibit the highest spatial resolution down to $\simeq$ 0.3 µm pixel size, as limited by visible light diffraction. However, many of the advantages of the scintillation counter are lost, at the expense of spatial resolution. For clinical applications, the low efficiency is prohibitive in view of dose limitations.

Ionization chambers

After chemical and optical detection output, we now turn to electronic output. The historically first electronic radiation detector was the Geiger–Müller counter, which uses a gas for ionization and charge separation by an external high voltage. To achieve high enough fields for charge separation, a thin wire is used as the anode. In the operation mode of an ionization chamber, the voltage applied is just high enough to prevent recombination of charges. In this way, the collected charge corresponds to the charge initially generated by ionization, and the current recorded between the electrodes is, hence, proportional to the incoming intensity. For the standard operation mode of the Geiger–Müller counter, the applied voltage is higher, resulting in a kinetic energy of the accelerated electrons, which is now high enough to ionize further gas atoms, leading to an electron avalanche. Due to the relatively low mobility of the cations, as well as screening effects, the dead time is considerable and limits the possible count rate. The Geiger–Müller counter is typically employed for small count rates of ionizing radiation, with the Poisson distributed events of single photon counting indicated by a characteristic acoustic click.

Semiconductor detectors

The principle of ionization and subsequent charge separation first realized in gas based counters has been transferred to solid state detection by semiconductor detectors, which are based on the generation of electron–hole pairs and subsequent separation in a bias voltage. Semiconductor detectors are presumingly the most advanced detectors for biomedical imaging. As an array of photo diodes, charge coupled devices (CCD), which are well known from visible light, can also be adapted to detect x-ray photons. To avoid absorption in nonactive layers on top of the chip (surface of chip fabrication), x-rays are often made to enter a thinned back side of the chip. For detection of hard x-rays, the thickness of the silicon sensor is important, i.e., absorption in the active layer has to be maximized. Upon illumination, the generated charge is stored first locally in each pixel (typically of 5–20 μm size), before being transferred pixel-by-pixel to an ADC. The different modes to achieve the transfer known from visible light cannot all be used for x-rays, since it is much more difficult to transfer and to park the charges in a 'shaded' area, where subsequent radiation cannot enter. Therefore, triggered shutters which close the CCD chip to exposure are used. The problem of shifting the charges across the chip to the ADC involves significant noise uptake and also induces idle readout time, which can easily dominate over the active exposure.

These problems are circumvented in pixel detectors, where each pixel has a separate ADC. These detectors are essentially noise free single photon counters with a high dynamic range. The disadvantages are a smaller number of pixels and a larger pixel size. Insensitive intermodule gaps also create a problem in large FOV recordings. It is only recently that pixel detectors have become available also for CdTe and GaAs, thus opening up a higher energy range.

For clinical radiography, the best solution is currently the flat panel detector, which offers a large pixel number and area. An array of thin film transistors (TFT) and photodiodes fabricated from amorphous semiconductors can be used to detect x-ray photons either directly, or after conversion into visible light, in most cases an amorphous gadolinium oxysulfide (Gadox) or cesium iodide (CsI), which is directly attached to the TFT layer. Compared to photon counting, flat panel detectors are energy integrating, which biases the detection towards high photon energy, an effect which is often favorable in radiography (to suppress the signal of low photon energy background). Flat panel detectors have a significantly reduced dose in clinical radiography.

3.5 Statistics of counting x-ray and gamma quanta

Following our brief presentation of x-ray generation and detection, we take note of the fundamental statistical properties of photon counting. In biomedical imaging, this is relevant for x-ray radiography, x-ray CT as well as the nuclear methods SPECT and

PET, which are based on radioactive tracers. The issue is most familiar from studying radioactive decay, which shall be briefly recalled here.

Let the probability that a radioactive nucleus decays within the (infinitesimal) time interval Δt be given by $p = a\Delta t$. For N identical radioactive atoms, this gives a probability

$$P(n) = \binom{N}{n} p^n (1-p)^{N-n} \tag{3.125}$$

to have exactly n decays in Δt. Equation (3.125) denotes the binomial distribution and can be understood as follows: If one arbitrarily picks n out of N nuclei, the factor p^n denotes the probability that these n nuclei decay and the factor $(1-p)^{N-n}$ the probability that the remaining $(N-n)$ nuclei do not decay in Δt. Since the atoms are indistinguishable, this is multiplied by the binomial coefficient $\binom{N}{n} = \frac{N!}{(N-n)!\,n!}$, which gives the number of possibilities to pick n out of N nuclei.

Now, let the mean number of decays in Δt be $Np = Na\Delta t = \mu$. We consider the limit $N \to \infty$ (since N is usually very large) and $p \to 0$, such that $\mu = Np = \text{const}$. Replacement of $p = \frac{\mu}{N}$ and some rearrangements yield

$$\begin{aligned} P(n) &= \frac{N!}{(N-n)!\,n!}\left(\frac{\mu}{N}\right)^n \left(1-\frac{\mu}{N}\right)^{N-n} \\ &= \frac{N(N-1)(N-2)\dots(N-n+1)}{n!}\frac{\mu^n}{N^n}\left(1-\frac{\mu}{N}\right)^{N-n} \\ &= \frac{N}{N}\cdot\frac{N-1}{N}\cdot\frac{N-2}{N}\cdots\frac{N-n+1}{N}\cdot\frac{\mu^n}{n!}\left(1-\frac{\mu}{N}\right)^N\left(1-\frac{\mu}{N}\right)^{-n}. \end{aligned} \tag{3.126}$$

Keeping n and μ fixed, the first n factors as well as the last factor approach unity for $N \to \infty$. Together with one possible definition of the exponential function, $\lim_{N\to\infty}(1-\frac{\mu}{N})^N = e^{-\mu}$, this yields the Poisson distribution

$$\lim_{N\to\infty} P(n) = \frac{\mu^n}{n!}e^{-\mu}\,. \tag{3.127}$$

In our example, it describes the probability of exactly n nuclear decays, given that the mean number of decays in the chosen interval is μ, the decay probability for each nucleus in Δt is very small and the total number of nuclei N is very large. It can be shown that the sum of two Poisson distributed random variables is again Poisson distributed (this does not hold, for example, for a division), with parameter $\mu = \mu_1 + \mu_2$.

This formulation used for radioactive decay (i.e., the generation of a photon) can also be applied to detection (i.e., the annihilation of a photon). Let p' denote the probability to detect a photon in a time interval Δt, using a detector of sensitivity d. Then p is described by $p = a\Delta t d = a'\Delta t$. Hence, Poisson statistics applies equally to photon generation and detection, including data recorded in the presence of additional absorption, scattering, imperfect detectors with low quantum efficiency, etc. The Poisson distribution is, therefore, the most important fundamental statistical property of medical physics and imaging. Let us note a few of its properties. With $\langle \dots \rangle$ denoting the mean, the moments of the Poisson distribution are

$$\langle n^k \rangle = \sum_{n=0}^{\infty} n^k P(n) = e^{-\mu} \sum_{n=0}^{\infty} n^k \frac{\mu^n}{n!} = e^{-\mu} \left(\mu \frac{\partial}{\partial \mu}\right)^k \sum_{n=0}^{\infty} \frac{\mu^n}{n!} = e^{-\mu} \left(\mu \frac{\partial}{\partial \mu}\right)^k e^{\mu} . \quad (3.128)$$

We conclude that $\langle n^k \rangle$ is a polynomial of degree k in μ, since all derivatives of the exponential leave the exponential unaffected. Since the exponential cancels with $e^{-\mu}$, only powers of μ remain. These polynomials are called Touchard polynomials $T_k(\mu)$. The first ones are easy to calculate:

$$\langle n^0 \rangle = 1, \qquad \langle n^1 \rangle = e^{-\mu} \mu \frac{\partial}{\partial \mu} e^{\mu} = \mu, \qquad \langle n^2 \rangle = e^{-\mu} \mu \frac{\partial}{\partial \mu} \mu \frac{\partial}{\partial \mu} e^{\mu} = \mu + \mu^2 . \quad (3.129)$$

One can show that the following recursion formula holds

$$T_{k+1}(\mu) = \mu\Big(T_k(\mu) + T_k'(\mu)\Big) . \quad (3.130)$$

From this we can deduce the variance of the Poisson distribution

$$\langle n^2 \rangle - \langle n \rangle^2 = \mu = \langle n \rangle , \quad (3.131)$$

which turns out to be identical to the mean, a well known characteristic property of this distribution. The standard deviation σ is hence

$$\sigma = \sqrt{\langle n^2 \rangle - \langle n \rangle^2} = \sqrt{\mu} . \quad (3.132)$$

This results in a relative uncertainty of a measurement of

$$\frac{\sqrt{\langle n^2 \rangle - \langle n \rangle^2}}{\langle n \rangle} = \frac{1}{\sqrt{\mu}} . \quad (3.133)$$

In other words, the higher the photon count (i.e., μ), the higher the absolute uncertainty, but the lower the relative uncertainty. This almost trivial statement has significant consequences, since it sets a limit to the possible reduction in the x-ray dose. The smaller the dose, the smaller the information for diagnostics. Therefore, modern developments are directed at optimal exploitation of the Poisson data, taking the knowledge of its statistical properties into account.

For large μ, it can be shown that the Poisson distribution (in the vicinity of μ) can be approximated by a Gaussian (normal) distribution centered at μ and with variance μ

$$P(x) = \frac{1}{\sqrt{2\pi\mu}} \exp\left(-\frac{(x-\mu)^2}{2\mu}\right) . \quad (3.134)$$

3.6 Summary and implications for clinical applications

The different interaction mechanisms between photons and matter determine many technological aspects of the medical application of x-rays. Here, we briefly summarize the most relevant results obtained so far using the example of recording a classical radiograph as sketched in Figure 3.20 (right), before we turn towards more advanced methods in subsequent chapters.

The most important characteristics of the different interaction channels between x-rays and matter and their dependencies on material and photon energy E are summarized in Table 3.2. In addition, the relative contributions to the total linear attenuation coefficient μ_{H_2O} of water, which can be used as a good approximation for most soft tissues, are plotted in Figure 3.20 (left). In an ideal energy range for diagnostic purposes based on absorption contrast, the attenuation is low enough for a sufficiently large fraction of photons to pass through the patient, but the photoelectric effect still contributes strongly to the total absorption. Due to the strong variation of photoelectric absorption with Z^4, it offers good contrast, e.g., between bone and soft tissue. In addition, since the incident photons are completely absorbed, no scattering reduces the image quality.

Let us estimate the photon energy needed to meet the first requirement. Assuming a typical 'sample' thickness of $d = 10\,\text{cm}$ and using water ($\rho = 1\,\text{g}\cdot\text{cm}^{-3}$) as a good substitute for most soft tissues, we have

$$\frac{I}{I_0} = e^{-\mu d} \overset{!}{=} \mathcal{O}\left(\frac{1}{e}\right) \quad \Rightarrow \quad \left(\frac{\mu}{\rho}\right)\rho d \overset{!}{=} \mathcal{O}(1) \tag{3.135}$$

$$\Rightarrow \quad \frac{\mu(E)}{\rho} \overset{!}{\approx} 0.1\,\text{cm}^2/\text{g} \quad \overset{\text{Figure 3.3}}{\Rightarrow} \quad E \approx 100\,\text{keV}\,. \tag{3.136}$$

This estimate yields photon energies on the order of 100 keV. Indeed, photon energies of up to about 140 kV are typically used to record a diagnostic radiograph, as well as a CT scan (note that this determines the maximum energy of a continuous bremsstrahlung spectrum, the mean photon energy is somewhat lower). However, above ~ 30 keV, photoelectric absorption is no longer the dominant interaction that

Table 3.2: Summary of the interactions of x-ray photons and matter: threshold energies, emitted particles and subsequent secondary processes, and variation of the linear attenuation coefficient μ with element and energy. Note that emission of charged particles is always accompanied by bremsstrahlung, which is not explicitly stated in the table.

	photoel. effect	**Thomson**	**Compton**	**pair production**
threshold energy	binding energies	—	—	$2m_e c^2(1 + \frac{m_e}{m_N})$
emitted particles	e^-	γ	γ, e^-	e^-, e^+
Secondary processes	charact. x-rays, Auger electrons	—	charact. x-rays, Auger electrons	annihilation radiation
element dep. of μ/ρ	$\propto Z^4/A$	$\propto Z^2/A$	$\propto Z/A \approx$ const.	approx. $\propto Z^2/A$
energy dep. of μ/ρ	$\propto 1/E^3$	$\propto 1/E^2$	complex	approx. $\propto \ln E$
type of scattering	—	coherent	incoherent	—

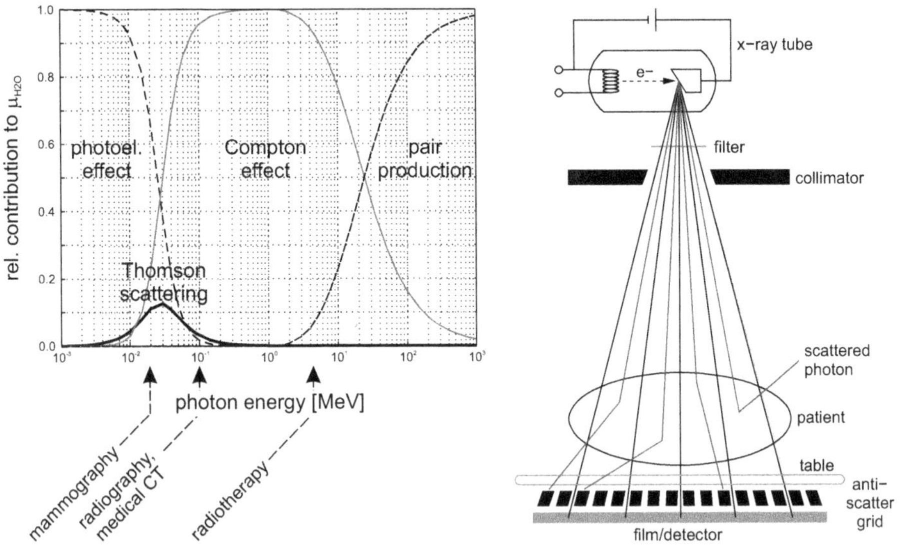

Fig. 3.20: *Left*: Relative contribution of the different primary interactions between photon and matter to the total linear attenuation coefficient μ of water, which is close to human soft tissues. *Right*: Scheme for recording a radiograph. A filter made of Al or Cu removes low energy components from the x-ray spectrum. Collimation ensures that radiation exposure is limited to the region of interest of the patient's body. An (optional) antiscatter grid reduces the amount of scattered photons and thus improves image quality [29].

leads to attenuation of the primary intensity. Compton scattering becomes more and more important and accounts for almost the entire attenuation above 100 keV, and μ is no longer very sensitive to Z but rather scales with the mass density (cf., Figure 3.3). Therefore, one could say that the energies used are a compromise between the acceptable dose exposition of the patient and sufficient image quality.[17]

The scattered photons generated in the patient have a broad angular distribution (Figure 3.12) and lead to a rather homogeneous illumination of the x-ray film that superimposes the "true" absorption contrast due to differences in μ and, thus, reduce the image quality. Therefore, antiscatter grids that allow only photons in the correct projection geometry to pass and remove scattered photons from the beam can be placed in front of the detector or film. Filters are employed to remove low energy photons ('beam hardening') that would not contribute to image formation, but lead to unnecessary and unacceptable doses at the entrance of the x-ray beam into the patient's body. Collimation is used to reduce the extent of the x-ray beam to the minimum re-

17 An exception with regard to photon energy is mammography, where voltages of about 25 kV are used (e.g., the characteristic K_α line of molybdenum), since the tissue layer to be traversed is only a few cm thick.

quired for a given diagnostic question. This reduces both the dose to the patient and the number of Compton-scattered photons that might compromise the image quality.

To record a radiograph, x-ray films have largely been replaced by digital detectors based on semiconductor technology. The image quality is determined by the three quantities contrast, resolution and noise discussed in Section 2.6. In order to maintain a good standard, quality assurance measurements on a regular basis are mandatory. For some indications, contrast agents are used. These are biocompatible chemical substances that contain an atomic species with a rather high atomic number, most often iodine ($Z = 53$) or barium ($Z = 56$), which strongly enhances photoelectric absorption. One example is the diagnosis of pathologies of blood vessels by digital subtraction angiography.

References

[1] J. Als-Nielsen and D. McMorrow. *Elements of Modern X-ray Physics.* John Wiley & Sons, 2nd edition, 2011.

[2] D. Attwood and A. Sakdinawat. *X-Rays and Extreme Ultraviolet Radiation: Principles and Applications.* Cambridge University Press, 2016.

[3] R. Behling. *Modern Diagnostic X-Ray Sources: Technology, Manufacturing, Reliability.* CRC Press, 2016.

[4] M. A. Blochin. *Physik der Röntgenstrahlen.* VEB Verlag Technik Berlin, 1957.

[5] International Commission on Radiation Units and Measurements. ICRU Report 44: Tissue Substitutes in Radiation Dosimetry and Measurement, 1988.

[6] A. H. Compton. A Quantum Theory of the Scattering of X-rays by Light Elements. *Phys. Rev.*, 21:483–502, 1923.

[7] W. Demtröder. *Experimentalphysik 2 – Elektrizität und Optik.* Springer, 2nd edition, 2002.

[8] N. A. Dyson. *X-rays in Atomic and Nuclear Physics.* Cambridge University Press, 2nd edition, 1990.

[9] A. Thompson et al. X-ray data booklet. Lawrence Berkeley National Laboratory, 2009. Available online at http://xdb.lbl.gov/.

[10] O. Glasser. *Wilhelm Conrad Röntgen und die Geschichte der Röntgenstrahlen.* Springer, 3rd edition, 1995.

[11] W.S. Graves, J. Bessuille, P. Brown, S. Carbajo, V. Dolgashev, K.-H. Hong, E. Ihloff, B. Khaykovich, H. Lin, K. Murari, E.A. Nanni, G. Resta, S. Tantawi, L.E. Zapata, F.X. Kärtner, and D.E. Moncton. Compact x-ray source based on burst-mode inverse compton scattering at 100 khz. *Phys. Rev. ST Accel. Beams*, 17:120701, 2014.

[12] D. J. Griffith. *Introduction to Electrodynamics.* Pearson, 4th edition, 2013.

[13] H. Haken and H. C. Wolf. *Atom- und Quantenphysik.* Springer Verlag, 8th edition, 2003.

[14] S. P. Hau-Riege. *Nonrelativistic Quantum X-Ray Physics.* Wiley-VCH, Berlin, 1st edition, 2014.

[15] W. Heitler. *The Quantum Theory of Radiation.* Oxford University Press, 3rd edition, 1954.

[16] Z. Huang, M. Bartels, R. Xu, M. Osterhoff, S. Kalbfleisch, M. Sprung, A. Suzuki, Y. Takahashi, T. N. Blanton, T. Salditt, and J. Miao. Grain rotation and lattice deformation during photoinduced chemical reactions revealed by in situ x-ray nanodiffraction. *Nature Materials*, 14:691–695, 2015.

[17] Z. Huang and R. D. Ruth. Laser-Electron Storage Ring. *Phys. Rev. Lett.*, 80:976–979, 1998.

[18] J. D. Jackson. *Klassische Elektrodynamik*. De Gruyter, 3rd edition, 2002.
[19] G. F. Knoll. *Radiation Detection and Measurement*. John Wiley & Sons, 1986.
[20] H. A. Kramers. On the theory of x-ray absorption and of the continuous x-ray spectrum. *Phil. Mag.*, 6:836–871, 1923.
[21] H. Krieger. *Grundlagen der Strahlungsphysik und des Strahlenschutzes*. Springer Spektrum, 4th edition, 2012.
[22] H. Krieger. *Strahlungsquellen für Technik und Medizin*. Springer Spektrum, 2nd edition, 2013.
[23] C. Leroy and P. G. Rancoita. *Priciples of Radiation Interaction in Matter and Detection*. World Scientific, 3rd edition, 2012.
[24] J. S. Maltz. Compton Emission Tomography. In A. Brahme, editor, *Comprehensive Biomedical Physics. Volume 1: Nuclear Medicine and Molecular Imaging*, chapter 1.05, pages 103–121. Elsevier, 2014.
[25] L. Marcu, E. Bezak, and B. Allen. *Biomedical Physics in Radiotherapy for Cancer*. Springer, 1st edition, 2012.
[26] M. Otendal, T. Tuohimaa, U. Vogt, and H. M. Hertz. A 9 kev electron-impact liquid-gallium-jet x-ray source. *Review of Scientific Instruments*, 2008.
[27] E. B. Podgorsak. *Radiation Oncology Physics: A Handbook for Teachers and Students*. International Atomic Energy Agency (IAEA), Vienna, 2005.
[28] G. G. Poludniowski and P. M. Evan. Calculation of x-ray spectra emerging from an x-ray tube. Part i. Electron penetration characteristics in x-ray targets. *Med. Phys.*, 34:2164–2174, 2007.
[29] M. Reiser, F.-P. Kuhn, and J. Debus. *Duale Reihe Radiologie*. Thieme, 3rd edition, 2011.
[30] S. Schleede, F. G. Meinel, M. Bech, J. Herzen, K. Achterhold, G. Potdevin, A. Malecki, S. Adam-Neumair, S. F. Thieme, F. Bamberg, K. Nikolaou, A. Bohla, A.Ö. Yildirim, R. Loewen, M. Gifford, R. Ruth, O. Eickelberg, M. Reiser, and F. Pfeiffer. Emphysema diagnosis using x-ray dark-field imaging at a laser-driven compact synchrotron light source. *Proc. Nat. Acad. Sci.*, 109(44):17880–17885, 2012.
[31] A. Sommerfeld. About the production of the continuous x-ray spectrum. *Proc. Nat. Acad. Sci. USA*, 15:393–400, 1929.

Symbols and abbreviations used in Chapter 3

$\underline{a}$	acceleration
A	atomic mass number (nucleon number)
$\underline{A}$	vector potential
ADC	analog-to-digital converter
α	fine structure constant
$\underline{B}$	magnetic induction
$\beta = \frac{v}{c}$	ratio of particle velocity and speed of light
c	speed of light in vacuum
e^-	electron
e^+	positron
e	elementary charge
E	energy
E_b	binding energy of an electron in an atom
$\underline{E}$	electric field
EM	electromagnetic

eV	electron volt
$\mathcal{E} = \frac{\hbar\omega}{m_e c^2}$	photon energy in terms of electron rest mass
ϵ_0	electric permittivity of vacuum
f	atomic form factor
f'	real part of atomic dispersion correction
f''	imaginary part of atomic dispersion correction
γ	Lorentz factor
h	Planck's constant
$\hbar = \frac{h}{2\pi}$	reduced Planck's constant
$\mathcal{H}'$	perturbation Hamiltonian
$i = \sqrt{-1}$	imaginary unit
I_0	incident photon intensity (photons per second)
$\underline{J}$	current density
K	undulator parameter
$\underline{k}, \lvert\underline{k}\rvert = \frac{2\pi}{\lambda}$	wavevector
λ	wavelength
$\lambda_C = \frac{h}{m_e c}$	Compton wavelength
m_e	electron rest mass
m_N	mass of atomic nucleus
μ	linear attenuation coefficient
μ_0	magnetic permeability of vacuum
n	neutron
N_A	Avogadro's number
∇	nabla operator
ν	frequency
$\omega = 2\pi\nu$	angular frequency
p	proton
P	polarization factor
$\underline{q}$	scattering vector
ρ	mass density
ρ_a	atomic number density
ρ_e	electron density
$\underline{p}$	momentum
r_0	classical electron radius (Thomson scattering length)
$\underline{S}$	Poynting vector
SR	synchrotron radiation
σ_a	total photon cross section per atom
$(\frac{d\sigma}{d\Omega})_a$	differential cross section per atom
U	voltage
$\underline{v}$	velocity
Z	atomic charge number (proton number)

4 Tomography

This chapter presents the fundamental concepts of computerized tomography (CT). CT provides three-dimensional (3d) reconstructions of objects or bodies from a series of different views (exposures), each at a different angle. More precisely, the spatial distribution of an observable f is reconstructed in n-dimensional space from projections into $(n-1)$-dimensional subspaces. The observable f can, for example, represent the electron density, the mass density, the x-ray attenuation coefficient, a radioactive tracer density, a fluorescence intensity and so forth, depending on the type of tomographic recording. Importantly, each projection must be given by an integral over a set of parallel hyperplanes. In 2d, the data must, hence, be given as line integrals through the object. Tomography is based on the mathematical properties of the Radon transform (RT). Important tomographic imaging modalities in medical applications are x-ray CT and 3d nuclear diagnostic imaging, namely single photon emission CT (SPECT) or positron emission tomography (PET). The concepts of tomography also apply to many other types of recording data, as long as they can be traced back to projections or line integrals. Sometimes tomography is used synonymously with 3d reconstruction, even if the spatial distribution of the signal is encoded differently, such as, for example, in MRI. Tomography was first considered by the Bohemian–Austrian mathematician Johann Radon in his famous 1917 publication [40], long before any practical use was expected. The pioneering work of medical CT by Allan Cormack and Sir Godfrey Hounsfield in the 1960s and 1970s was honored with the Nobel Prize for Physiology and Medicine in 1979 [1, 20]. Today, tomography is an entire topic of its own, both in applied mathematics, as well as in several fields of science and engineering, and above all in medical imaging. In compiling this chapter, we found a number of seminal monographs and textbooks to be particularly helpful and recommendable for further reading, namely the classic textbook by A. C. Kak and M. Slaney [23], the modern textbook by T. Buzug [12] covering in useful detail all aspects from the fundamental level all the way to medical applications, and for the mathematical aspects of tomography in particular the treatments of F. Natterer [33, 34].

4.1 Tomography in a nutshell

Before addressing tomography in a more general setting and notation, and with more mathematical depth, we start by a simple straightforward introduction, formulated in Cartesian coordinates, following the derivation in [23], and in particular the short presentation in [6], the author of which has also kindly provided didactic graphics. As stated above, the task of tomography is to reconstruct the 3d distribution of the physical quantity $f(x, y, z)$ from its projections. Consider, for example, the projection along y, corresponding to the line integral over y for all (x, z), forming a projection

DOI 10.1515/9783110426694-004

image $\bar{f}$

$$\bar{f}(x,z) = \int_{-\infty}^{\infty} f(x,y,z)\,dy\,. \tag{4.1}$$

We assume that f is bounded and has a compact support, such that this integral exists. Again, f can, in general, be any physical quantity (x-ray absorption coefficient, tracer concentration, fluorescence or photoelectron yield per unit volume, ...), which is accessible by a line integral. For simplicity, we assume f to be a scalar quantity,[1] and for concreteness, we consider x-ray CT. We describe the x-ray beam as a plane monochromatic EM wave with vacuum wavenumber $k = 2\pi/\lambda$ propagating in the y direction. After penetration of an object characterized by $n(x,y,z) = 1 - \delta(x,y,z) + i\beta(x,y,z)$, amplitude and intensity recorded in the (x,z) plane can be written as

$$A(x,z) = A_0(x,z)\exp\left[ik\int_{-\infty}^{\infty} n(x,y,z)\,dy\right] \tag{4.2}$$

$$= A_0(x,z)\exp\left[ik\int_{-\infty}^{\infty}(1-\delta(x,y,z))\,dy - k\int_{-\infty}^{\infty}\beta(x,y,z)\,dy\right]\,, \tag{4.3}$$

$$I(x,z) = |A(x,z)|^2 = \underbrace{|A_0(x,z)|^2}_{=I_0(x,z)}\exp\left[-2k\int_{-\infty}^{\infty}\beta(x,y,z)\,dy\right]\,. \tag{4.4}$$

The attenuation and accumulated phase shift after traversing the object are, thus, determined by the projection integrals

$$\bar{\beta}(x,z) = -\frac{\ln\left(I(x,z)/I_0(x,z)\right)}{2k} = \int_{-\infty}^{\infty}\beta(x,y,z)\,dy\,, \tag{4.5}$$

$$\bar{\delta}(x,z) = -\frac{\varphi(x,z)}{k} = \int_{-\infty}^{\infty}\delta(x,y,z)\,dy\,. \tag{4.6}$$

The first integral is often written in terms of the linear attenuation coefficient $\bar{\mu}(x,z) = 2k\bar{\beta}(x,z)$, and describes the interaction mechanism exploited in conventional CT, while the second accounts for phase contrast, which will be treated in detail in Chapter 6. Importantly, for absorption contrast, a linear line integral only arises after taking the logarithm of the recorded intensity $I(x,z)$. Figure 4.1 illustrates this simple tomographic geometry, based on a parallel beam and linear line integrals of $f(x,y,z)$, following suitable discretization of the object into volume elements (voxels).

1 Generalizations to vector tomography can, of course, also be relevant in biomedical imaging, for example, a vector field of collagen fiber orientation, of cerebral blood flow or of bone crystallite orientation.

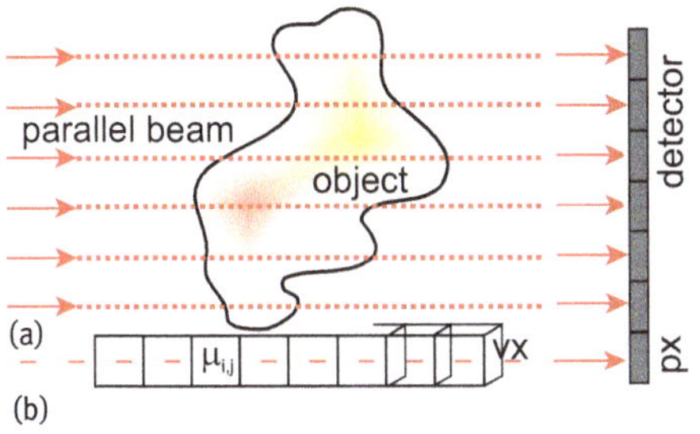

Fig. 4.1: (a) The simplest implementation of tomography is based on line integrals in a parallel beam setting. (b) A given detector pixel (px) records the object function f integrated over a line, i.e., a sum over the discrete voxel (vx) line. Instead of preparing a parallel (collimated) beam by suitable combinations of source, slits, monochromator, etc., the collimation could also be implemented by a collimator array in front of the detector selecting only perpendicular rays to the detector entrance face.

In order to reconstruct the full 3d distribution, multiple 'views' of the object are required. The corresponding set of projection images can be recorded with the beam and detector rotating around the object, or the object rotating in a fixed beam, or a suitable combination of both. In fact, rotations of a source detector gantry around a patient are typical for medical CT, while rotations of the object are typical for analytical micro CT applications. Let (x, y, z) denote the coordinate frame of the laboratory system, comprising source and detector, and (x', y', z') the coordinates of the object, rotated by an angle θ with respect to the laboratory frame around the shared axis $z = z'$

$$\begin{bmatrix} x' \\ y' \\ z' \end{bmatrix} = \begin{bmatrix} \cos\theta & -\sin\theta & 0 \\ \sin\theta & \cos\theta & 0 \\ 0 & 0 & 1 \end{bmatrix} \begin{bmatrix} x \\ y \\ z \end{bmatrix} . \tag{4.7}$$

Let the beam direction in the laboratory frame be along y, and each detector pixel (x, z) record the line integral of $f(x, y, z)$ over y. The xy planes then form a stack of parallel (tomographic) planes, where each plane can be treated separately in a 2d space. In each plane, the detector data is a 1d curve obtained from the 2d function $f_z(x, y) := f(x, y, z)$ by a line integral along y, for $\theta = 0$ simply given by

$$P_z(\theta = 0, x) = \int_{-\infty}^{\infty} f_z(x, y)\, dy . \tag{4.8}$$

For arbitrary angle θ, we have

$$P_z(\theta, x) = \mathcal{R}f = \int f_z(x\cos\theta - y\sin\theta, x\sin\theta + y\cos\theta)\, dy , \tag{4.9}$$

where $\mathcal{R}$ denotes the RT [40], here in 2d (but for a whole stack of planes). Figure 4.2 shows the 2d geometry in Cartesian coordinates. Image formation is described by $\mathcal{R}$,

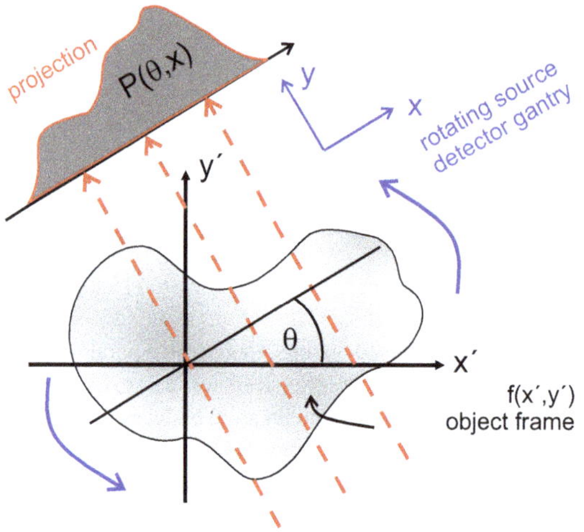

Fig. 4.2: The projection geometry in 2d based on line integrals, with the source detector gantry (x, y) rotated with respect to the Cartesian object frame (x', y') by θ.

which is a linear operator. Assuming that the corresponding inverse operator $\mathcal{R}^{-1}$ required for object reconstruction exists, one can formally write

$$f_z = \mathcal{R}^{-1} P_z \,. \tag{4.10}$$

Most of this chapter is devoted to finding explicit expressions and approximations of this inverse operator, for both continuous and discrete functions f. If a suitable representation of the 2d inverse operator $\mathcal{R}^{-1}$ for reconstruction of the function in each plane $f_z(x', y')$ is found, the 3d function $f(x', y', z')$ can be reconstructed plane by plane for all z.

If we plot the projection data $P_z(\theta, x)$ according to equation (4.9) for each plane as a 2d array, with $\theta \in [0 \ldots 180°)$ along the horizontal and the x dependence along the vertical axis, we obtain a so called sinogram. We drop the index z for a given plane and denote this 2d function by $P(\theta, x)$. It represents the function $f(x', y')$ in the projection space (or Radon space) and is equivalent to the real space representation $f(x', y')$. Consider, for example, a point like density distribution in the object, described mathematically by a set of Dirac δ functions. Since the RT is linear, it is sufficient to consider a single point. For concreteness, let us assume $f(x', y') = \delta_{2d}(x' - x'_0, y' - y'_0)$, where δ_{2d} denotes the 2d delta function. Then, the sinogram becomes

$$P(\theta, x) = \delta(x - r_0 \sin(\theta + \psi)) \,, \tag{4.11}$$

with $r_0 = (x'^2_0 + y'^2_0)^{1/2}$ and $\psi = \arctan(x'_0 / y'_0)$. In other words, a single point in the object maps to a sinusoidal curve in the projection space, see the illustration in Figure 4.3. In practice, inspection of recorded data in the sinogram is very useful to detect artifacts, e.g., due to faulty movement of the object.

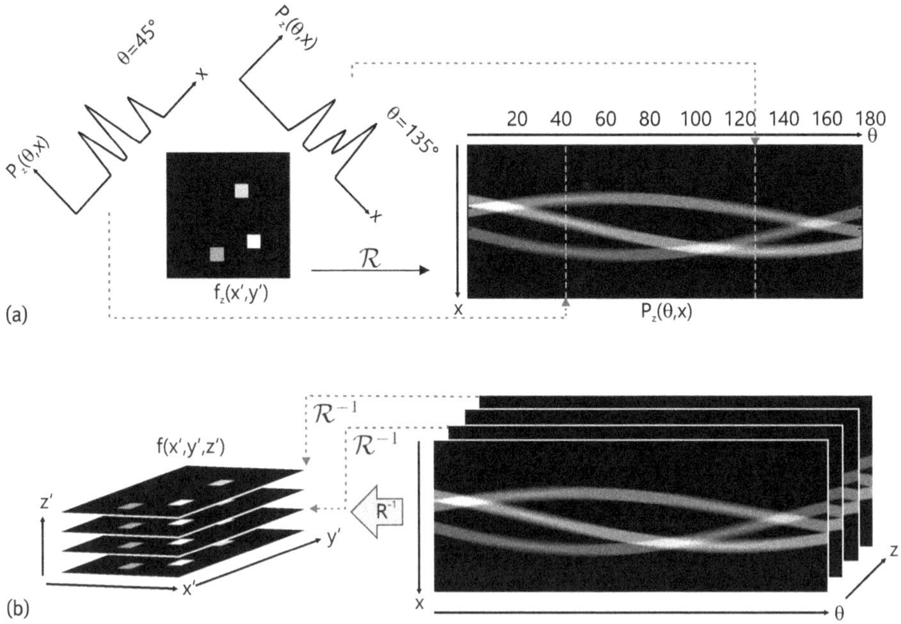

Fig. 4.3: (a) Visualization of a sinogram obtained for three squares of different density (gray value) in a plane. The projections shown are indicated by the *dashed vertical lines in the sinogram.* (b) 3d volumes are reconstructed by filtered backprojection (FBP) of the data stored in the sinogram, separately for each plane in a stack. From [6].

Next, we consider the Fourier transform $\tilde{P}_z(\theta, k)$ of the projection integral $P_z(\theta, x)$ (equation 4.9) with respect to x, and then transform to the object coordinates (x', y')

$$\begin{aligned} \tilde{P}_z(\theta, k) &= \frac{1}{\sqrt{2\pi}} \int P_z(\theta, x) \exp[-ikx]\, dx \\ &= \frac{1}{\sqrt{2\pi}} \iint f_z(x\cos\theta - y\sin\theta, x\sin\theta + y\cos\theta) \exp[-ikx]\, dx\, dy \\ &= \frac{1}{\sqrt{2\pi}} \iint f_z(x', y') \exp[-ik(x'\cos\theta + y'\sin\theta)]\, dx'\, dy' \\ &= \frac{1}{\sqrt{2\pi}} \iint f_z(x', y') \exp[-i(k\cos\theta)x'] \exp[-i(k\sin\theta)y']\, dx'\, dy' \,. \end{aligned} \tag{4.12}$$

In the last expression, we can recognize the 2d Fourier transform $\tilde{f}_z = \mathcal{F}[f_z]$, evaluated at the point (k_x, k_y), with $k_x = k\cos\theta$ and $k_y = k\sin\theta$. Since k is real and arbitrary, $\tilde{f}$ is evaluated, for each fixed θ, along a radial line in Fourier space. In other words, the 1d Fourier transform of a projection corresponds to a line cut (or slice) through the object's 2d Fourier transform,

$$\tilde{P}_z(\theta, k) = \sqrt{2\pi}\, \tilde{f}_z(k\cos\theta, k\sin\theta) \,. \tag{4.13}$$

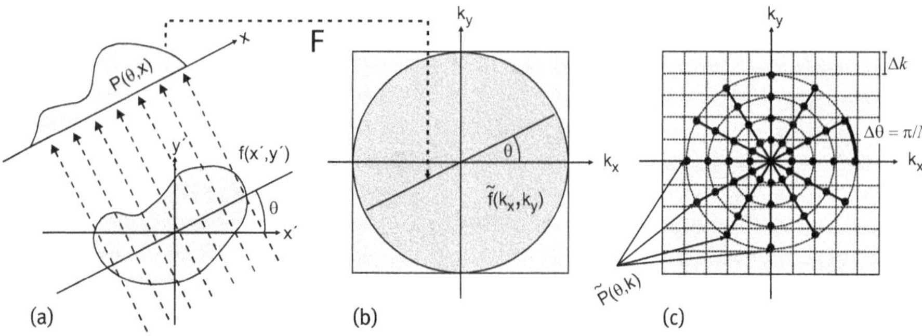

Fig. 4.4: Visualization of the Fourier slice theorem. The 1d Fourier transform $\tilde{P}_z(\theta, k)$ of a projection $P_z(\theta, x)$ obtained as shown in (a) corresponds to a line cut in the 2d Fourier transform $\tilde{f}_z(k\cos\theta, k\sin\theta)$ of the object plane $f_z(x', y')$, as sketched in (b). (c) The density of (discrete) data points falls off as $1/k$, posing a challenge to sample the high spatial frequencies. Provided there is sufficient number of projections and suitable regridding, 2d fast Fourier transform (FFT) algorithms can be used to reconstruct $f_z(x', z')$ in each plane. Adapted from [6].

This relationship is known as the Fourier slice theorem (FST), here stated in 2d. It is clear that it is valid separately for each plane z of the object. In each plane, the lines go through the origin, and the line density decreases as $1/k$, see the sketch in Figure 4.4. In practice, the vast majority of tomographic reconstruction is not carried out by an inverse 2d FFT as one might suspect based on equation (4.13), but by a very robust and efficient operation known as filtered backprojection (FBP). Before presenting this and other forms of reconstruction in more detail in the next section, we present a simple derivation [23], well suited for the nutshell treatment of this introduction. To stress the fact that this is a purely 2d framework of tomography, we drop the subscript z for the plane, and write the object function $f(x', y')$ in terms of its Fourier representation as

$$
\begin{aligned}
f(x', y') &= \frac{1}{2\pi} \int_{-\infty}^{\infty} \int_{-\infty}^{\infty} \tilde{f}(k_x, k_y) \exp[i(k_x x' + k_y y')]\, dk_x\, dk_y \\
&= \frac{1}{2\pi} \int_0^{2\pi} \left[\int_0^{\infty} \tilde{f}(\theta, k) \exp[ik(x' \cos\theta + y' \sin\theta)]\, k\, dk \right] d\theta \\
&= \frac{1}{2\pi} \int_0^{\pi} \left[\int_{-\infty}^{\infty} \tilde{f}(\theta, k)\, |k|\, \exp[ikx]\, dk \right] d\theta \\
&= \frac{1}{2\pi} \int_0^{\pi} \left[\frac{1}{\sqrt{2\pi}} \int_{-\infty}^{\infty} \tilde{P}(\theta, k)\, |k| \exp[ikx]\, dk \right] d\theta
\end{aligned}
$$

where we first switched from Cartesian to polar coordinates, then exploited centrosymmetry $\tilde{f}(\theta + \pi, k) = \tilde{f}(\theta, -k)$ to rewrite the integration limits and finally used the FST to replace $\tilde{f}$ by the projection data $\tilde{P}(\theta, k)$. The last line now instructs us how to treat the data:

1. Perform a filtering step of the projection data $P(\theta, x)$ in Fourier space with a ramp filter, i.e., a weight factor increasing linearly with k, yielding filtered projections

$$Q_y(\theta, x) = \frac{1}{\sqrt{2\pi}} \int_{-\infty}^{\infty} \tilde{P}(\theta, k)\, |k|\, \exp[ikx]\, dk\,. \tag{4.14}$$

2. The second step is the actual backprojection. For a given angle θ, the value of the filtered projection $Q_y(\theta, x)$ is 'smeared out' in the y direction, i.e., added to all points along a straight line parameterized by $x = x' \cos\theta + y' \sin\theta$. This is repeated for all projection angles, i.e., by integration over θ

$$f(x', y') = \frac{1}{2\pi} \int_0^{\pi} Q_y(\theta, x' \cos\theta + y' \sin\theta)\, d\theta\,. \tag{4.15}$$

Backprojection and the importance of filtering are illustrated in Figure 4.5. Multiplication with $|k|$ can be interpreted as a compensation for the decreasing line density.

For discrete data, filtering and backprojection both deserve careful consideration, and reconstruction quality is typically very sensitive to the numerical implementation. This sensitivity can be lifted to some extent if the data is not oversampled, i.e., if the size of the voxel (i.e., the discrete volume element) is much smaller than the desired resolution, and if the number of projections is much larger than required, but this is always at the cost of something – for example detector pixel number, measurement time and dose. This brings us to the important question of how many projections $N_\theta = \pi/\Delta\theta$ need to be recorded to obtain a desired reconstruction quality. Consider the discretized object grid in reciprocal space, as shown in Figure 4.4 (c). Towards high spatial frequencies, the data go out to the spatial frequency $k_{\max} = \pi/\Delta x$, where Δx denotes the pixel size, which is also taken to be equal to the resolution for the sake of simplicity. The sampling intervals in real and reciprocal space in the discrete Fourier transformation are related via $\Delta x \Delta k = 2\pi/N$, where N denotes the number of pixels along x (cf., equation 2.147). The angular sampling condition is fulfilled, if

$$k_{\max}\Delta\theta \leq \Delta k\,, \tag{4.16}$$

which leads to the condition

$$N_\theta := \frac{\pi}{\Delta\theta} \geq N\frac{\pi}{2}\,, \tag{4.17}$$

i.e., the number of projections N_θ has to be larger than the number of resolution elements N along a given direction times $\pi/2$.

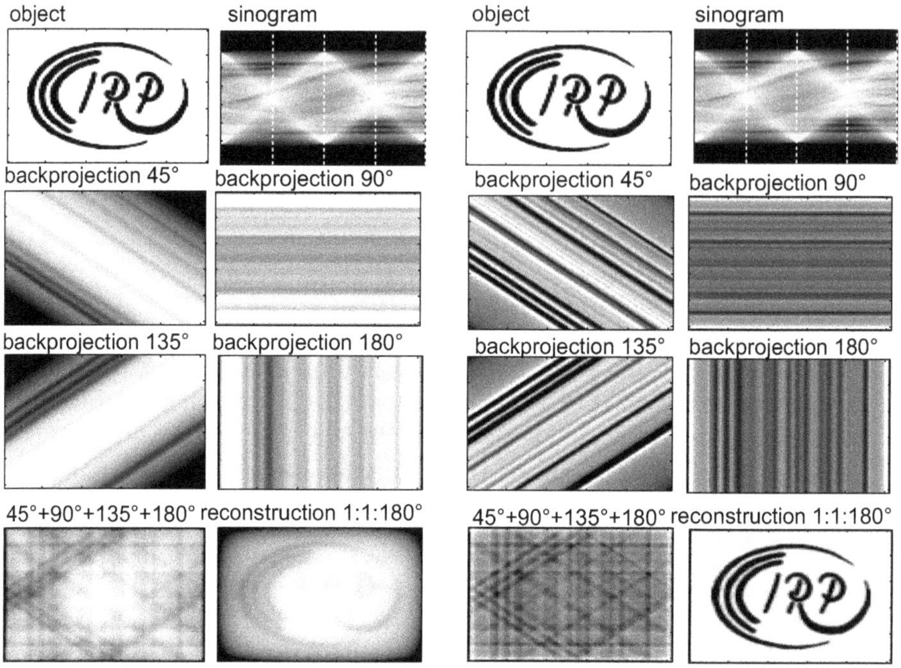

Fig. 4.5: Backprojection of the logo of Institut für Röntgenphysik, without filtering (*left*) and with filtering (*right*). Four projection angles do not yield any recognizable shape of the object, while 180 projections in steps of 1° give a reasonable approximation. However, without filtering the gray values are flawed and the reconstruction is blurred. In contrast to the unfiltered data, the filtered signal can also become negative, such that the reconstruction volume can not only be 'filled up' but also 'emptied' by backprojection. Adapted from [5].

We close this introduction with a note on the generality of tomographic concepts. Indeed, the concept of tomographic imaging has proved to be applicable to many more types of experiments than x-ray radiography. Suitable generalization to other radiation, either electromagnetic or particulate, is conceptually straightforward. The challenges are more on the experimental side, for example, the strong absorption and inelastic scattering of visible light photons. Absorption and scattering is even more prohibitive for electrons, which prevents the use of electron tomography of biomolecular matter for objects larger than a virus. Contrarily, low absorption and, hence, contrast limit the use of thermal neutrons, apart from the lack of suitable sources. Ultrasound would also be a good candidate for tomographic 3d inspection, but again strong interaction leads to diffraction of the wave in the body. Apart from the nature of the interactions, suitable sources and detectors are an important issue for all forms of tomography. Finally, for medical diagnostics and imaging of humans, the diagnostic benefit has to be weighted against the risks associated with a certain radiation probe.

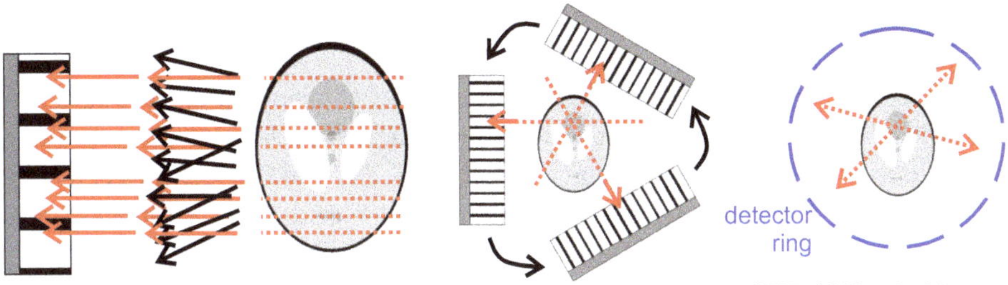

Fig. 4.6: The concept of tomography is very general. For example, the source of radiation can also be within the body, such as for the nuclear diagnostic imaging techniques SPECT and PET. (a) SPECT: The directional information is based on collimators in front of the detector, so that a single detection element receives only signals from a line (or plane) perpendicular to the detector. (b) A full dataset is obtained by rotating one or more detectors around the body. (c) PET: Signal reducing collimation is not required in PET, since the emission process itself with two simultaneously emitted γ quanta indicates the line of response, see Section 3.2.6.

Nuclear diagnostic tomography techniques are, in fact, very useful due to their specific labeling capabilities. In this case, the radiation sources are brought into the body itself, which does not change the mathematical description of the tomographic problem, at least not on the rudimentary level (neglecting self absorption, for example). The principles are sketched in Figure 4.6 for single photon emission computed tomography (SPECT) and positron emission tomography (PET). The radiation sources consist of well designed radioisotope compounds, emitting γ radiation in the case of SPECT and positrons in the case of PET. In SPECT, the directional information is then achieved by a set of collimators, so that the registration of a γ quantum in a detector array can be traced back to an emission line (line of response) going through the body. As sketched in Figure 4.6 (a), the collimation leads to a large loss in signal. Most γ quanta do not contribute to the image, in fact, and the radioisotope concentration and, thus also the dose to the patient, must be ramped up accordingly. Importantly, the signal in one detection element again amounts to a line integral of the radioisotope density. As shown in (b), the image is reconstructed from several projections, recorded by rotating one or more detection arrays. In PET, single events of β^+ decay in the body are detected by a coincidence measurement of γ quanta emitted diametrically at an angle of 180° by electron/positron annihilation (cf., Section 3.2.6). To sum up, the coincident events recorded in opposing detector modules yield the integral of the radioactive tracer concentration along lines in the body. A collimator as required in SPECT is not necessary in PET, since the 180° emission of photons accounts for the line of response. In this way, a significantly larger fraction of events (radioactive decays) can be recorded.

4.2 Mathematical fundamentals of tomography

In this section, we present the mathematical concepts of tomographic reconstruction under ideal conditions of noise and artifact free data. We start by addressing the same issues as in the nutshell presentation above, i.e., RT, FST and FBP, but now in a more general notation (in particular for arbitrary dimension n) and with more mathematical background. Further, we describe different (alternative) reconstruction methods.

4.2.1 Radon transformation (RT)

We are interested in a function $f(\underline{x})$, describing a certain physical property of an object, in particular its spatial distribution. The function f could denote electron density, tracer concentration, attenuation coefficient for a certain type of radiation, or any other scalar quantity amenable to a tomographic recording. It could also be the component of a vector. The important point is that the recorded data originate from projection integrals over f, written in the notation of Figure 4.7 for 2d as

$$g(\theta, s) := \int_{-\infty}^{\infty} f(s\underline{n}_\theta + r\underline{n}_\theta^\perp)\,\mathrm{d}r\,, \tag{4.18}$$

where $\underline{n}_\theta^\perp$ is the unit vector with angle θ with respect to the x axis, and $\underline{n}_\theta$ is a unit vector perpendicular to $\underline{n}_\theta^\perp$. Throughout this chapter we assume that all functions are in the Schwartz space $\mathcal{S}(\mathbb{R}^n)$, i.e., the space of infinitely often differentiable functions that, along with all their derivatives, fall off towards infinity faster than any inverse power of the argument. This ensures, for example, that all Fourier transforms exist and are unique.

More generally, the n-dimensional RT of the function $f(\underline{x})$, $\underline{x} \in \mathbb{R}^n$ is given by

$$(\mathcal{R}f)\,(\underline{n}_\theta, s) := \underbrace{\int f(s\underline{n}_\theta + \underline{y})\,\mathrm{d}^{n-1}y}_{\text{integration over a hyperplane}} \qquad \text{with} \quad \underline{y} \in \theta^\perp$$
$$= \int f(\underline{x})\,\delta(\underline{x}\cdot\underline{n}_\theta - s)\,\mathrm{d}^n x\,. \tag{4.19}$$

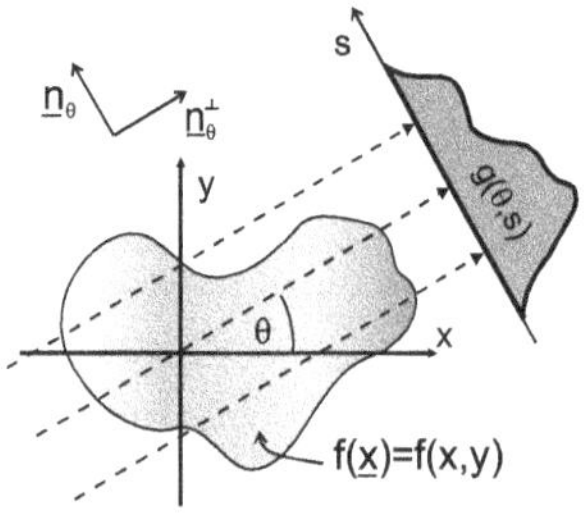

Fig. 4.7: Visualization of equation (4.18). Parallel line integrals through a sample are recorded for a projection angle θ. Note that in this notation θ is shifted by $\pi/2$ with respect to the definition in Figure 4.2.

Here $\theta^\perp$ is the subspace of $\mathbb{R}^n$ perpendicular to $\underline{n}_\theta$. For $n = 2$ this coincides with the definition in equation (4.18). For $n > 2$, the RT does not correspond to an integration over lines, but over hyperplanes.[2]

4.2.2 Fourier slice theorem (FST)

The most important question is whether the RT is invertible in arbitrary dimension n. In other words, we would like to know if it is possible to reconstruct the function $f(\underline{x})$ from $(\mathcal{R}f)(\underline{n}_\theta, s)$. The answer is again given by the Fourier slice theorem (FST)

$$(\mathcal{F}_s\mathcal{R}f)\,(\underline{n}_\theta, t) = (2\pi)^{\frac{n-1}{2}}\,(\mathcal{F}f)\,(t\underline{n}_\theta)\,. \tag{4.20}$$

The Fourier transform on the left hand side is over the variable s only, such that the result still depends on $\underline{n}_\theta$ but also on the new Fourier variable t. The Fourier transform on the right hand side is taken over the full $\mathbb{R}^n$. To see this, consider the (one dimensional) Fourier transform $\mathcal{F}_s$ with respect to the variable s of the RT $(\mathcal{R}f)(\underline{n}_\theta, s)$:

$$\begin{aligned}
(\mathcal{F}_s\mathcal{R}f)\,(\underline{n}_\theta, t) &= (2\pi)^{-\frac{1}{2}}\int \mathrm{d}s\, e^{-its}\,(\mathcal{R}f)\,(\underline{n}_\theta, s) \\
&= (2\pi)^{-\frac{1}{2}}\int \mathrm{d}s\, e^{-its}\int d^n x\, f(\underline{x})\,\delta(\underline{x}\cdot\underline{n}_\theta - s) \\
&= (2\pi)^{-\frac{1}{2}}\int \mathrm{d}^n x\, e^{-it\underline{x}\cdot\underline{n}_\theta}\, f(\underline{x}) \\
&= (2\pi)^{-\frac{1}{2}}(2\pi)^{\frac{n}{2}}\,(\mathcal{F}f)\,(t\underline{n}_\theta) \\
&= (2\pi)^{\frac{n-1}{2}}\,(\mathcal{F}f)\,(t\underline{n}_\theta)
\end{aligned} \tag{4.21}$$

Note that the expression $\mathcal{F}[f(\underline{x})](t\underline{n}_\theta)$ is a complete n-dimensional Fourier transform evaluated along a slice $t\underline{n}_\theta$. Hence, *the Fourier transform (1d) of $\mathcal{R}f$ with respect to s corresponds to a line cut through $\mathcal{F}f$ (n-dimensional) along $\underline{n}_\theta$*. Thus, knowledge of $(\mathcal{R}f)$ allows for the computation of the full Fourier transform of f, and by inverse Fourier transformation hence also of f. In other words, the RT can be inverted, and this holds for all dimensions n.

The FST already provides a reconstruction method by inverse Fourier transformation of equation (4.21):

$$(2\pi)^{\frac{n-1}{2}}\left(\mathcal{F}^{-1}\mathcal{F}f\right)(\underline{x}) = (2\pi)^{-\frac{n}{2}}\int \mathrm{d}^n k\, e^{i\underline{k}\cdot\underline{x}}\,(\mathcal{F}_s\mathcal{R}f)\left(\frac{\underline{k}}{|\underline{k}|}, |\underline{k}|\right) \tag{4.22}$$

$$= (2\pi)^{-\frac{n}{2}}\int_0^\infty \mathrm{d}t\, t^{n-1}\int \mathrm{d}\underline{n}_\theta\, e^{i\underline{x}\cdot\underline{n}_\theta t}\,(\mathcal{F}_s\mathcal{R}f)(\underline{n}_\theta, t)\,, \tag{4.23}$$

2 In $n > 2$, the RT is, hence, *not* the same as the related x-ray transform, as defined by equation (4.137) in the section on projection geometries further below.

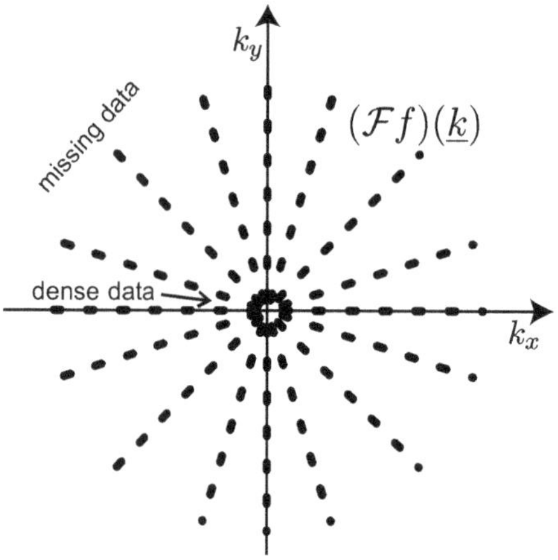

Fig. 4.8: The information recorded by tomography is distributed along radial rods (spikes) in Fourier space, with increasing blank areas between the spikes as we increase the distance to the origin, i.e., the spatial frequency.

where we have used an n-dimensional transform into polar coordinates: $\underline{k} \rightarrow t\underline{n}_\theta$. The expression $(\mathcal{F}_S\mathcal{R}f)(\frac{\underline{k}}{|\underline{k}|}, |\underline{k}|)$ corresponds to the Fourier space filled by the spikes (rods) visualized in Figure 4.8.[3]

In a more concise form, the function $f(\underline{x})$ can be reconstructed by calculation of

$$f(\underline{x}) = (2\pi)^{-n+\frac{1}{2}} \int \mathrm{d}t\, t^{n-1} \int \mathrm{d}\underline{n}_\theta\, e^{i\underline{x}\cdot\underline{n}_\theta t}\, (\mathcal{F}_S\mathcal{R}f)\,(\underline{n}_\theta, t)\,. \tag{4.24}$$

Hence, the function $f(\underline{x})$ can be obtained by Fourier transforming the one-dimensional, measured intensity profiles, arranging them in the n-dimensional Fourier space followed by an inverse Fourier transform, according to the FST.

However, FST based reconstruction is not very suitable for incomplete data, when information is reduced to only a few lines, see Figure 4.8. When the number of projection angles is insufficient, or the angles are not evenly distributed (the sample holder, for example, can easily obstruct some angles creating a missing wedge, see Section 4.5.1), FST based reconstruction can yield strong artifacts. Moreover, even for complete data, the density of sampling points at high spatial frequency can pose a challenge, as well as the numerical regridding of data in order to apply the DFT with the usual Cartesian coordinates. Filling the blank (un-sampled) regions with zeros or preferably by interpolation, one can easily obtain strong artifacts, resulting in unacceptable image quality.[4] Therefore, one has to use the information (data) more

3 The prefactor on the left hand side of equation (4.22) $(2\pi)^{\frac{n-1}{2}}$ is merely included for convenience because in the second step, it is absorbed in the expression $(\mathcal{F}_S\mathcal{R}f)(\frac{\underline{k}}{|\underline{k}|}, |\underline{k}|)$ according to the FST. The remaining prefactor $(2\pi)^{-\frac{n}{2}}$ is part of the n-dimensional inverse Fourier transform.

4 If the data in Fourier space is properly convolved with a function corresponding to the computational domain, i.e., by a sinc function, satisfactory results can be obtained. More recently, FST based reconstruction has become an active topic of research again, resulting from progress in implementing the radial DFT (on a polar grid) [4].

cleverly. To this end, compressive sensing and algebraic reconstruction techniques (ART) are suitable approaches. These more advanced techniques will be addressed below, after we have considered the conventional reconstruction approaches based on backprojection.

In the following paragraphs, we first develop the treatment by further definition of operators and by expanding the notation. This may seem confusing and overloaded at first, but on longer term provides more leverage.

4.2.3 Backprojection operator

For fixed angle $\underline{n}_\theta$, the operator $\mathcal{R}_{\underline{n}_\theta}$ defined by its action on a function f as $(\mathcal{R}_{\underline{n}_\theta} f)(s) := (\mathcal{R}f)(\underline{n}_\theta, s)$ defines a mapping $\mathcal{S}(\mathbb{R}^n) \to \mathcal{S}(\mathbb{R})$. A corresponding dual operator $R^{\#}_{\underline{n}_\theta}$ (as shown below) is given by

$$\left(R^{\#}_{\underline{n}_\theta} g\right)(\underline{x}) := g(\underbrace{\underline{n}_\theta \cdot \underline{x}}_{=s}) , \tag{4.25}$$

and represents a mapping $\mathcal{S}(\mathbb{R}) \to \mathcal{S}(\mathbb{R}^n)$. We also anticipate that g may, in addition to s, depend on $\underline{n}_\theta$ explicitly, in which case we write $(R^{\#}_{\underline{n}_\theta} g)(\underline{x}) := g(\underline{n}_\theta, \underline{n}_\theta \cdot \underline{x})$. Algebraically, this operator corresponds to the transpose of $\mathcal{R}_{\underline{n}_\theta}$

$$\left(\mathcal{R}_{\underline{n}_\theta} f\right)(s) = \int \mathrm{d}^n x\, \delta(\underline{x} \cdot \underline{n}_\theta - s)\, f(\underline{x}) , \quad \text{and} \tag{4.26}$$

$$\left(\mathcal{R}^{\#}_{\underline{n}_\theta} g\right)(\underline{x}) = \int \mathrm{d}s\, \delta(\underline{x} \cdot \underline{n}_\theta - s)\, g(\underline{n}_\theta, s) . \tag{4.27}$$

Next, we show that $R^{\#}_{\underline{n}_\theta}$ as defined above is, indeed, the dual operator of $R_{\underline{n}_\theta}$. For constant $\underline{n}_\theta$ we have

$$\begin{aligned}\int \mathrm{d}s \left(\mathcal{R}_{\underline{n}_\theta} f\right)(s)\, g(\underline{n}_\theta, s) &= \int \mathrm{d}s \int \mathrm{d}^n x\, \delta(\underline{x} \cdot \underline{n}_\theta - s)\, f(\underline{x})\, g(\underline{n}_\theta, s) \\ &= \int \mathrm{d}^n x\, f(\underline{x}) \int \mathrm{d}s\, \delta(\underline{x} \cdot \underline{n}_\theta - s)\, g(\underline{n}_\theta, s) \\ &= \int \mathrm{d}^n x\, f(\underline{x}) \left(\mathcal{R}^{\#}_{\underline{n}_\theta} g\right)(\underline{x}) ,\end{aligned} \tag{4.28}$$

and after integration over the unit sphere ($\underline{n}_\theta$), we obtain

$$\begin{aligned}\int \mathrm{d}\underline{n}_\theta\, \mathrm{d}s\, (\mathcal{R}f)(\underline{n}_\theta, s)\, g(\underline{n}_\theta, s) &= \int \mathrm{d}^n x\, f(\underline{x}) \int \mathrm{d}\underline{n}_\theta \left(\mathcal{R}^{\#}_{\underline{n}_\theta} g(\underline{n}_\theta, \cdot)\right)(\underline{x}) \\ &= \int \mathrm{d}^n x\, f(\underline{x}) \int \mathrm{d}\underline{n}_\theta\, g(\underline{n}_\theta, \underline{x} \cdot \underline{n}_\theta) \\ &= \int \mathrm{d}^n x\, f(\underline{x}) \left(\mathcal{R}^{\#} g\right)(x) .\end{aligned} \tag{4.29}$$

$\mathcal{R}^{\#}$ is, hence, the dual operator to the full (“all angles”) Radon operator $\mathcal{R}$.

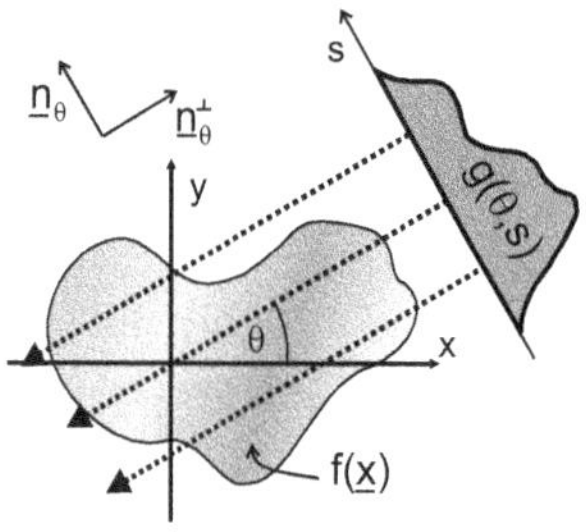

Fig. 4.9: The backprojection operator fills the measured intensity evenly along the projection line. Each pixel in the plane, which is reached from the detector pixel under an angle θ receives the value of that detector pixel.

4.2.4 General inversion formula

To derive a general inversion formula based on backprojection, we also need the so called Riesz operator I^α (also denoted as the Riesz potential), defined by

$$\mathcal{F}(I^\alpha f)(\underline{k}) = |\underline{k}|^{-\alpha}(\mathcal{F}f)(\underline{k}) \tag{4.30}$$

for all real $\alpha < n$. Computing the inverse Fourier transform, adopting a somewhat sloppy notation for ease of reading, we have

$$(I^\alpha f)(\underline{x}) = \left(\mathcal{F}^{-1}|\underline{k}|^{-\alpha}(\mathcal{F}f)(k)\right)(\underline{x}) , \tag{4.31}$$

which is well defined (square integrable) for $(\mathcal{F}f)(\underline{x})$ sufficiently smooth at $\underline{x} = 0$ and for $\alpha < n$. From this definition it follows directly that $I^{-\alpha}$ is the inverse of I^α

$$\left(I^{-\alpha}I^\alpha f\right)(\underline{x}) = \left(\mathcal{F}^{-1}|\underline{k}|^{\alpha}\mathcal{F}\mathcal{F}^{-1}|\underline{k}|^{-\alpha}\mathcal{F}f\right)(\underline{x}) = f(\underline{x}) . \tag{4.32}$$

Finally, we find that $I^0 = 1$. We now continue by considering the application of the Riesz operator to a function f that is to be reconstructed from projection data

$$\left(I^\alpha f\right)(\underline{x}) = \left(\mathcal{F}^{-1}|\underline{k}|^{-\alpha}\mathcal{F}f\right)(\underline{x}) = (2\pi)^{-\frac{n}{2}} \int d^n k\, e^{i\underline{k}\cdot\underline{x}}\, |\underline{k}|^{-\alpha}\, (\mathcal{F}f)\,(\underline{k}) \tag{4.33}$$

Transformation of equation (4.33) to polar coordinates of the form $\underline{k} = t \cdot \underline{n}_\theta$ gives

$$(2\pi)^{-\frac{n}{2}} \int d^n k\, e^{i\underline{k}\cdot\underline{x}}\, |\underline{k}|^{-\alpha}\, (\mathcal{F}f)\,(\underline{k}) = (2\pi)^{-\frac{n}{2}} \int_0^\infty dt\, t^{n-1} \int d\underline{n}_\theta\, e^{i\underline{x}\cdot\underline{n}_\theta t}\, t^{-\alpha}\, (\mathcal{F}f)\,(t\cdot\underline{n}_\theta) . \tag{4.34}$$

By using the FST given in equation (4.20) we can replace $(\mathcal{F}f)(t\cdot\underline{n}_\theta)$

$$\begin{aligned} &(2\pi)^{-\frac{n}{2}} \int_0^\infty dt\, t^{n-1} \int d\underline{n}_\theta\, e^{i\underline{x}\cdot\underline{n}_\theta t}\, t^{-\alpha}\, (\mathcal{F}f)\,(t\cdot\underline{n}_\theta) \\ &= (2\pi)^{-\frac{n}{2}} \int_0^\infty dt\, t^{n-1} \int d\underline{n}_\theta\, e^{i\underline{x}\cdot\underline{n}_\theta t}\, t^{-\alpha} (2\pi)^{\frac{-n+1}{2}} \left(\mathcal{F}_s \mathcal{R}_{\underline{n}_\theta} f\right)(t) . \end{aligned} \tag{4.35}$$

By substitution of $\underline{n}_\theta \to -\underline{n}_\theta$ and $t \to -t$, we see that the integration kernel $(\mathcal{F}_s\mathcal{R}_{\underline{n}_\theta}f)(t)$ is even

$$\left(\mathcal{F}_s\mathcal{R}_{\underline{n}_\theta}f\right)(t) = (2\pi)^{\frac{n-1}{2}}\left(\mathcal{F}f\right)(t\cdot\underline{n}_\theta) = (2\pi)^{\frac{n-1}{2}}\left(\mathcal{F}f\right)(-t\cdot(-\underline{n}_\theta)) = \left(\mathcal{F}_s\mathcal{R}_{-\underline{n}_\theta}f\right)(-t) \quad (4.36)$$

so that we can extend the integral over the full range in t compensated by the additional prefactor $1/2$. Thus, we have

$$\begin{aligned}\left(I^\alpha f\right)(\underline{x}) &= \frac{1}{2}(2\pi)^{-n+\frac{1}{2}}\int_{-\infty}^{\infty} dt\,|t|^{n-1-\alpha}\int d\underline{n}_\theta\, e^{i\underline{x}\cdot\underline{n}_\theta t}\left(\mathcal{F}_s\mathcal{R}_{\underline{n}_\theta}f\right)(t)\\ &= \frac{1}{2}(2\pi)^{-n+\frac{1}{2}}\int d\underline{n}_\theta\left(I^{\alpha+1-n}\mathcal{R}_{\underline{n}_\theta}f\right)(\underline{x}\cdot\underline{n}_\theta)\,,\end{aligned} \quad (4.37)$$

where the definition of the Riesz operator (4.31) was used. With the definition of the full backprojection

$$\left(\mathcal{R}^\# g\right)(\underline{x}) = \int d\underline{n}_\theta\, g(\underline{n}_\theta, \underline{x}\cdot\underline{n}_\theta)\,, \quad (4.38)$$

we find that

$$\left(I^\alpha f\right)(\underline{x}) = \frac{1}{2}(2\pi)^{-n+\frac{1}{2}}\left(\mathcal{R}^\# I^{\alpha+1-n}\mathcal{R}f\right)(\underline{x})\,. \quad (4.39)$$

Hence, with the inverse operator $I^{-\alpha}$, we conclude that

$$f(\underline{x}) = \left(I^{-\alpha}I^\alpha f\right)(\underline{x}) = \frac{1}{2}(2\pi)^{-n+\frac{1}{2}}\left(I^{-\alpha}\mathcal{R}^\# I^{\alpha+1-n}\mathcal{R}f\right)(\underline{x})\,. \quad (4.40)$$

With this general inversion formula (4.40) we have obtained a family of reconstruction formulas distinguished by the parameter α. For all dimensions n and for all allowable choices $\alpha < n$, we thus have a formula to reconstruct f from its RT. The choice in α gives us quite a bit of diversity in algorithmic design. Next, we discuss the most popular examples. Notably, we can choose between applying the Riesz operator in the lower dimensions of the projection space, or in the higher dimensional embedding space, or in both. This corresponds essentially to choices in the sequences of backprojection and filtering operations.

4.2.5 Different reconstruction methods by choice of α

Different choices for α in the general inversion formula (4.40) result in different reconstruction methods. In particular, one can select between a filtering step before backprojection, a filtering step after backprojection, or a combination of both. For $\alpha = n-1$, for example, the general inversion formula becomes

$$f(x) = \frac{1}{2}(2\pi)^{-n+\frac{1}{2}}\left(I^{-n+1}\mathcal{R}^\#\mathcal{R}f\right)(\underline{x})\,, \quad (4.41)$$

and, hence, by a Fourier transform, we see that

$$\left(\mathcal{F}f\right)(\underline{k}) = \frac{1}{2}(2\pi)^{-n+\frac{1}{2}}\,|\underline{k}|^{n-1}\mathcal{F}\left(\mathcal{R}^\#\mathcal{R}f\right)(\underline{k})\,. \quad (4.42)$$

The Riesz operator thus acts like a high pass filter applied to the backprojection. This choice of performing the backprojection first and then the filtering operation in the n-dimensional space is known as filtered layergram reconstruction. The realization of the filtered layergram operation is simple, as no filtering is required before backprojection. However, the method is susceptible to perturbations (noise, blurring, data inconsistencies), at least if not cleverly implemented numerically. It is, therefore, not much used in practice. This may change for special applications. For experimental situations where filtering is required in higher dimensional space, because, for example, the object has been vibrating along a certain (known) direction, it may be of advantage to combine such additional filters with the filters needed for reconstruction and use a single filtering step in n dimensions.

For $\alpha = 0$, the reconstruction becomes equivalent to the Fourier method given in equation (4.24), which we addressed already in the paragraph on the FST. In this case, ($I^0 = 1$),

$$f(\underline{x}) = \frac{1}{2}\,(2\pi)^{-n+\frac{1}{2}}\left(\mathcal{R}^{\#} I^{1-n} \mathcal{R} f\right)(\underline{x})\,, \tag{4.43}$$

and the filter I^{1-n} can be regarded as correcting for the decreasing density of sampling points with increasing spatial frequencies, see Figure 4.8.

4.2.6 Dimensionality of reconstruction and relation to the Hilbert transform

Next, we address the dimensionality n of the Radon and reconstruction spaces. We show that, depending on whether n is even or odd, the RT in some sense has nonlocal or local properties, respectively. This is important in view of local tomography (ROI tomography), where the object is larger than the reconstruction volume, see Figure 4.10. To this end, we reformulate the reconstruction equation, based on the Hilbert transform for the case n even. Using these relations, we can also briefly present the original formulation by J. Radon.

We start by considering the case of $n \geq 3$ and odd and $\alpha = 0$. Then, $n - 1$ is even and, hence,

$$\mathcal{F}(I^{1-n} g)(t) = |t|^{n-1} (\mathcal{F} g)(t) = t^{n-1} (\mathcal{F} g)(t)\,. \tag{4.44}$$

Inverse Fourier transformation and applying the differentiation rule (2.120) gives

$$\begin{aligned}
\left(\mathcal{F}^{-1} \mathcal{F} I^{1-n} g\right)(s) &= (2\pi)^{-\frac{1}{2}} \int dt\, e^{ist}\, t^{n-1} (2\pi)^{-\frac{1}{2}} \int dz\, e^{-izt} g(z) \\
&= (2\pi)^{-1} \int dt \left(-i\frac{\partial}{\partial s}\right)^{n-1} e^{ist} \int dz\, e^{-izt} g(z) \\
&= (2\pi)^{-1} \left(-i\frac{\partial}{\partial s}\right)^{n-1} \int dz\, g(z) \int dt\, e^{it(s-z)} \\
&= \left(-i\frac{\partial}{\partial s}\right)^{n-1} \int dz\, g(z)\, \delta(s - z)\,.
\end{aligned} \tag{4.45}$$

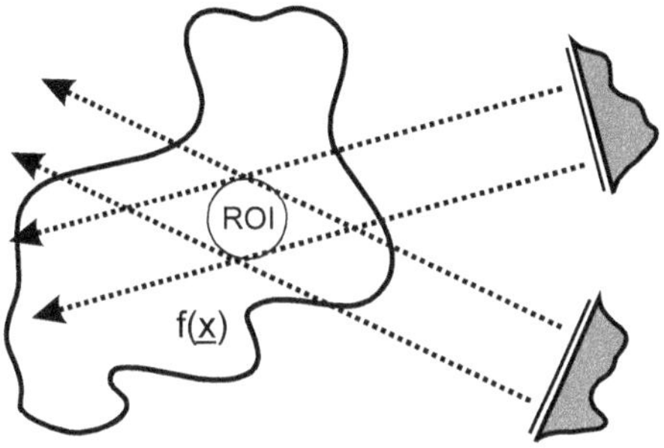

Fig. 4.10: The geometry of local tomography (ROI tomography). Here, e.g., a high resolution region is reconstructed within a larger body, which cannot, however, be contained in the reconstruction volume, e.g., since the number of voxels would be prohibitive. Therefore the recorded data is incomplete. By considering fundamental properties of the Radon transformation in an n-dimensional space, an interesting observation can be made. For n odd, the RT at a given position depends only on the local environment of the reconstruction point. Artifacts that can easily occur for reconstruction with n even, could hence be avoided. For example, replacing a conventional reconstruction of a 3d object based on the 2d RT in parallel planes by a truly 3d RT promises significant improvements for local tomography.

Hence, it follows that

$$I^{1-n}g(s) = \left(-i\frac{\partial}{\partial s}\right)^{n-1} g(s)\,. \tag{4.46}$$

The general inversion formula (4.40) for $\alpha = 0$ is

$$\begin{aligned} f(\underline{x}) &= \frac{1}{2}(2\pi)^{-n+\frac{1}{2}}\left(\mathcal{R}^{\#}I^{1-n}\mathcal{R}f\right)(\underline{x}) \\ &= \frac{1}{2}(2\pi)^{-n+\frac{1}{2}}\int d\underline{n}_\theta\, I^{1-n}(\mathcal{R}f)(\underline{n}_\theta, s)\Big|_{s=\underline{x}\cdot\underline{n}_\theta}\,. \end{aligned} \tag{4.47}$$

We finally use equation (4.46) to replace $I^{1-n}(\mathcal{R}f)(\underline{n}_\theta, s)$

$$f(\underline{x}) = \frac{1}{2}(2\pi)^{-n+\frac{1}{2}}(-i)^{n-1}\left(\mathcal{R}^{\#}\left(\frac{\partial}{\partial s}\right)^{n-1}(\mathcal{R}f)\right)(\underline{x})\,. \tag{4.48}$$

Thus for n odd, it is sufficient for reconstruction of $f(x)$ at position $\underline{x}$ to know the RT $(\mathcal{R}f)(\underline{n}_\theta, s)$ at those positions $(\underline{n}_\theta, s)$ where the hyperplane defined by $\underline{n}_\theta$ and s contains $\underline{x}$, and the corresponding derivatives with respect to s. Hence, the problem in some sense becomes "local" in Radon space. This enables reconstruction of subregions of $f(\underline{x})$ from local data of $\mathcal{R}f$ by simple backprojection without the artifacts expected for an even dimension n.

When the dimension n is even, this is not the case. Then we have instead

$$|t|^{n-1} = t^{n-1}\operatorname{sgn}(t)\,. \tag{4.49}$$

The sign function can be expressed as

$$\operatorname{sgn}(t) = \underbrace{\frac{1}{\pi}\int_{-\infty}^{+\infty} dr\,\frac{\sin(tr)}{r}}_{=\pi\,\operatorname{sgn}(t)} = \frac{1}{\pi}\int_{-\infty}^{\infty} dr\,\frac{1}{2ir}\left(e^{itr} - e^{-itr}\right). \tag{4.50}$$

The representation of $\operatorname{sgn}(t)$ can be proved by complex analysis and is used here to study the action of the Riesz operator with $\alpha = 1 - n$ on g

$$\begin{aligned}
\left(I^{1-n}g\right)(s) &= \left(\mathcal{F}^{-1}t^{n-1}\operatorname{sgn}(t)\mathcal{F}g\right) \\
&= (2\pi)^{-\frac{1}{2}}\int dt\, e^{ist}t^{n-1}\frac{1}{\pi}\int_{-\infty}^{\infty} dr\,\frac{1}{2ir}\left(e^{itr} - e^{-itr}\right)(2\pi)^{-\frac{1}{2}}\int dz\, e^{-itz}g(z) \\
&= \frac{1}{2\pi^2}\left(-i\frac{\partial}{\partial s}\right)^{n-1}\left[\int dr\int dz\left(\int dt\, e^{it(s+r-z)} - e^{it(s-r-z)}\right)\frac{1}{2ir}\,g(z)\right] \\
&= (\pi)^{-1}\left(-i\frac{\partial}{\partial s}\right)^{n-1}\int dr\left[\int dz\,(\delta(s+r-z) - \delta(s-r-z))\,\frac{1}{2ir}\,g(z)\right] \qquad (4.51) \\
&= (\pi)^{-1}\left(-i\frac{\partial}{\partial s}\right)^{n-1}\int dr\,\frac{1}{2ir}\,(g(s+r) - g(s-r)) \\
&= (\pi)^{-1}\left(-i\frac{\partial}{\partial s}\right)^{n-1} 2\int dr\,\frac{1}{2ir}g(s+r) \\
&= i\left(-i\frac{\partial}{\partial s}\right)^{n-1}\underbrace{\frac{1}{\pi}\int dz\,\frac{g(z)}{s-z}}_{\substack{:=(\mathcal{H}g)(s)\\ \text{Hilbert transform}}}\,, \qquad (4.52)
\end{aligned}$$

where we have used the definition of the Hilbert transform in the last step. The Hilbert transform has the following property

$$\left(\frac{\partial}{\partial s}\right)^{n-1}(\mathcal{H}g(z))(s) = \left(\mathcal{H}\left(\frac{\partial}{\partial z}\right)^{n-1} g(z)\right)(s)\,. \tag{4.53}$$

Plugging the last expression for $(I^{1-n}g)(s)$ as well as the last mentioned property of the Hilbert transform into the general inversion formula (4.40) and with $s = \underline{x}\cdot\underline{n}_\theta$ we find that

$$\begin{aligned}
f(\underline{x}) &= \frac{i}{2}(2\pi)^{-n+1}(-i)^{n-1}\int \mathrm{d}\underline{n}_\theta\frac{1}{\pi}\int dz\,\frac{1}{s-z}\left(\frac{\partial}{\partial z}\right)^{n-1}\left.\frac{(\mathcal{R}f)(\underline{n}_\theta, z)}{s-z}\right|_{s=\underline{x}\cdot\underline{n}_\theta} \\
&= \frac{i}{2}(2\pi)^{-n+1}(-i)^{n-1}\left(\mathcal{R}^{\#}\mathcal{H}\left[\left(\frac{\partial}{\partial z}\right)^{n-1}(\mathcal{R}f)(z)\right]\right)(\underline{x}) \qquad (4.54)
\end{aligned}$$

Therefore, the function $\mathcal{R}f(\underline{n}_\theta, z)$ is required over its entire domain, for reconstruction at position $\underline{x}$, not just for hyperplanes containing $\underline{x}$, since the Hilbert transform is non-local. This makes the 2dRT problematic for local tomography applications. Expressing the Hilbert transformation $\mathcal{H}$ by two integrals as in equation (4.51) and writing the $(n-1)$st derivative of $(\mathcal{R}f)(\underline{n}_\theta, z)$ with respect to z by the superscript $(n-1)$, we have

$$
\begin{aligned}
f(\underline{x}) &= \frac{1}{\pi}\int \frac{dz}{2z}\Big[(\mathcal{R}f)^{n-1}(\underline{n}_\theta, \underline{x}\cdot\underline{n}_\theta + z) - (\mathcal{R}f)^{n-1}(\underline{n}_\theta, \underline{x}\cdot\underline{n}_\theta - z)\Big]\cdot(-i)^{n-1} \\
&= i(2\pi)^{-n+1}\frac{1}{\pi}\int \frac{dz}{2z}\int d\underline{n}_\theta\,(\mathcal{R}f)^{n-1}(\underline{n}_\theta, \underline{x}\cdot\underline{n}_\theta + z)\cdot(-i)^{n-1}\,,
\end{aligned} \tag{4.55}
$$

where the last line follows using symmetry of the integrand under the transformation $\underline{n}_\theta \to -\underline{n}_\theta$ and $z \to -z$. If we now set

$$
F_{\underline{x}}(z) := \int d\underline{n}_\theta\,(\mathcal{R}f)(\underline{n}_\theta, \underline{x}\cdot\underline{n}_\theta + z)\,, \tag{4.56}
$$

we finally obtain the original formulation of J. Radon (1917)

$$
f(\underline{x}) = \text{const.}\cdot\int \frac{dz}{z}\,F_{\underline{x}}^{(n-1)}(z)\,. \tag{4.57}
$$

4.2.7 Reconstruction as a convolution problem

The most commonly used algorithm in particular for x-ray CT is the so called filtered backprojection (FBP). We have already learned that simple backprojection does not yield a satisfactory reconstruction. However, in the last section we found that for n even

$$
f(\underline{x}) = -\frac{1}{2}(2\pi)^{-n-1}(-i)^{n-1}\left(\mathcal{R}^{\#}\mathcal{H}(\mathcal{R}f)^{(n-1)}\right)(\underline{x})\,, \tag{4.58}
$$

where we have expressed the $(n-1)$st derivative of $(\mathcal{R}f)(\underline{n}_\theta, s)$ with respect to s by the superscript $(n-1)$. Now, we can consider the operator $\mathcal{H}\cdot(\frac{\partial}{\partial s})^{n-1}$ as a "filter", notably a convolution filter. For reconstruction, this filter must be applied to the projections, before the result is backprojected. The challenge is to implement the filter in a practically robust way, suitable for the discrete setting, stable against noise, and with appropriate treatment of the principal value integral underlying the Hilbert transform.

To this end, in view of its practical relevance, we need one further result, concerning the sequence of convolution and backprojection. We consider the convolution of the RT with a function $h = h(\underline{n}_\theta, s)$:

$$
\begin{aligned}
(h * \mathcal{R}_{\underline{n}_\theta}f)(r) &= \int ds\, h(\underline{n}_\theta, r-s)\,(\mathcal{R}_{\underline{n}_\theta}f)(s) \\
&= \int ds\, h(\underline{n}_\theta, r-s)\int d^n z\,\delta(\underline{z}\cdot\underline{n}_\theta - s)\,f(\underline{z}) \\
&= \int d^n z\, h(\underline{n}_\theta, r-\underline{z}\cdot\underline{n}_\theta)\,f(\underline{z})
\end{aligned} \tag{4.59}
$$

We then apply the backprojection operator $\mathcal{R}^\#$ on both sides, which yields

$$\begin{aligned}\left(\mathcal{R}^\#(h * \mathcal{R}_{\underline{n}_\theta} f)\right)(\underline{x}) &= \int d\underline{n}_\theta \int d^n z\, h\left(\underline{n}_\theta, \underline{x}\cdot\underline{n}_\theta - \underline{z}\cdot\underline{n}_\theta\right) f(\underline{z}) \\ &= \int d^n z\, f(\underline{z}) \int d\underline{n}_\theta\, h\left(\underline{n}_\theta, (\underline{x}-\underline{z})\cdot\underline{n}_\theta\right) \\ &= (f * \mathcal{R}^\# h)(\underline{x})\,. \end{aligned} \tag{4.60}$$

This is an interesting result. The convolution of a function with the RT $\mathcal{R}f$ followed by backprojection is identical to the convolution of the backprojected filter kernel with the original function f.[5] The relevance of this identity is the following: We can choose $h(\underline{n}_\theta, s)$, such that $\mathcal{R}^\# h$ approximates a δ function. In view of the fact that we have discrete voxels, the "δ function" does not need to be sharper than the voxel size. Thus, we can choose $\mathcal{R}^\# h$ accordingly, for example, by

$$\left(\mathcal{F}\mathcal{R}^\# h\right)(\underline{k}) = (2\pi)^{-\frac{n}{2}} \Phi\left(\frac{|\underline{k}|}{c}\right), \tag{4.61}$$

where c denotes a limit frequency and Φ is a function with $\Phi(0) = 1$ and $\Phi(x) \to 0$ for $x \to \infty$ (a sufficiently fast decay has to be assumed). The limit $c \to \infty$ then corresponds to $(\mathcal{R}^\# h)(\underline{x}) = \delta(\underline{x})$ (a constant in Fourier space corresponds to a δ function in real space). The question arises how exactly $h(\underline{n}_\theta, s)$ should be chosen. To this end, we compute by a long but straightforward calculation

$$\begin{aligned}\left(\mathcal{F}\mathcal{R}^\# h\right)(\underline{k}) &= (2\pi)^{-\frac{n}{2}} \int d^n x\, e^{-i\underline{k}\cdot\underline{x}} \int d\underline{n}_\theta\, h(\underline{n}_\theta, \underline{x}\cdot\underline{n}_\theta) \\ &= (2\pi)^{\frac{n-1}{2}} |\underline{k}|^{1-n} \left((\mathcal{F}_s h)\left(\frac{\underline{k}}{|\underline{k}|}, |\underline{k}|\right) + (\mathcal{F}_s h)\left(-\frac{\underline{k}}{|\underline{k}|}, -|\underline{k}|\right)\right). \end{aligned} \tag{4.62}$$

Since

$$(\mathcal{F}\mathcal{R}^\# h)(|\underline{k}|) = (2\pi)^{-\frac{n}{2}} \Phi\left(\frac{|\underline{k}|}{c}\right) \tag{4.63}$$

does not depend on the direction of $\underline{k}$, also $\mathcal{F}_s h$ is independent of the direction. Therefore, it follows that $(\mathcal{F}_s h)(\underline{n}_\theta, s) = (\mathcal{F}_s h)(|s|)$ and, hence, also

$$(\mathcal{F}_s h)(s) = \frac{1}{2} \cdot (2\pi)^{-\frac{n-1}{2}} |s|^{n-1} \cdot (2\pi)^{-\frac{n}{2}} \Phi\left(\frac{|s|}{c}\right). \tag{4.64}$$

Thus, one can use the same filter kernel for each projection $\underline{n}_\theta$. The filter is a bandpass, since small spatial frequencies are suppressed by $|s|^{n-1}$, while high frequencies are suppressed by $\Phi(\frac{|s|}{c})$. In summary, we have the following reconstruction result:

$$f(\underline{x}) \approx \left(f * \mathcal{R}^\# h\right)(\underline{x}) = \left(\mathcal{R}^\#(h * \mathcal{R}_{\underline{n}_\theta} f)\right)(\underline{x}), \tag{4.65}$$

$$h(s) = (2\pi)^{\frac{1}{2}-n} |s|^{n-1} \Phi\left(\frac{|s|}{c}\right). \tag{4.66}$$

5 This can also be inferred directly from the FST.

Hence, first a convolution is required for each projection $\underline{n}_\theta$ and this convolved – or better, filtered – data then has to be backprojected. Both are simple and fast operations. We do, however, have to live with the disadvantage that $\mathcal{R}^\# h$ is only approximately a δ function.

4.2.8 Problems of discretization and speed

In practice $(\mathcal{R}f)(\underline{n}_\theta, s)$ is only given for discrete angles θ_k and values s_l. Therefore, one needs to discretize the inversion methods. In this respect, numerical accuracy and speed are key properties. These issues do not matter in the idealized mathematical treatment, but are highly relevant in practice. After considerable progress in computer performance over the last decades, one might argue that tomographic reconstruction no longer poses a computational challenge. At the same time, however, the requirements have also increased, notably the reconstruction volume in terms of the number of voxels. At present, the reconstruction volume is often more limited by issues of memory size than by numerical speed, and reading the data is often more time consuming than the actual computation. Techniques requiring many iterations of reconstruction and "real-time" reconstruction parallel to data recording remain particularly challenging. It is, therefore, still important to consider the numerical complexity.

Let us start by analyzing the Fourier reconstruction for the case $n = 2$, i.e., the direct implementation of the Fourier slice theorem

$$(2\pi)^{\frac{n-1}{2}} (\mathcal{F}f)(t\underline{n}_\theta) = \left(\mathcal{F}_s \mathcal{R}_{\underline{n}_\theta} f\right)(t) . \tag{4.67}$$

For this, we need the one-dimensional Fourier transform of Radon data for each angle θ_k. We demand that for $|s| \geq R$ the data are zero, $(\mathcal{R}f)(\underline{n}_\theta, s) = 0$, i.e., the object to be reconstructed is fully contained in a circle of radius R. The interval $-R \leq s \leq +R$ can be split into $2q$ subintervals of equal length

$$s_l = R\frac{l}{q} , \quad l = -q, \dots , q . \tag{4.68}$$

The DFT is then

$$\tilde{g}_{kr} := \frac{1}{\sqrt{2q}} \sum_{l=-q}^{q} e^{-2\pi i \frac{rl}{2q}} (\mathcal{R}f) \left(\underline{n}_{\theta k}, R\frac{l}{q}\right) . \tag{4.69}$$

According to the sampling theorem, the discrete transform allows for the exact reconstruction only if there is a limit spatial frequency (band limit), otherwise we will encounter aliasing problems. This DFT amounts to $\mathcal{O}(q \log q)$ calculation steps per angle, giving $\mathcal{O}(p\, q \log q)$ total operations for p angles, and with $p \approx q$, therefore, $\mathcal{O}(q^2 \log q)$. Next, we have to consider interpolation in Fourier space. For simple interpolation, we have a computational complexity of $\mathcal{O}(q^2)$. Note that the resolution of the data in Fourier space is $\varrho = \frac{\pi}{R}$, independent of p and q, i.e., it does not increase with higher sampling. Finally, a 2d inverse Fourier transform has to be applied, requiring a computation time of $\mathcal{O}(q^2 \log q)$. We thus also have for the entire reconstruction $\mathcal{O}(q^2 \log q)$. As we will see, this is relatively fast compared to other algorithms.

As the next reconstruction method, we consider FBP, which was given by

$$f(\underline{x}) \approx \mathcal{R}^{\#}(h * \mathcal{R}_{\underline{n}_\theta} f)(\underline{x}) , \tag{4.70}$$

for a suitable filter $h(s)$. In discrete form this corresponds to

$$f(\underline{x}) \approx \sum_{i=0}^{p} \frac{2\pi}{p} \sum_{l=-q}^{q} \Delta s \, h(\underline{x} \cdot \underline{n}_{\theta i} - s_l) \, r(\underline{n}_{\theta i}, s_l) . \tag{4.71}$$

Naively computing this expression requires $\mathcal{O}(pq)$ operations for each $\underline{x}$. Since $h(s)$ is a filter with limit frequency c, we must have $\frac{R}{2q} \leq \Delta s \leq \frac{1}{c}$ (i.e., c has to be chosen such that this is fulfilled). In this way, we obtain a reasonable voxel size $\Delta x \approx \frac{1}{c}$, i.e., the number of voxels in one direction is $M = Rc$. With $\frac{R}{2q} \leq \frac{1}{c}$ we have $M \approx 2q$ and, hence, the complexity $\mathcal{O}(M^2 pq) = \mathcal{O}(pq^3) = \mathcal{O}(q^4)$ for $p = \mathcal{O}(q)$. Using FFT methods for the filtering, one can reduce the run time to $\mathcal{O}(q^3)$, since precomputing the convolutions with FFT has complexity $\mathcal{O}(q \log q)$ instead of the naive $\mathcal{O}(pq)$, resulting in a total of $\mathcal{O}(M^2 p) = \mathcal{O}(q^3)$ operations.

FBP is, hence, slower than Fourier reconstruction, but since the errors go to zero linearly in Δs and $\frac{2\pi}{p}$ (because of the approximation of the integrals by Riemann sums), the artifacts are better controlled. However, one has to note that the finite sampling creates aliasing effects for both methods, when the reconstructed objects are not band limited. Consider for comparison the sampling of an audio signal. The signal can be filtered in an analog way before digital sampling. This prevents aliasing effects and assures that the signal played corresponds exactly to the low pass filtered input signal. Without the analog prefiltering step, high frequencies beyond the hearing threshold would leak into the audible low frequency range. Unfortunately, there is no such analog prefiltering step possible in tomography,[6] so that high spatial frequencies in the object can lead to artifacts in reconstruction.

4.3 Reconstruction based on Chebyshev series and orthogonal polynomial expansion

The following section is included as an advanced topic, which may be skipped in an introductory reading or course. Reconstruction based on the Chebyshev series is presented not primarily in view of a practical reconstruction method, but to investigate

6 This might not be entirely true, since we expect that a finite source size in CT could provide a kind of analog filter, removing the high spatial frequencies.

what information we have at hand when probing discrete angles and data points. To this end, we consider an approach proposed in [10, 50]. This method reconstructs a certain class of functions *exactly*, from finite data, and is based on a decomposition of the projections into Chebyshev polynomials of the second kind.

4.3.1 Chebyshev polynomials

First, a short excursion to Chebyshev polynomials is required. Generally formulated, the Gegenbauer polynomials $G_n^\lambda(s)$ are those polynomials that are orthogonal on the interval $[-1, 1]$ with the weight function $W(s) = (1 - s^2)^{\lambda - \frac{1}{2}}$

$$\int_{-1}^{1} \mathrm{d}s\,(1 - s^2)^{\lambda-\frac{1}{2}}\, G_n^\lambda(s)\, G_m^\lambda(s) = \text{const.} \cdot \delta_{mn}\,. \tag{4.72}$$

The special case $\lambda = 0$ is known as Chebyshev polynomials of the first kind, and the case $\lambda = 1$ as Chebyshev polynomials of the second kind

$$T_n(s) = G_n^0(s)\,, \tag{4.73}$$

$$U_n(s) = G_n^1(s)\,. \tag{4.74}$$

For Chebyshev polynomials of the second kind, we have the following representation

$$U_n(s) = \frac{\sin((n+1)\arccos s)}{\sin(\arccos s)}\,, \tag{4.75}$$

$$U_n(\cos\theta) = \frac{\sin((n+1)\theta)}{\sin\theta}\,. \tag{4.76}$$

First, we show that these are, indeed, polynomials. With the addition theorems

$$\sin(\alpha+\beta) = \sin\alpha\,\cos\beta + \cos\alpha\,\sin\beta \tag{4.77}$$

$$\cos(\alpha+\beta) = \cos\alpha\,\cos\beta - \sin\alpha\,\sin\beta\,, \tag{4.78}$$

and by induction we can write

$$U_n(\cos\theta)$$

$$\overset{(4.77)}{=} \frac{\sin\theta\cos n\theta + \sin n\theta\cos\theta}{\sin\theta} \tag{4.79}$$

$$= \cos((n-1+1)\theta) + \cos\theta\,\underbrace{\frac{\sin n\theta}{\sin\theta}}_{U_{n-1}(\cos\theta)} \tag{4.80}$$

$$\overset{(4.78)}{=} \cos\theta\cos((n-1)\theta) - \underbrace{\sin^2\theta}_{1-\cos^2\theta}\,\underbrace{\frac{\sin((n-1)\theta)}{\sin\theta}}_{U_{n-2}(\cos\theta)} + \cos\theta\, U_{n-1}(\cos\theta) \tag{4.81}$$

$$= \underbrace{\cos\theta\Big(\cos((n-1)\theta) + \cos\theta\, U_{n-2}(\cos\theta)\Big)}_{U_{n-1}(\cos\theta)\text{ by (4.80)}} - U_{n-2}(\cos\theta) + \cos\theta\, U_{n-1}(\cos\theta) \tag{4.82}$$

$$= 2\cos\theta\, U_{n-1}(\cos\theta) - U_{n-2}(\cos\theta)\,, \tag{4.83}$$

or alternatively

$$U_n(s) = 2s\,U_{n-1}(s) - U_{n-2}(s)\,. \tag{4.84}$$

As the base clause, we can use

$$U_0(s) = 1\,, \tag{4.85}$$

$$U_1(s) = \frac{\sin 2\theta}{\sin\theta} = \frac{2\sin\theta\cos\theta}{\sin\theta} = 2\cos\theta = 2s\,. \tag{4.86}$$

Since these two expressions are polynomials in s, this is also true for all other $U_n(s)$. The orthogonality of Chebyshev polynomials is shown with equation (4.72)

$$\int_{-1}^{1} \mathrm{d}s\,\sqrt{1-s^2}\,U_n(s)U_m(s) \tag{4.87}$$

$$\overset{s\to\cos\theta}{=} \int_0^{\pi} \mathrm{d}\theta\,\sin\theta\,\underbrace{\sin\theta}_{\sqrt{1-s^2}}\,U_n(\cos\theta)U_m(\cos\theta) \tag{4.88}$$

$$\overset{(4.76)}{=} \int_0^{\pi} \mathrm{d}\theta\,\sin\left((n+1)\theta\right)\sin\left((m+1)\theta\right) \tag{4.89}$$

$$= \int_0^{\pi} \mathrm{d}\theta\,\frac{1}{2i}\left(e^{i(n+1)\theta} - e^{-i(n+1)\theta}\right)\frac{1}{2i}\left(e^{i(m+1)\theta} - e^{-i(m+1)\theta}\right) \tag{4.90}$$

$$= -\frac{1}{4}\int_0^{\pi} \mathrm{d}\theta\,\left[e^{i(n+m+2)\theta} + e^{-i(n+m+2)\theta} - e^{i(n-m)\theta} - e^{-i(n-m)\theta}\right] \tag{4.91}$$

$$= -\frac{1}{4}\left[\underbrace{\frac{e^{i(n+m+2)\theta}}{i(n+m+2)}\bigg|_0^{\pi} - \frac{e^{-i(n+m+2)\theta}}{i(n+m+2)}\bigg|_0^{\pi}}_{=0} - 2\delta_{nm}\pi\right] \tag{4.92}$$

$$= \frac{\pi}{2}\delta_{nm}\,. \tag{4.93}$$

A last property of Chebyshev polynomials that we will need is

$$U_m(\cos\theta) = \sum_{\nu=-m}^{m} e^{i\nu\theta}\,, \tag{4.94}$$

which is easy to derive from equation (4.76) and which we, therefore, state here without proof.

4.3.2 Chebyshev backprojection

This function class is useful because it has a very important property: The $U_n(s)$ are eigenfunctions of the operator $R_{\underline{n}_\theta} R^{\#}_{\underline{n}_\phi}$ when the reconstruction volume is limited to a radius 1. In this case, we have

$$\left(R_{\underline{n}_\theta} R^{\#}_{\underline{n}_\phi} U_m\right)(t) = \int\limits_{|x|\le 1} \mathrm{d}^2x\, \delta(\underline{x}\cdot\underline{n}_\theta - t) U_m(\underline{x}\cdot\underline{n}_\phi) \tag{4.95}$$

$$= \int\limits_{-\sqrt{1-t^2}}^{\sqrt{1-t^2}} \mathrm{d}t'\, U_m\left((t'\underline{n}_\theta^\perp + t\underline{n}_\theta)\cdot\underline{n}_\phi\right) \tag{4.96}$$

$$= \sqrt{1-t^2}\int\limits_{-1}^{1} \mathrm{d}t''\, U_m\underbrace{\left(t''\sqrt{1-t^2}\underbrace{\underline{n}_\theta^\perp\underline{n}_\phi}_{=\sin(\phi-\theta)} + t\underbrace{\underline{n}_\theta\underline{n}_\phi}_{=\cos(\phi-\theta)}\right)}_{s}, \tag{4.97}$$

where we have used the substitution $t'' = t'\sqrt{1-t^2}$. Since $U_m(s)$ is a polynomial in $s = t''\sqrt{1-t^2}\sin(\phi-\theta) + t\cos(\phi-\theta)$, each polynomial term s^k with $k \le m$ consists of powers

$$\left(t''\sqrt{1-t^2}\sin(\theta-\phi)\right)^j\left(t\cos(\theta-\phi)\right)^{k-j}, \tag{4.98}$$

with $j = 0, \ldots, k$. For j odd, the integral over t'' is 0, since t'' appears as an odd power. Therefore, only the even j remain, i.e. also only even powers of $\sqrt{1-t^2}$. The integral is hence a polynomial in t of order m. Hence, we have

$$\left(R_{\underline{n}_\theta} R^{\#}_{\underline{n}_\phi} U_m\right)(t) = \sqrt{1-t^2} P_m(t) \tag{4.99}$$

with polynomial $P_m(t)$ of order m. Since the Chebyshev polynomials are orthogonal, we have

$$\int \mathrm{d}t \sqrt{1-t^2} P_m(t) U_{m'}(t) = 0 \qquad \text{for} \qquad m' > m, \tag{4.100}$$

because P_m can be expressed as a linear combination of U_k with $k = 0, \ldots, m$, to which all $U_{m'}$ are orthogonal.

Consider once more the operator $R_{\underline{n}_\theta} R^{\#}_{\underline{n}_\phi}$:

$$\left(R_{\underline{n}_\theta} R^{\#}_{\underline{n}_\phi} f\right)(t) = \int\limits_{-\sqrt{1-t^2}}^{\sqrt{1-t^2}} \mathrm{d}t'\, f\left(\underbrace{t'\sin(\phi-\theta) + t\cos(\phi-\theta)}_{=:s}\right) \tag{4.101}$$

$$= \int \frac{\mathrm{d}s}{\sin(\phi-\theta)} f(s)\, K(t,s) \tag{4.102}$$

with a symmetric integration kernel $K(t,s) = K(s,t)$. Hence, also $R_{\underline{n}_\theta} R^{\#}_{\underline{n}_\phi}$ is symmetric. Since the operator is symmetric, we can apply it to either function in the scalar product. Hence, for all $m' > m$, we have

$$0 = \int\limits_{-1}^{1} \mathrm{d}t \left(R_{\underline{n}_\theta} R^{\#}_{\underline{n}_\phi} U_m\right)(t)\, U_{m'}(t) \tag{4.103}$$

$$= \int\limits_{-1}^{1} \mathrm{d}t\, U_m(t) \left(R_{\underline{n}_\theta} R^{\#}_{\underline{n}_\phi} U_{m'}\right)(t). \tag{4.104}$$

Since, on the other hand, $(R_{\underline{n}_\theta} R^{\#}_{\underline{n}_\phi} U_{m'})(t) = \sqrt{1-t^2} P_{m'}(t)$ holds, and with relabeling $m \to m'$ and $m' \to m$, we obtain the orthogonality relation for all $m \neq m$

$$0 = \int_{-1}^{1} \mathrm{d}t \sqrt{1-t^2} P_m(t) U_{m'}(t) \qquad \forall\, m' \neq m\,. \tag{4.105}$$

This means that $P_m(t)$ is identical to $U_m(t)$ up to a prefactor

$$P_m(t) = c \cdot U_m(t)\,. \tag{4.106}$$

Since $P_m(1) = 2U_m(\cos(\phi - \theta))$ (see equation 4.97), and

$$U_m(1) = \lim_{\theta \to 0} \frac{\sin(m+1)\theta}{\sin\theta} = m + 1\,, \tag{4.107}$$

we hence have

$$P_m(t) = \frac{\left(R_{\underline{n}_\theta} R^{\#}_{\underline{n}_\phi} U_m\right)(t)}{\sqrt{1-t^2}} = \underbrace{\frac{2U_m\left(\cos(\phi-\theta)\right)}{m+1}}_{\text{„eigenvalue"}} U_m(t)\,. \tag{4.108}$$

4.3.3 Chebyshev expansion

Now we will see how this can be used. We consider a function $f(\underline{x})$, which we want to reconstruct from its discrete RT. From the exact inversion formula, we know that

$$f(\underline{x}) = \text{const.} \times \left(R^{\#}\mathcal{H}(Rf)^{n-1}\right)(\underline{x}) \tag{4.109}$$

$$\overset{n=2}{=} \text{const.} \times \int_0^{2\pi} \mathrm{d}\phi \, \left(\mathcal{H}(R_{\underline{n}_\theta} f)'\right)(\underline{x} \cdot \underline{n}_\phi)\,, \tag{4.110}$$

where $(Rf)^{n-1}$ again denotes the $(n-1)$st derivative of $(\mathcal{R}f)$ with respect to s. We hence know that $f(\underline{x})$ can be written as

$$f(\underline{x}) = \frac{1}{2} \int_0^{2\pi} \mathrm{d}\phi \; g(\phi, \underline{x} \cdot \underline{n}_\phi) = \int_0^{\pi} \mathrm{d}\phi \; h(\phi, \underline{x} \cdot \underline{n}_\phi) \tag{4.111}$$

with some function $h(\phi, s)$. We restrict ourselves to functions $f(\underline{x})$, with support on the unit disk. Then also $h(\phi, s)$ is nonzero only for $-1 \leq x \leq 1$. Moreover, for each ϕ the function $h(\phi, s)$ can be expanded into Chebyshev polynomials

$$g(\phi, s) = \sum_{m=0}^{\infty} w_m(\phi) U_m(s)\,, \qquad -1 \leq s \leq 1\,, \tag{4.112}$$

where $w_m(\phi)$ are the coefficients and for $f(\underline{x})$ we have

$$f(\underline{x}) = \sum_{m=0}^{\infty} \int_0^{\pi} \mathrm{d}\phi \, w_m(\phi) \, U_m \left(\underline{x} \cdot \underline{n}_{\phi}\right) . \tag{4.113}$$

Next, we consider the RT of $f(\underline{x})$:

$$(\mathcal{R}f)\,(\underline{n}_{\theta}, s) = \int_{|x|\leq 1} \mathrm{d}^2 x \, \delta(\underline{x} \cdot \underline{n}_{\theta} - s) \int_0^{\pi} \mathrm{d}\phi \, g(\phi, \underline{x} \cdot \underline{n}_{\phi}) \tag{4.114}$$

$$= \sum_{m=0}^{\infty} \int_0^{\pi} \mathrm{d}\phi \, w_m(\phi) \int_{|x|\leq 1} \mathrm{d}^2 x \, \delta(\underline{x} \cdot \underline{n}_{\theta} - s) U_m(\underline{x} \cdot \underline{n}_{\phi}) \tag{4.115}$$

$$= \sum_{m=0}^{\infty} \int_0^{\pi} \mathrm{d}\phi \, w_m(\phi) \frac{2}{m+1} U_m \left(\cos(\phi - \theta)\right) U_m(s) \sqrt{1 - s^2} \tag{4.116}$$

$$= \sqrt{1 - s^2} \sum_{m=0}^{\infty} \underbrace{\int_0^{\pi} \mathrm{d}\phi \, w_m(\phi) \frac{2}{m+1} U_m \left(\cos(\phi - \theta)\right)}_{\text{coefficients } c_m(\theta)} U_m(s) . \tag{4.117}$$

We will use these $c_m(\theta)$ later. For now, we note that we can obtain them directly from the RT as coefficients of the Chebyshev polynomial expansion. If we multiply the above equation with $\frac{2}{\pi} U_{m'}(s)$ and integrate, we obtain

$$c_{m'}(\theta) = \frac{2}{\pi} \int_{-1}^{1} \mathrm{d}s \, (\mathcal{R}f)\,(\underline{n}_{\theta}, s) U_{m'}(s) . \tag{4.118}$$

So far, everything has been continuous. Next, we introduce the discrete and equidistant angles $\phi_k = \frac{\pi k}{K}$ with $k = 1, \ldots, K$. Then we have

$$\frac{1}{K} \sum_{k=1}^{K} U_m(\underline{x} \cdot \underline{n}_{\phi_k}) U_m \left(\cos(\phi_k - \phi)\right) \tag{4.119}$$

$$= \frac{1}{2K} \sum_{k=1}^{2K} U_m(\underline{x} \cdot \underline{n}_{\phi_k}) U_m \left(\cos(\phi_k - \phi)\right) , \tag{4.120}$$

since for $\phi_k \to \phi_k + \pi$ the sign of both U_m changes, and so, writing out the first polynomial U_m with explicit polynomial coefficients u_{ml} and the second one with equation (4.94)

$$= \frac{1}{2K} \sum_{k=1}^{2K} \underbrace{\sum_{l=0}^{m} u_{ml} (\underline{x} \cdot \underline{n}_{\phi_k})^l}_{U_m(\underline{x} \cdot \underline{n}_{\phi_k})} \sum_{\nu=-m}^{m} e^{i\nu(\phi_k - \phi)} . \tag{4.121}$$

We now rewrite

$$(\underline{x} \cdot \underline{n}_{\phi_k})^l = r^l (\underline{n}_{\theta} \cdot \underline{n}_{\phi_k})^l \tag{4.122}$$

$$= \left(\frac{r}{2}\right)^l \left(e^{i(\theta - \phi_k)} + e^{-i(\theta - \phi_k)}\right)^l , \tag{4.123}$$

and use the binomial formula

$$(a+b)^l = \sum_{p=0}^{l} \binom{l}{p} a^{(l-p)} b^p \qquad l \in \mathbb{N} \tag{4.124}$$

to evaluate the sum over k. This results in $\delta_{\nu,2p-l}$, which we now use to compute the sum over ν. The result is

$$\frac{1}{K}\sum_{k=1}^{K} U_m(\underline{x}\cdot\underline{n}_{\phi_k})U_m\left(\cos(\phi_k-\phi)\right) = U_m(\underline{x}\cdot\underline{n}_{\phi}) \qquad \forall\, m < K\,. \tag{4.125}$$

This means that the Chebyshev polynomial $U_m(\underline{x}\cdot\underline{n}_\phi)$ backprojected under any angle ϕ can be represented *exactly* by backprojections of the same U_m under K discrete angles ϕ_k, if $m < K$, i.e., if the expansion is sufficiently accurate. In this way, we now have a result for the exact reconstruction of all functions that can be expressed as a linear combination of $U_m(\underline{x}\cdot\underline{n}_\phi)$ with $m < K$. For this class, we only need to know the RT under K angles ϕ_k, since the $c_m(\phi_k)$ can be computed from these by a Chebyshev expansion. These functions have the following form:

$$f(\underline{x}) = \sum_{m=0}^{K-1}\int_0^{\pi} \mathrm{d}\phi\, w_m(\phi)U_m(\underline{x}\cdot\underline{n}_\phi) \tag{4.126}$$

$$= \sum_{m=0}^{K-1}\int_0^{\pi} \mathrm{d}\phi\, w_m(\phi)\frac{1}{K}\sum_{k=1}^{K} U_m(\underline{x}\cdot\underline{n}_{\phi_k})U_m\left(\cos(\phi_k-\phi)\right) \tag{4.127}$$

$$= \frac{1}{K}\sum_{k=1}^{K}\sum_{m=0}^{K-1} U_m(\underline{x}\cdot\underline{n}_{\phi_k})\underbrace{\int_0^{\pi} \mathrm{d}\phi\, w_m(\phi)U_m\left(\cos(\phi_k-\phi)\right)}_{=\frac{m+1}{2}c_m(\phi_k)}\,. \tag{4.128}$$

In this way, we have been able to discretize the angle, but it seems that we still need the full – i.e., continuous – information $(\mathcal{R}f)(\phi_k, s)$ for each angle ϕ_k. On the other hand, we also know that the RT equals a polynomial of at most order $K-1$ in s, up to a prefactor of $\sqrt{1-s^2}$, see equation (4.116). Therefore, the right hand side of

$$c_m(\phi_k) = \frac{2}{\pi}\int_{-1}^{1} \mathrm{d}s(\mathcal{R}f)(\phi_k, s)U_m(s)\,, \tag{4.129}$$

is an integral over a polynomial of at most order $2(K-1)$, multiplied with $\sqrt{1-s^2}$. Fortunately, it is possible to compute such an integral exactly from a discrete data set, using the so called quadrature formula

$$\int_{-1}^{1} \mathrm{d}s\,\sqrt{1-s^2} + p(s) = \sum_{i=1}^{K} w_i p(s_i)\,, \tag{4.130}$$

with suitably chosen w_i and locations s_i. For Chebyshev polynomials of the second kind, this is provided by Gaussian quadrature, (see, for example, *Numerical Recipes* [39]). Gaussian quadrature uses the locations and weights

$$s_i = \cos\frac{i\pi}{K+1}\,, \tag{4.131}$$

$$w_i = \frac{\pi}{K+1}\left(1 - s_i^2\right)\,, \tag{4.132}$$

and for the coefficients one obtains

$$c_m(\phi_k) = \sum_{i=1}^{K} \frac{\pi}{K+1}(1+s_i^2)\frac{\mathcal{R}f(\phi_k, s_i)}{\sqrt{1-s_i^2}} U_m(s_i) \tag{4.133}$$

$$= \sum_{i=1}^{K} \frac{\pi}{K+1} \sin^2\theta_i \frac{\mathcal{R}f(\phi_k, s_i)}{\sin\theta_i} \frac{\sin((m+1)\theta_i)}{\sin\theta_i} \quad \text{with } \theta_i := \frac{i\pi}{K+1} \tag{4.134}$$

$$= \frac{\pi}{K+1}\sum_{i=1}^{K}(\mathcal{R}f)(\phi_k, s_i)\,\sin((m+1)\theta_i)\,. \tag{4.135}$$

We see that all functions of type

$$f(\underline{x}) = \sum_{m=0}^{K-1}\int_0^{\pi} \mathrm{d}\phi\; w_m(\phi)\; U_m(\underline{x}\cdot\underline{n}_\phi) \tag{4.136}$$

can be reconstructed *exactly* for all $\underline{x}$ with $|\underline{x}| \leq 1$ from the $K \times K$ data points $(\mathcal{R}f)(\phi_k, s_i)$. However, the data points are not equidistant, but with $s_i = \cos\theta_i$ are more dense at the boundaries $s \to \pm 1$. This is intuitively understandable, since at the boundary the Radon projections contain less and less information, since the values stem from integrations over smaller segments. Therefore, a denser sampling is required. However, the required locations s_i are not necessarily where the detector pixels are located, which in practice, are often equidistant. In this case, some accuracy is lost, since the values of the RT at s_i have to be computed from interpolation. A noteworthy exception is PET, where the pixels are positioned on a circular detector ring; these are exactly the correct locations for this method.

So indeed, a certain function class can be reconstructed exactly from finite data. Note the complete analogy with the sampling theorem. When we are sure to have a band limited signal (in this case Chebyshev limited), the (continuous!) signal can be reconstructed exactly from a sufficiently large discrete number of sampled data points. However, as noted before, unlike with audio sampling, it is impossible to apply a Chebyshev low pass filter to a real object before measurement, hence aliasing artifacts may appear.

In summary, the reconstruction algorithm presented in this section consists of the following steps:

1. Compute $c_m(\phi_k)$ from the data using equation (4.135).
2. Multiply $c_m(\phi_k)$ with $\frac{m+1}{2K}$, see equation (4.128).
3. Compute the "filtered projections" (cf., equation (4.128))

$$\sum_{m=0}^{K-1} \underbrace{\frac{m+1}{2K} c_m(\phi_k)}_{c_m^*(\phi_k)} U_m(s) =: \lambda_k(s) .$$

4. Compute the backprojection $f(\underline{x}) = \frac{1}{K} \sum_{k=1}^{K} \lambda_k(\underline{x} \cdot \underline{n}_{\phi_k})$ (again from equation (4.128)).

Step 1 is the most interesting, requiring only $\mathcal{O}(K \log K)$ operations, since it can be handled by a discrete sin transformation (FFT). Thus, only a total of $\mathcal{O}(K^2 \log K)$ steps is needed (including all angles). Step 3 can also be reduced to a sin transformation (with interpolation). Step 4 is the limiting step with $\mathcal{O}(M^2 K)$ operations (for a grid of size M^2). Since polynomials of degree $K-1$ have only limited resolution, it is appropriate to choose $M \approx K$ (a Chebyshev polynomial of degree m has m roots in the interval $[-1, 1]$, to resolve these one needs at least m voxels). Hence the method is of similarly favorable complexity as FBP, but offers the potential to be exact.

4.4 Tomographic projection geometries

In the previous sections, we considered the mathematical foundations of tomography in the geometry of parallel projections and under ideal recording conditions. In this section, we address the experimentally very relevant configuration of cone beam projection. As we saw above, the simplest implementation of tomography is based on parallel projections and in the framework of the 2dRT. Parallel projections are acquired by rotating the object around a single rotation axis, or equivalently by rotating the source detector unit around the sample. In these cases, both image recording and reconstructions involve only the 2dRT and inverse 2dRT, applied to a set of parallel planes, with a normal vector along the rotation axis. Importantly, the 2dRT is applied independently to a set of parallel planes through the object or body. Due to its simplicity regarding both the geometry of data acquisition and the reconstruction, this reduction of a volumetric problem to the 2dRT is usually desired. The underlying assumptions, however, are restrictive: One assumes ideal parallel projection images of objects rotated around a stable axis, perfectly aligned with respect to the optical axis and the detector. When scans or the recording geometry deviate from this ideal parallel beam setting, the situation may become more complicated. This section addresses more general projection geometries, while the next section deals with artifacts arising from nonideal data acquisition.

Figure 4.11 presents an example of a high resolution mouse thorax micro-CT, recorded in cone beam geometry with a high brilliance liquid metal jet anode [26]. While a simple geometric variable transformation can be used to convert a cone

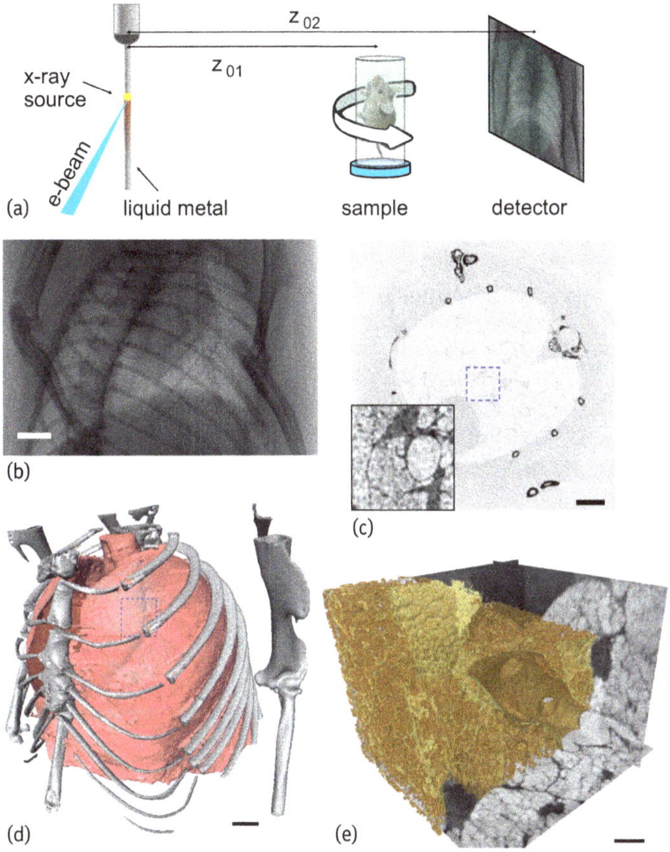

Fig. 4.11: (a) Schematic view of a modern micro-CT setup. Electrons are focused onto a liquid anode source, in this case with spot sizes typically in the range ∈ [4 . . . 40 μm], generating almost isotropic emission of x-rays, from which a cone beam is selected by apertures. By adjustment of source-to-object distance z_{01}, and the object detector distance z_{02}, different geometric magnifications, FOV, flux densities, and – as we will see in Chapter 6 – also contrast mechanisms are chosen. (b) Projection with large FOV covering the entire thorax of the mouse. (c) Slice perpendicular to the rotation axis. The inset shows a zoom into the lung area marked by the dashed rectangle with adjusted contrast. (d) 3d rendering showing the thorax, containing automatically segmented bones (*gray*), the heart (*red*) and lung tissue (*pink*). The *dashed lines* indicate the zoom region shown in (e). (e) 3d rendering of the "zoomed" reconstruction volume recorded at higher magnification by moving the detector downstream. In this setting phase contrast is exploited, as detailed in Chapter 6. The subfigure shows orthogonal slices (orthoslices) and renderings of the soft tissue structure. Scale bars denote 2 mm in (b,c,d) and 500 μm in (e). Adapted from [26].

beam recording into an effective parallel, as discussed below, cone beam recordings of thicker objects or with larger cone angle require a more involved reconstruction scheme, denoted as fan beam reconstruction and cone beam reconstruction, for cases of fan shaped or conical beams, respectively. In addition, the angular range to be covered has to be larger than the typical 180° (roughly by the cone angle).

4.4.1 X-ray transform, Fresnel scaling theorem and cone beam reconstruction

The RT does not actually describe the radiographic image in 3d. For this purpose, we need the x-ray transform (sometimes also simply called the ray transform), defined for parallel beams as

$$\mathcal{P}[f]\,(\hat{\theta},\underline{x}) := \int_0^\infty f(\underline{x} + t\hat{\theta})\,dt\,. \tag{4.137}$$

Here, $\hat{\theta} \in \Omega \subseteq S^2$ denotes a unit vector contained in a domain of direction vectors of the unit sphere S^2, under which the object is sampled, i.e., a projection angle. The points $\underline{x}$ are in the subspace $\hat{\theta}^{\perp}$ perpendicular to the projection direction $\hat{\theta}$, i.e., in a plane perpendicular to the projection axis, say the source or the detection plane. For $n = 2$, this is equivalent to the RT. However, for $n > 2$ the definition of the x-ray transform differs. While the RT is based on integrals over $(n-1)$-dimensional hyperplanes, the value of the x-ray transform is given by the integral over a line going through $\underline{x}$ in the direction of $\hat{\theta}$. The x-ray transform as defined by equation (4.137) describes the formation of the radiographic image in a parallel beam. Its mathematical properties are well studied, see, for example, [44].

For cone beam illumination with a central ray $\hat{\theta}$ (see Figure 4.15), the x-ray transform can be written as

$$\mathcal{C}_{\underline{s}}[f]\,(\hat{\theta},\underline{y}) := \int_0^\infty f(\underline{s} + r\,\hat{r}(\underline{y}))\,dr\,, \tag{4.138}$$

where $\underline{s}$ denotes the (fixed) source position and $\underline{y} = (y_2, y_3)^{\mathrm{T}}$ a position on the detector. The detector normal is along $\hat{\theta}$. The object function f is mapped onto its projection values along lines from the source to a detector position $\underline{y}$. The tangent unit vector of the ray from the source to $\underline{y}$ is denoted by $\hat{r}(\underline{y})$. With the notation specified in Figure 4.15 (a), we can write

$$\hat{r}(\underline{y}) := \frac{y_2\hat{y}_2 + y_3\hat{y}_3 + z_{12}\hat{\theta} - \underline{s}}{|y_2\hat{y}_2 + y_3\hat{y}_3 + z_{12}\hat{\theta} - \underline{s}|}\,, \tag{4.139}$$

where z_{12} is the detector distance from the origin assumed to be located in the object, for example, in the center of rotation. Correspondingly, the source position is written as $\underline{s} = -z_{01}\hat{\theta}$, with z_{01} denoting the distance between source and object. For each source position, the optical axis is given by the central ray of the cone with unit vector $\hat{\theta}$. The source detector gantry can then be moved, as in helical CT, for example, with the source trajectory parameterized by a curve Γ with $\underline{s} \in \Gamma$, or equivalently, the object can be rotated as in analytical CT. The set of recorded central ray directions $\hat{\theta}$ forms the so called orbit on the unit sphere of directions S^2.

A range of projection angles of 180° is sufficient for reconstruction of parallel beam analytical CT data. We already know that, in this case, the 3d object can be decomposed in parallel planes, with each plane reconstructed by the 2dRT. The 2dRT is

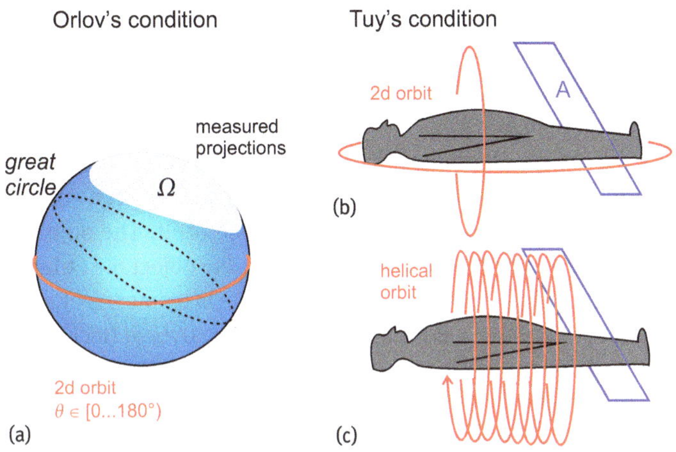

Fig. 4.12: Geometric sampling criteria for projection directions and orbits of the source. (a) The parallel beam projection geometry has to satisfy Orlov's completeness condition. The sketch shows the unit sphere of directions S^2 (*blue*), and the subset of measured projections Ω (*gray*). According to Orlov, every great circle on S^2 has to cut the domain of measured projections, which is not the case for Ω, but for the standard $\theta \in [0 \dots 180^\circ)$ scan shown in *red*. (b) In contrast to projections based on a parallel beam geometry, point source projections where the source emits a cone of radiation have to satisfy Tuy's condition. Accordingly, each plane crossing the reconstruction volume has to intersect with the orbit. As an example, the plane A is shown in *blue*. Tuy's condition is clearly not fulfilled for any of the two (standard) orbits shown, but would be satisfied, if both orbits were combined.

invertible, if $\theta \in [0 \dots 180^\circ)$ is sufficiently sampled. One may wonder which other scans could potentially yield a complete set of data. For example, when rotations around a first axis are limited to a smaller range, giving rise to a missing wedge, a rotation around a second (also only partially accessible) axis may be used for compensation. Let $\Omega \subset S^2$ denote the set of directions measured, and S^2 the unit sphere of directions. According to Orlov's completeness condition, which was initially stated in the context of electron microscopy and tomography of biomolecules [35], the object is completely sampled if every equatorial circle (great circle) in S^2 crosses Ω [17, 34, 35]. For example, this is the case for a single great semicircle on S^2, i.e., the standard tomographic scan over 180°, see Figure 4.12. In fact, Orlov's condition can easily be understood based on the Fourier slice theorem. Accordingly, each measured projection direction $\hat{\theta} \in S^2$ corresponds to a plane in reciprocal space going through the origin and defined by $k_x(\hat{\theta}\cdot\hat{e}_x)+k_y(\hat{\theta}\cdot\hat{e}_y)+k_z(\hat{\theta}\cdot\hat{e}_z) = 0$. The measurement planes in reciprocal space form a set of all such planes in $\tilde{f}$, each corresponding to a projection angle. Now consider a point $\underline{k} = (k_x, k_y, k_z)^{\mathrm{T}}$ in the object's reciprocal space $\tilde{f}(k_x, k_y, k_z)$. For this Fourier component to be contained in the data, the vector (k_x, k_y, k_z) has to be in one of the measurement planes, i.e., it has to be normal to a projection direction $\hat{\theta}$. Stated differently, there has to be at least one projection $\hat{\theta}$, for which $\hat{\theta} \cdot \underline{k} = 0$. On the unit sphere of directions S^2, the points corresponding to a given $\underline{k}$ form a great circle. Therefore, every great circle has to intersect the set of measured projections. In most cases, this set will be realized by a continuous scan of the source with respect to the object, see the red orbit shown in Figure 4.12.

While Orlov's completeness condition applies to parallel projection, Tuy's completeness condition is formulated for the case of a point source moving on a curve Γ around the object f. In tomography, Γ is also called the source orbit. The object is sampled completely, if the orbit Γ intersects all planes through the object, i.e., planes cutting its support *supp(f)* [34, 46]. A simple way to achieve this is by means of a helical orbit, see Figure 4.12 (b,c). Contrarily, the standard scan of analytical (technical) CTs consisting of a 180° rotational range – strictly speaking – does not yield complete data, since it does not satisfy Tuy's condition. However, as we will see next, such a scan will be still sufficient for sufficiently small cone angles γ.

To address the cone beam projection geometry, we consider the situation sketched in Figure 4.13. The object is positioned at distance z_{01} behind the source, and its projection image is recorded at distance z_{12} behind the object. Accordingly, the projection image appears geometrically magnified by a factor of

$$M = \frac{z_{01} + z_{12}}{z_{01}} . \tag{4.140}$$

This corresponds to the simple geometrical magnification expected from geometrical optics. According to the full wave optical treatment presented in Chapter 6, we also have to transform the propagation distance, i.e., in addition to the lateral length scales, also the longitudinal length scale, according to [36]

$$z_{\text{eff}} = \frac{z_{12}}{M} = \frac{z_{01}\, z_{12}}{z_{01} + z_{12}} . \tag{4.141}$$

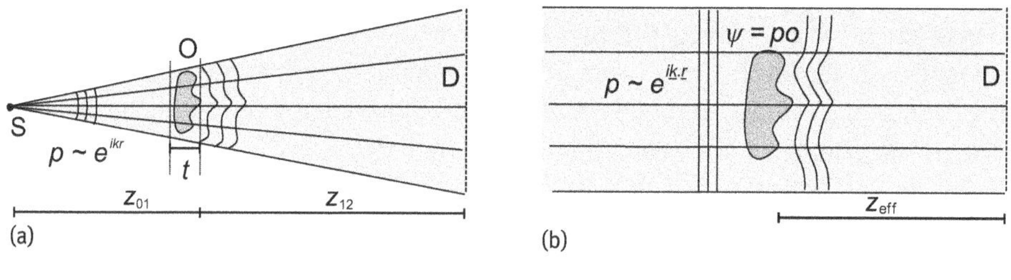

Fig. 4.13: According to the Fresnel scaling theorem, the wavefields in the detection plane behind an object are equal for the cone beam and parallel beam case, up to a simple variable transformation. As expected from ray optics, the data on the detector (D) are expanded in the cone beam case corresponding to the geometric magnification factor $M = (z_{01} + z_{12})/z_{01}$. In addition, the propagation distance between the object's exit plane and the detector has to be transformed according to $z_{\text{eff}} = z_{12}/M$. For the theorem to hold, we must assume a sufficiently thin object, where the exit wave $\psi = po$ is given by multiplication of the probe p (illumination) and the complex valued object transmission function o.

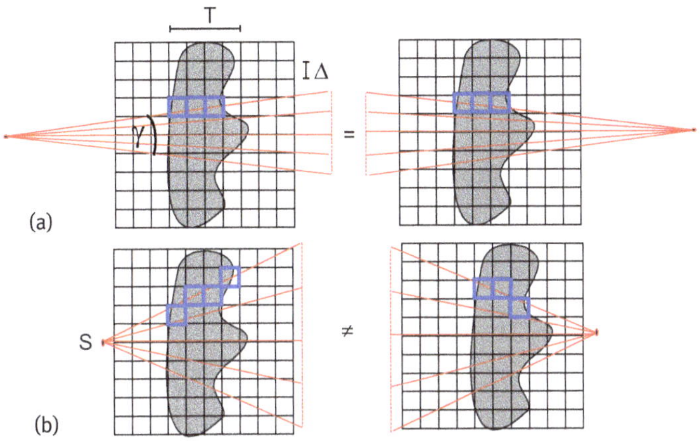

Fig. 4.14: Illustration of cone beam projections. (a) For small cone beam angle $\gamma \leq \arctan(\Delta/T)$), each ray intersects voxels only of a single plane. In this case, two projections separated by a rotation of 180° are approximately equal, as in parallel beam projection. (b) For larger cone angles, this is no longer the case. Hence, cone beam data is, in general, not part of the Radon space. Projection and backprojection algorithms have to take the cone beam path into account.

This is the Fresnel scaling theorem [36]. It states that the field measured behind an object that is illuminated by a point source is equivalent to that of an object illuminated by a parallel beam, if we shrink the detector pixel size by a factor of M, and adapt the distance between object and detector to $z_{12} \rightarrow z_{\text{eff}}$. However, if the point source illumination can be mapped onto the parallel case, how can the cone beam situation be different from a parallel projection, regarding completeness of tomographic scans? The answer is simple: The Fresnel scaling theorem can only be applied if the object is sufficiently thin, essentially 2d. If this is not the case, the magnification M is different for the front and the back of the object, which already hints at a problem. How thin is sufficiently thin? This can be readily answered based on Figure 4.14. In traversing the object, a ray should not intersect more than a single row of voxels, or more precisely, the resolution element Δ, i.e., the object thickness T must be smaller than

$$T \leq \Delta/\tan(\gamma)\,, \tag{4.142}$$

where γ denotes the opening angle of the cone beam, as sketched in Fig.4.14 If this is not the case, independent reconstruction of 2d slices as in parallel projection is no longer possible. Instead, projection and backprojection must be adapted for the oblique nature of the rays, and operations acting simultaneously on a subset of the volume become necessary. Worse than that is the fact that in cone beam geometry, the density of rays is inhomogeneous. It is higher towards the source than towards the detector. This way, the part of the object close to the source has 'higher weight', and two projections of the same object spaced by 180° are, in general, not equal. For these

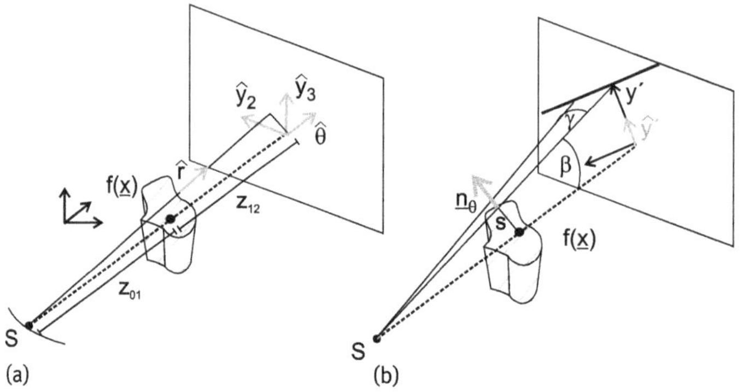

Fig. 4.15: (a) Cone beam geometry. For a given source position $\underline{s}$, the unit vector $\hat{\theta}$ indicates the direction of the central ray and the detector normal, with z_{01} and z_{12} the source object and the object detector distances, respectively. The coordinate system of the sample is denoted by $\underline{x}$ and the detector frame by $\underline{y}$. Each detector position corresponds to a line integral through the object, according to $\mathcal{C}_{\underline{s}}[f] = \int f(\underline{s} + r\hat{r}(\underline{y}))\, dr$. (b) Geometry and definitions required for Grangeat's formula. Each line on the detector defines a plane through the object, which can either be parameterized by its $\underline{n}_\theta$ and distance to the origin S, or by the unit vector $\hat{y}'$ and distance y' on the detector. The data is integrated with proper weights along lines (see the *black line*, integral over γ), and differentiated along $y' = z_{02} \tan\beta$.

reasons, the measured data does not form a Radon space, and – strictly speaking – all concepts presented so far would fail. However, luckily a few clever generalizations and observations have been made in view of generalizing tomographic concepts to cone beam geometry. Most importantly, according to Grangeat's formula [16], the first derivative of the 3dRT can be derived from a weighted cone beam transform $\mathcal{C}f$ according to [16, 34]

$$\frac{1}{\cos^2\beta}\frac{\partial}{\partial y'}\mathcal{C}f(y') = \frac{\partial}{\partial s}\mathcal{R}_{\underline{n}_\theta}f(s)\,, \tag{4.143}$$

with angles and coordinates defined in Figure 4.15 following [30], and the weighted cone beam transform given by

$$\mathcal{C}f(y') = \int_{-\pi/2}^{\pi/2}\int_{\mathbb{R}^+} f(r, \beta(y'), \gamma)\,\frac{z_{01} + z_{12}}{\cos(\gamma)}\,dr\,d\gamma\,. \tag{4.144}$$

Here, $\mathcal{R}_{\underline{n}_\theta}f(s)$ denotes the 3dRT, corresponding to planes normal to the direction vector $\underline{n}_\theta$, see Figure 4.15. Grangeat's formula provides the starting point for an exact cone beam method involving the following steps [12]: Weighting of the data recorded with the planar detector, computation of the partial derivatives with respect to the principal axes, computation of the line integrals, and interpolation in Radon space. The significance of this approach is the fact that it enables reconstruction of cone beam data even at large opening angle γ. The exact method based on Grangeat's differentiation trick as described in [12] is quite involved, and in practice the vast majority of

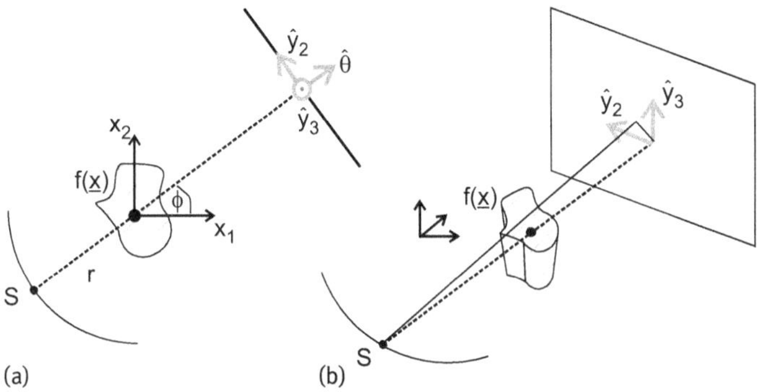

Fig. 4.16: Geometry and definitions of unit vectors required for the FDK reconstruction formula.

cone beam data is instead reconstructed by an approximative cone beam algorithm proposed by Feldkamp, Davis and Kress in 1984 [14], the famous FDK algorithm. The FDK algorithm is derived under the assumption of small cone beam angle $\gamma \ll \pi$. Of course, it can be larger than for the flat object approximation to hold, otherwise the algorithm would be obsolete. We consider a point source S on a circular orbit with radius r and data recorded by a planar detector, i.e., the standard geometry of analytical CT. Despite the fact that this orbit strictly speaking does not fulfill Tuy's condition, FDK reconstruction works well, again if γ is not too large. The point of normal incidence on the detector defines the projection direction $\hat{\theta}$ and the origin of the detector coordinate system $\underline{y} = (y_2, y_3)^T$. The coordinate system of the object is denoted by $\underline{x}$ with its origin in the center of rotation. The plane defined by $(y_2, \hat{\theta})$ is called the central slice. Data in this slice can be reconstructed exactly by fan beam reconstruction, i.e., without any approximations. With the geometric definitions as given in Figure 4.16, this results in the reconstruction formula [34]

$$f(\underline{x}) \simeq \int_{S^1} \frac{z_{01}^2}{z_{01} - \underline{x} \cdot \hat{\theta}} \int_{-\rho}^{\rho} dy_2' \, h(y_2 - y_2') \, g(\hat{\theta}, y_2' \hat{y}_2 + y_3 \hat{y}_3) \frac{z_{01}}{\sqrt{z_{01}^2 + {y'_2}^2 + y_3^2}} , \tag{4.145}$$

where the projection data at position $\underline{y} = y_2 \hat{y}_2 + y_3 \hat{y}_3$ in the detection plane is $g(\hat{\theta}, \underline{y})$. The position on the detector $\underline{y}$ is geometrically related to the object $\underline{x}$ by

$$y_2 = \frac{z_{01}}{z_{01} - \underline{x} \cdot \hat{\theta}} \underline{x} \cdot \hat{y}_2 , \qquad y_3 = \frac{z_{01}}{z_{01} - \underline{x} \cdot \hat{\theta}} , \tag{4.146}$$

and $h(s)$ is the usual backpropagation filter (4.66)

$$h = \frac{1}{2} (2\pi)^{1/2-n} |s|^{n-1} \, \Phi\left(\frac{|s|}{c}\right) . \tag{4.147}$$

The FDK algorithm can be efficiently implemented numerically. Figure 4.17 shows an FDK reconstruction obtained for an excised mouse cochlea [7], using commercially

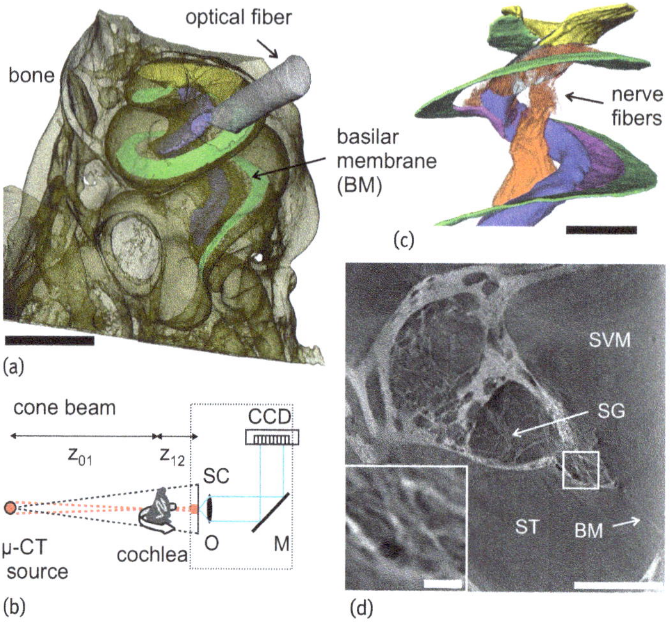

Fig. 4.17: High resolution tomography of a mouse cochlea, obtained at a liquid anode cone beam micro-CT setup [7], showing (a) a 3d rendering of the excised cochlea with bone (*brown, semi-transparent*), surrounding soft tissues such as the basilar membrane (*green*), Reissner's membrane (*yellow*), Rosenthal's canal (*blue*), and an optical fiber (*gray*), as part of research directed at novel optogenetic cochlea implants. (b) In contrast to the previous example of a mouse lung (Figure 4.11), an inverse geometry with $z_{12} \ll z_{01}$ is used, in combination with a high resolution detector system, consisting of a scintillator (SC), objective (O), mirror (M) and an optical CCD. (c) Magnified view showing nerve tissue (*orange*). (d) Orthoslice through a ROI, revealing fine anatomical details, such as the scala tympani (ST), basilar membrane (BM), scala vestibuli et media (SVM) and spiral ganglion (SG) for inverse geometry. The data was obtained for $z_1 = 65$ mm, $z_2 = 20$ mm, and effective pixel size $p_{\text{eff}} = 0.57$ μm. The spatial resolution is on the order of 3 voxels, i.e., 1.75 μm. The image quality and sensitivity to soft tissues is achieved by exploiting phase contrast and phase retrieval algorithms, see Chapter 6. Scale bars are (a) 800 μm, (c) 400 μm,(d) 200 μm and 20 μm (*inset*). Adapted from [7]. Multimedia views are available online.

available software (Bronnikov algorithms, Arnhem) [11]. FDK reconstruction is also provided by the freely available ASTRA toolbox [37]. Important in data recording is the fact that the rotational range should be enlarged by the cone beam angle, i.e., one should scan $\theta \in [0 \ldots (180^\circ + \gamma)]$, instead of $\theta \in [0 \ldots 180^\circ]$ as in the parallel beam case.

4.4.2 Tomography based on the 3d Radon transform

As mentioned above, Grangeat's solution to cone beam tomography is based on the 3dRT. We now consider reconstruction by 3dRT in more detail. The required area integrals can be obtained by postintegration on the planar detector, i.e., integration of the detector data along one direction. Let us for now assume a parallel beam or a sufficiently small cone angle α, neglecting the weighting and filtering steps according to Grangeat, which have received much attention in the literature. Instead we have the geometric parameters of analytical CT in mind, where the object is much smaller than the source object distance z_{01}. For such conditions of typical micro-CT setups, it was shown recently that the area integrals required by reconstruction according to the 3dRT can be efficiently implemented, and that the stringent conditions on source size for high resolution and partial coherence can be relaxed in one direction [48]. This way, tomography applications, for example, with nanoscale resolution, which today require highly brilliant synchrotron radiation become accessible for laboratory micro-CT setups. Moreover, the local properties of reconstruction by 3dRT can be exploited for so called local or ROI tomography. Here we present the underlying concept of these ideas, closely following the original work [47].

We recall that tomography in the framework of 2dRT is based on recording a set of signal curves $g_z(\theta, s) := g(z, \theta, s)$, for all projection angles θ of a single rotation axis (z axis) and for a set of parallel planes indexed by z. For given θ, each point s of a signal curve $g_z(\theta, s)$ results from a line integral. Contrarily, the 3dRT of a function $f(\underline{x})$, $\underline{x} \in \mathbb{R}^3$ requires integrals over planes according to

$$(\mathcal{R}f)(\underline{n}_{\theta,\phi}, s) := \int\limits_{\underline{y}\in(\theta,\phi)^\perp} f(s\,\underline{n}_{\theta,\phi} + \underline{y})\,\mathrm{d}^2 y = \int \mathrm{d}^3 x\, f(\underline{x})\,\delta(\underline{x}\cdot\underline{n}_{\theta,\phi} - s)\,, \tag{4.148}$$

where $(\theta, \phi)^\perp$ is the subspace of $\mathbb{R}^3$ perpendicular to $\underline{n}_{\theta,\phi}$. For a given point $\underline{n}_{\theta,\phi}$ on the unit sphere, the function f has to be integrated over a set of parallel planes intersecting the object, each plane contributing a data point $g(\theta, \phi, s) = g_{\theta,\phi}(s)$ or simply $g(s)$ for one projection direction $\underline{n}_{\theta,\phi}$. Figure 4.18 illustrates the 2d integration of the object function over such a plane. Complete data collection for 3dRT now requires rotations by two angles, for example, parameterized by θ and ϕ. The angles can be written in vector notation as $\underline{n}_{\theta,\phi}$ and, hence, more explicitly for 3dRT than the notation $\underline{n}_\theta$ used for arbitrary dimension before. Here, we simply parameterize the directions on the unit sphere S^2. To obtain a complete data set, the entire unit sphere must then be covered, i.e., sets of planes intersecting the object have to be probed at all $\underline{n}_{\theta,\phi}$. Contrarily, the generation of a complete 2dRT data set only requires rotation around one axis.

As was demonstrated experimentally in [47], this concept is compatible with anisotropic sources. That work was motivated by the idea that area integrals can be efficiently computed with sources extended in the direction over which the data is integrated out. The entire 3d reconstruction volume still inherits an isotropic resolution corresponding to the high resolution direction. In this way, instrumental constraints (on source, focusing or collimation optics) can be relaxed, and more photons contribute to image formation. In the case of isotropic illumination, the 1d signal curves for the 3dRT can be simply generated from conventional 2d projections, by integra-

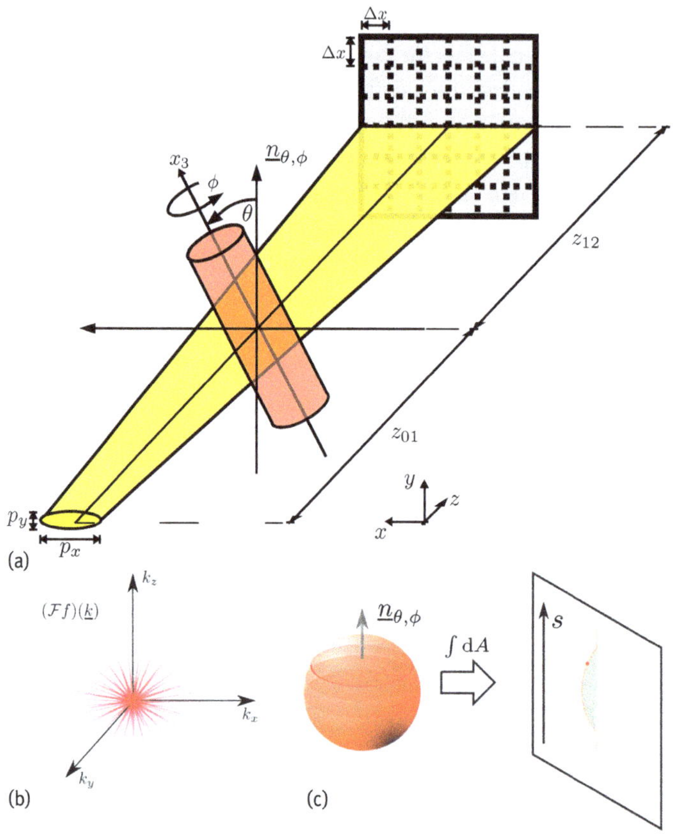

Fig. 4.18: Illustration of the 3d RT. (a) The tomographic rotation axis is successively tilted by an angle θ with respect to the laboratory y axis, while the object is rotated around this axis by an angle ϕ. 2d projection data is acquired and integrated along x. In this direction, the source size can be relaxed, resulting in higher usable flux for given resolution or partial coherence. (b) Sketch of the signal measured for a 3dRT data set in 3d Fourier space resulting in a "hedgehog like" structure. (c) Every object plane is mapped to one point of the 3d RT by area integration. The normal vector to the plane $(\theta, \phi)^\perp$ defines a family of parallel planes, mapping onto the 1d signal curve $g_{\theta,\phi}(s)$; s denotes the Radon coordinate. From [48].

tion of the projection images along different directions in the detection plane, see Figure 4.19. This requires that the entire object 'fits' the detector size in the lateral (integration) direction. The procedure can be denoted as 'postintegration', or simply 'detector integration', and yields one-dimensional profiles corresponding to the area integrals for all required angles. In this case, the data recording is still identical to the conventional 2dRT with one physical rotation, and only the inversion is based on the 3dRT, offering certain advantages for local tomography, as will be addressed below.

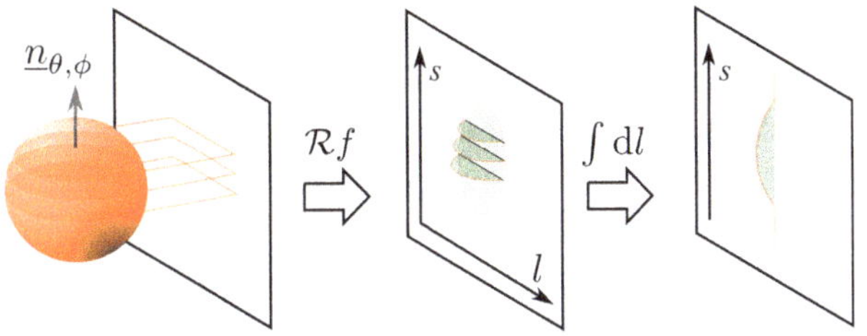

Fig. 4.19: Relation between the 2dRT and 3dRT. The conventional projection image is formed by the integral through the object along the direction perpendicular to the detector. By integrating along a direction parallel to the detector (here denoted by coordinate *l*), integration over one plane through the object is performed, and hence the prerequisite for the 3dRT.

3dRT reconstruction follows the general inversion formula equation (4.40) [41] to reconstruct f from $\mathcal{R}f$, formulated for $n = 3$

$$f(\underline{x}) = \frac{1}{2}(2\pi)^{-5/2}\left(I^{-\alpha}\mathcal{R}^{\#}I^{\alpha-2}\mathcal{R}f\right)(\underline{x}) . \tag{4.149}$$

Different choices for the parameter $\alpha < 3$ can be compared in view of their reconstruction properties for 3dRT, since robustness against noise and other imperfections in data acquisition could very well be different from the 2dRT case. Here, we consider only FBP, i.e., $\alpha = 0$, yielding (4.46) [47]

$$f(\underline{x}) = \frac{i^2}{2}(2\pi)^{-5/2}\left(\mathcal{R}^{\#}\left(\frac{\partial}{\partial s}\right)^2(\mathcal{R}f)(\underline{n}_\theta, s)\right)(\underline{x}) . \tag{4.150}$$

Importantly, the 3dRT has local properties, i.e., for reconstruction of $f(\underline{x})$ at the position $\underline{x}$, as only the RT $(\mathcal{R}f)(s, \underline{n}_\theta)$ and its second derivative at $s = \underline{x}\cdot\underline{n}_\theta$ enters. Contrarily, for 2dRT as for all even n, reconstruction at a given point is influenced by the global object structure, as expressed by the algebraic kernel of the Hilbert integral transform. In order to compute the required area integral for 3dRT, the object needs to be fully illuminated along the lateral direction, but detection points have to be 'dense' only along the normal of the integration plane, and only around a single point, where the rotation axes intersect. Further, the number of angular projections N_p can be significantly relaxed, as N_p decreases for smaller FOV.

This brings us to the interesting question of the angular sampling required for 3dRT. Using the same arguments of Nyquist sampling like previously for 2dRT, the necessary number of projections can be estimated to [47]

$$N_\text{p} \simeq \frac{\pi}{\sqrt{3}} N_\text{r}^2 , \tag{4.151}$$

where N_r is the number of resolution elements in the reconstruction volume. In contrast to 2dRT, where the number of projections scales linearly with N_r, the relation is quadratic for 3dRT. However, as the 3dRT scheme integrates over planes, more signals are summed up, and the accumulation time per projection in 3dRT can be reduced accordingly. Provided there will be future technical improvements, in particular sufficiently fast motor movement and short detector readout, the total accumulation time can be decreased significantly for anisotropic sources with higher permissible flux.

However, it is important to control the errors in approximating area integrals by ways of extended sources and detector integration. The error $\mathcal{E} = A/A_0$ defined as the ratio between the measured area integral A and the true area integral A_0 can be decomposed into [30]

$$\mathcal{E} = (1 + \mathcal{E}_{\mathrm{cb}})(1 + \mathcal{E}_{\mathrm{src}})(1 + \mathcal{E}_{\mathrm{nl}}) \approx (1 + \mathcal{E}_{\mathrm{cb}} + \mathcal{E}_{\mathrm{src}} + \mathcal{E}_{\mathrm{nl}}) , \tag{4.152}$$

where $\mathcal{E}_{\mathrm{cb}}$ is the so called cone beam error, related to the fact that the rays are denser in those parts of the object facing the source than in those facing the detector. $\mathcal{E}_{\mathrm{nl}}$ denotes the nonlinear error, which arises from taking the exponential mean (for absorption CT). In other words, the error that applies to all situations where the function f appears in the exponent of the integration kernel. In this case, taking the logarithm of the detector pixel is required, but it is not exact if the object function varies over the size of a voxel. $\mathcal{E}_{\mathrm{src}}$ is the error related to the fact that an extended source of width $2S$ has been used rather than a point source. If A_0 and A_s denote the area integrals of the point and extended sources, respectively, we hence have

$$\mathcal{E}_{\mathrm{src}} := \frac{A_s - A_0}{A_0} . \tag{4.153}$$

This error can be estimated based on the geometric parameters [30]

$$\mathcal{E}_{\mathrm{src}} \approx \frac{1}{6}\left(\frac{z_{01} + z_{12}}{z_{01}}\right)^3 \left(\frac{S}{z_{01} + z_{12}}\right)^2 . \tag{4.154}$$

Using this estimation with typical parameters of analytical CT, the equation above shows that area integrals can be perfectly carried out with anisotropic sources, and hence profit from an increased flux, without compromising resolution. Figure 4.20 shows projections and tomography results obtained for a hazelnut [48], using a standard fine focus source with anisotropic spot size, effectively 1 mm × 0.1 mm. Using the 3dRT scheme, the resolution does not suffer from the long extension in one direction. At the same time the flux is increased.

Finally, we note that a number of tomographic applications cannot be carried out with full beams and area detectors, where reconstruction based on 2dRT is the most efficient. Consider, for example, problems where the quantity f to be reconstructed is the x-ray fluorescence or a diffraction signal. In such cases, a pencil beam (point beam) must be used to record the signal in order to achieve spatial information, since each detection element collects signals from the entire illuminated volume, see [13, 29, 43] for recent examples. Data recording, then, always involves scanning three degrees of freedom, typically two translations (scanning a projection image) and one rotation (collecting a set of projection angles). For such applications, data acquisition

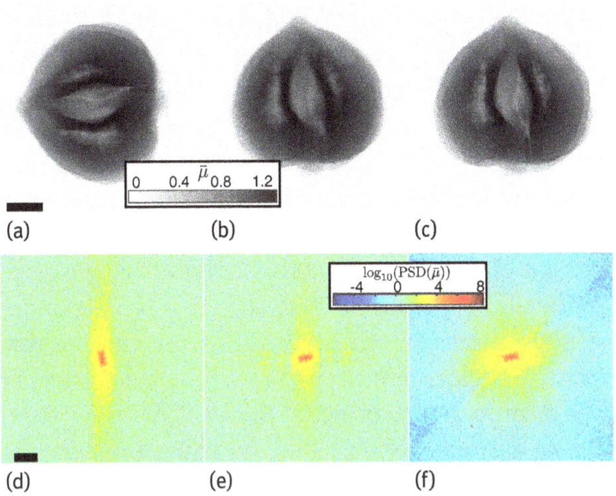

Fig. 4.20: Experimental results demonstrating the isotropic 'filling' of the 3d Fourier space by 3dRT. Projection images of a hazelnut are shown for $\theta = 90°$ (a) and $\theta = 0°$ (b), but constant ϕ. In (a) sharp edges are transferred at the side walls of the hazelnut, corresponding to the high resolution direction of the source, while the top and the bottom walls are blurred, and (b) shows opposite behavior. Reconstruction by 3*dRT* filtered backprojection (3dRT-FBP) yields isotropic resolution, despite the anisotropically blurred projection images, see the re-projection of the reconstructed volume (c) (for the same ϕ as (a,b)). The PSD (d–f) corresponding to (a–c) quantify this effect. While in (d,e) the signal extends over a large range in the vertical direction but decays rapidly along the horizontal direction, the signal is isotropically distributed up to high Fourier components in (f). Scale bars: 3 mm in (a–c), 3 mm^{-1} in (d–f). From [48].

based on 3dRT could be particularly interesting. It would involve one translation and two rotations, in combination with a light sheet illumination, similar to optical microscopy [49].

4.5 Tomography artifacts and algebraic reconstruction

In any real tomographic application, the image quality will mostly be determined by the presence or relative absence of artifacts. It is certain that it is important to understand the properties of the Radon operator and its inverse in mathematical spaces of continuous functions, and under idealized conditions. Noise models are included in many theoretical treatments, and issues of robustness, regularization and numerical stability have received considerable attention. However, two different issues dominate the efforts and considerations of tomographic practice. The first is due to the discrete nature of tomographic data as addressed above; the second is due to the inconsistencies and errors in recording the data. Reasons are as diverse as bad pixels and sensitivity variation of the detector, rotation misalignment, wobble of the tomographic axis, interior movements of the object, broad spectrum of the radiation, scattering within the sample, deviations in the geometry from an idealized setting and so forth. However, such systematic errors as indicated above are often difficult to take into account in the reconstruction scheme. Indeed, these errors must be addressed primarily on the experimental level. Following proper identification, improvements may be sought in tomographic alignment, data recording and raw data correction. If not possible otherwise, also in postprocessing on the levels of the projections, the sinogram or finally the 3d reconstruction. In this section, we first present some of the more common types of artifacts, including possible correction schemes. We then address algebraic reconstruction techniques (ART). In ART algorithms, reconstruction is carried out iteratively with simulation of the forward problem (image formation). In this way, a much higher level of experimental detail can be incorporated than in FBP. Hence, one can overcome some of the typical idealizations, for example, regarding the recording geometry, the interaction of the radiation and the sample or the spectral content.

4.5.1 Tomography artifacts and corrections

Dark current, empty beam (flat field), and ring artifacts: The vast majority of tomography applications is based on full field recordings. In looking at a projection, we want to be sure to inspect the object, not the flaws of the beam or the detector. To this end, we must know the dark current I_{dark} of the detector and its response to uniform illumination (sensitivity). At the same time, we want to divide out any variations of the illumination I_{flat} (empty beam), as would result from an inhomogeneous radiation of the source, or residual elements in the beam path. In fact, sensitivity and empty beams can be accounted for simultaneously. A common well suited correction scheme is, therefore, the pixel wise correction according to

$$I_{\text{corr}} = \frac{I_{\text{raw}} - I_{\text{dark}}}{I_{\text{flat}} - I_{\text{dark}}} . \tag{4.155}$$

This correction requires a sufficient number of dark and empty beam recordings, as well as stable beam conditions (stationary empty beams). The empty beam and sensitivity correction is also denoted as flat field correction. If a pixel is defective, however, returning zero or extremely high values (hot pixel), division is not advisable. Before further processing, median filtering (cf., Section 2.3) can then be used to remove outliers. Correction will often work only up to a certain degree. Suppose, in each line of the detector, a few systematic intensity variations remain. They will appear as horizontal lines in the sinogram, since they are constant in θ. Wavelet filtering can be used to remove these horizontal image components [32]. In the reconstruction, one will then observe ringlike artifacts in planes perpendicular to the rotation axis. As illustrated in Figure 4.21, it is possible to correct these ring artifacts by suitable filtering.

Misalignment of the rotation axis and errors in rotation angles: In a reconstruction, we want to assume that the rotation axis is vertical to the beam and located in

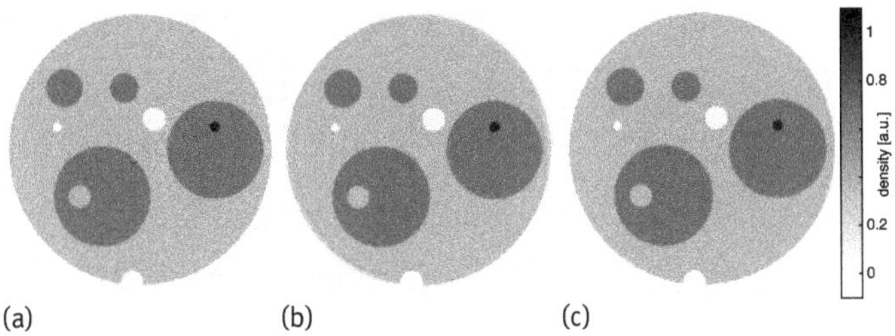

Fig. 4.21: Simulations illustrating so called ring artifacts, resulting, for example, from systematic variations in the detector sensitivity, in particular from bad pixels. A sinogram was first generated from a phantom consisting of spheres with density values between 0 and 1. The values in some horizontal lines of the sinogram were deliberately changed by 2%, to simulate a pixel wise variation in detector sensitivity or quantum efficiency. (a) Uncorrected reconstruction. (b) Reconstruction from the corrected sinogram, after application of a wavelet filter. (c) Ring artifact removal by a Fourier filter. The sinogram is summed over θ and smoothed by Fourier filtering. Subsequently, the difference between the filtered and unfiltered signal is computed and subtracted from the entire sinogram. From [25].

the central pixel column of the detector. In practice, we have finite deviation angles α and β, associated with the pitch and roll degrees of freedom of the tomographic rotation stage, respectively. The roll angle β of the tomographic axis describes a sideward tilt of the axis. It can later be corrected by a corresponding rotation of the detector images, but this requires interpolation and can be accompanied by a loss of resolution. Contrarily, the pitch of the axis α, i.e., the forward or backward tilt, may eventually be more problematic.[7]

Figure 4.22 (a–d) illustrates a proper alignment routine for the case of a micro-CT cone beam tomography, to prevent corresponding artifacts [6]. Suppose now that the tomographic axis has been properly aligned perpendicular to the optical axis, but that one notices after data recording that the axis was not centered in the detector but displaced laterally with respect to the optical axis. The projection images can be shifted to correct for this. In practice, a good strategy is to reconstruct 2d orthoslices for different positions of the rotation axis and then fix the axis for the position that yields the best image quality. Further, the rotational scan could have been subject to flaws. Figure 4.23 shows an example where cables have hindered a free rotation, resulting in hampered rotations. Nevertheless, the recorded data could partially be 'saved' by corrections performed in the sinogram [24]. Inspection of the sinogram is, hence, always recommended.

Missing wedge: We have seen that a full tomographic scan in analytical CT must cover $\theta \in [0\dots180^\circ]$, or eventually even a larger range in the case of cone beam

7 There are also software solutions to this form of misalignment, provided by advanced CT reconstruction packages, for example, by the CERA software package of Siemens.

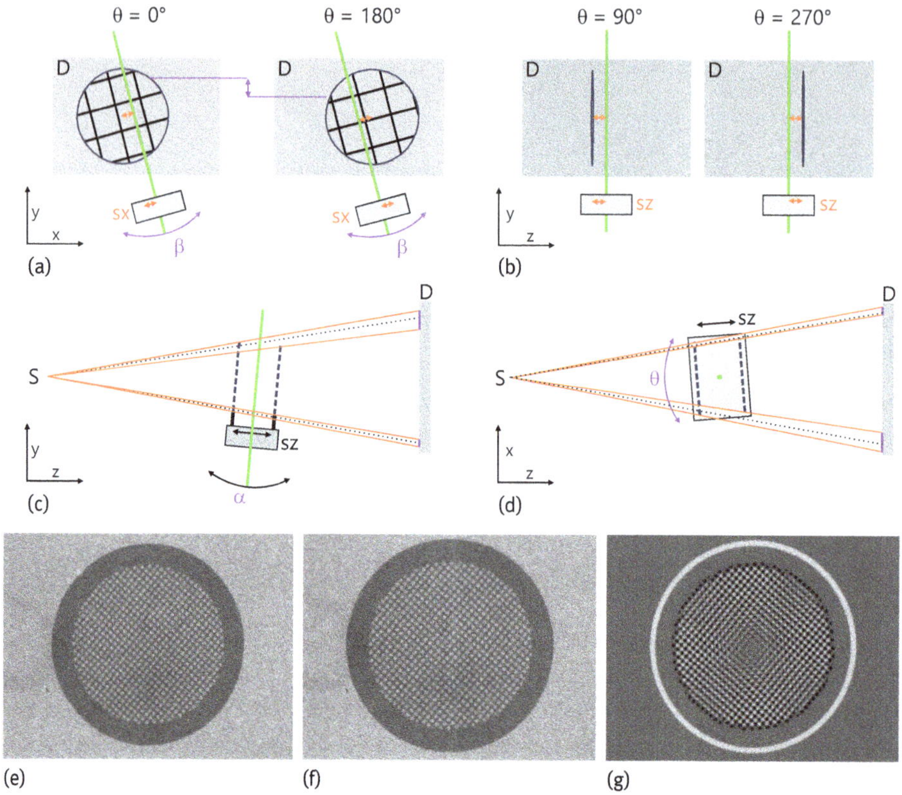

Fig. 4.22: Alignment of the tomographic axis for the micro-CT recordings shown in Figure 4.17 and Figure 4.11. (a,b) An alignment grid is used to center the object with respect to the rotation axis (positioners sx, sz above the rotation), as well as the roll β of the rotation axis. (c) Alignment of the pitch α. (d–f) Two images of the grid are recorded with a translational shift along the beam z (positioner sz), resulting in slightly different magnification. (g) Computation of a difference image indicates the fix point, to which the detector center can be moved. Note that the optical axis is given by the line between source and detector center. Alternatively, if the detector is not to be moved, α is varied. From [6].

recordings. However, if the object is not always accessible 'from all sides' due to steric constraints or opaque object chambers, artifacts are introduced. Figure 4.24 shows the example of a hydrated tissue slice from the central nervous system of a mouse. Clearly, such soft tissue is not self supporting. Neither can it easily be inserted in a transparent capillary (thin glass, plastic, etc.). Instead, the object is better sandwiched between thin foils. However, the metal ring providing stability obstructs the beam path for a certain angular range – the so called missing wedge. The lack of information in this missing wedge leads to streak artifacts around structures with a strong contrast.

Artifacts of local tomography: Suppose we want to reconstruct the fine structure of a Neanderthal tooth, say some microscale layering in the enamel, to investigate the

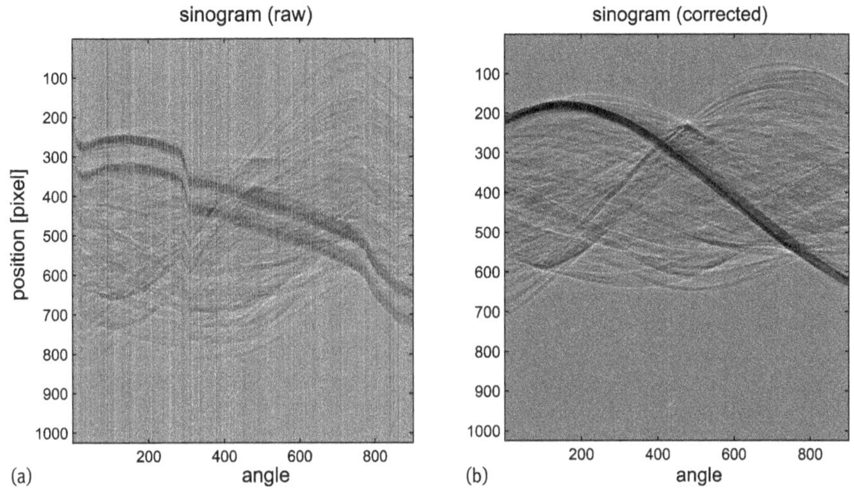

Fig. 4.23: Example illustrating the importance of inspecting the data in the sinogram. (a) Sinogram of an analytical CT recording, during which cables of the rotating sample stage have exerted forces and hampered the rotations. In addition, a doubling of features is observed, resulting from an empty beam recording by shifting the object in the FOV. Both effects, the intended shifts and the unintentional rotational inaccuracies due to the cabling can be accounted for a posteriori, yielding (b) a corrected sinogram, from which reconstruction was successful. From [24].

dental health of our ancestors. The anthropologist who found the skull does not give us permission to cut out the tooth. Reconstructing the entire head with a volume larger than $(0.1\,\mathrm{m})^3$ at the desired voxel size 10^{-6} m would entail 10^{15} voxels in the reconstruction volume, which is impossible for any kind of digital processing. Apart from memory and processing, also the number of projections $N_\mathrm{p} \simeq N_\mathrm{r}\pi/2 \simeq 1.57 \times 10^5$ needed to sample this volume at the desired resolution would be prohibitive. Instead, we must select a region of interest (ROI). This simple example shows that larger detectors, faster computers and more brilliant x-ray beams alone will not meet the challenge of reaching high resolution in large objects. Rather, we have to improve the ways in which we can perform local or ROI tomography. In fact, if strong objects such as the jaw enter in some projections, but are outside the reconstruction volume, we will face strong artifacts, denoted as artifacts of local tomography. One strategy to reduce such artifacts may be to use suitable postprocessing and filtering steps [18, 51, 52], or – on a more fundamental level – to change the recording geometry, for example, by using 3dRT (see Section 4.4.2), or by measuring spatial derivatives of the function f to be reconstructed [38]. Figure 4.25, reproduced from [25], illustrates a simple, yet effective way to suppress these artifacts, in particular on small spatial frequencies by suitable replicate padding of the ROI's boundary values, which is also described in [28].

Beam hardening artifacts: If the transmission of an object or within a certain part of the object becomes too low, CT data will become very noisy due to Poisson photon noise. The resulting noise artifacts in reconstruction are sometimes denoted as the photon starvation effect. Notwithstanding the noise, CT scans recorded with

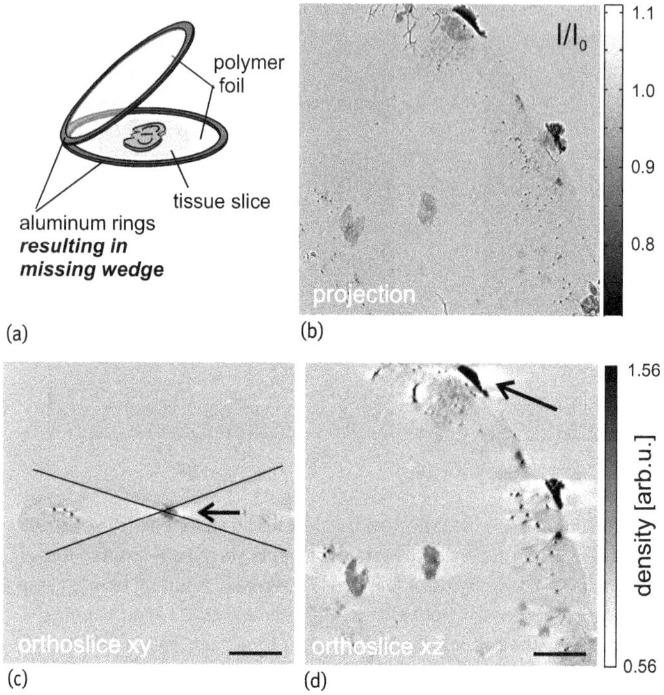

Fig. 4.24: Effects of a missing wedge in data recording. (a) The tissue is placed in between thin foils supported by metal rings, which are opaque for the photon energy used. (b) Projection perpendicular to the plane of the tissue, showing silver stained neurons and aggregates. The view is unobstructed. However, the reconstructed volume lacks the data of the missing wedge, see the indicated wedge in (c). (c) An orthoslice corresponding to the plane orthogonal to the rotation axis. Artifacts appear in the reconstruction, see *arrows* in (c,d). (d) An orthoslice with the same orientation as the projection in (b). From [45].

monochromatic radiation still give fairly faithful reconstruction values. This is no longer the case for broad bandpass sources such as most laboratory instruments, characterized by an extended emission spectrum $I_0(E)$, e.g., the bremsstrahlung spectrum of x-ray tubes. In contrast to synchrotron radiation, one is in desperate need of signal, which means that one is tempted to use this entire spectrum offered by the source. However, the attenuation of the x-rays through matter then differs for the different spectral components and, hence, the spectrum changes while the beam traverses the object; this is an effect known as beam hardening. To account for the broad spectrum, the Beer–Lambert law has to be integrated over the photon energy spectrum to obtain the total transmitted intensity

$$I(z) = \int_0^{E_{\max}} dE\, I_0(E)\, \exp\left[-\int_0^z dz'\mu(E,z')\right] . \tag{4.156}$$

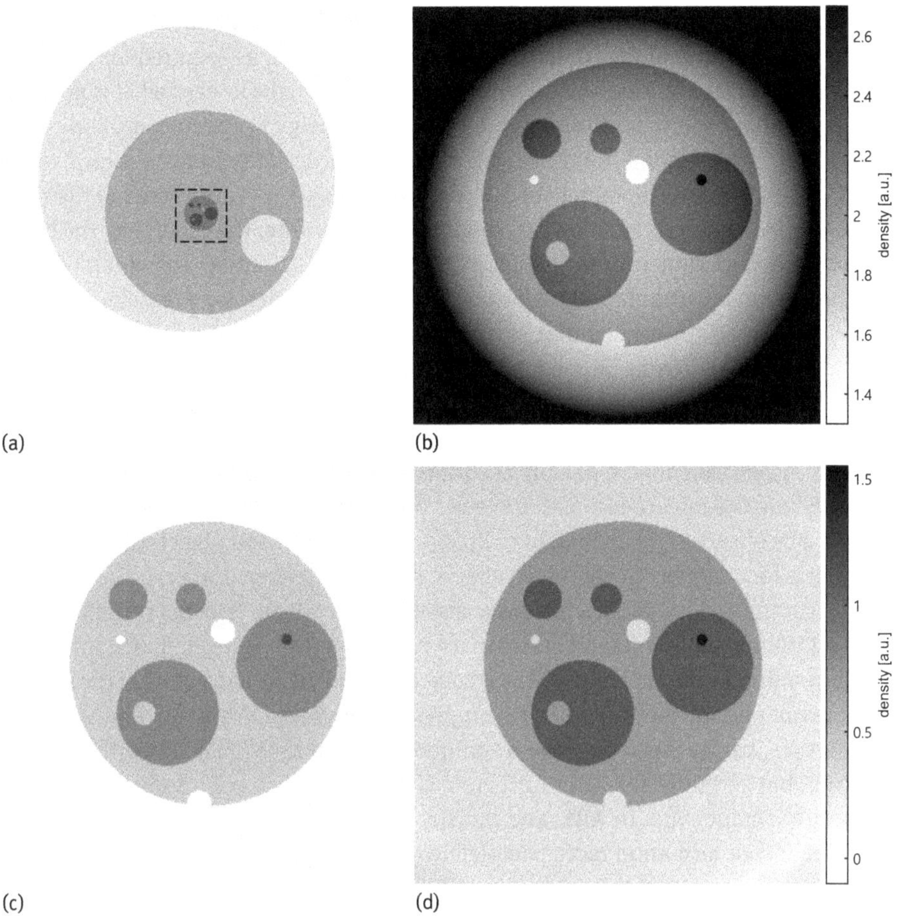

Fig. 4.25: Simulations illustrating the artifacts that occur when the object is larger than the reconstruction volume (local or ROI tomography). Note that ROI tomography can be easily simulated, for example, by cutting out a rectangular region in the sinogram, which covers the entire angular but not the spatial extent of the full sinogram. One then reconstructs from this subset and compares to the reconstruction of the full data set. (a) The ROI indicated by the *black square* is embedded in a large phantom consisting of 4096×4096 pixels, which cannot be completely probed (i.e., in experiments due to restricted detector or beam size). (b) Reconstruction of the 600 pixel ROI, from $N_p = 7200$ projections. Note that N_p fulfills the sampling criteria for the entire volume. Pronounced low frequency artifacts are observed in the reconstruction, when comparing to the enlarged ROI shown in (c). (d) Improved reconstruction, obtained from only $N_p = 360$ projections. The sinogram has been padded from 600 ('measured') pixels around the rotation axis to 1200, by replication of the boundary values. This way, the artifacts can be efficiently suppressed. The object's density values are reconstructed with fairly correct ratios, but are no longer on absolute scale. From [25].

Strictly speaking, this expression can no longer be reduced to a linear line integral by taking the logarithm of the transmitted intensity $-\ln(I/I_0)$, as required for tomography (in absorption contrast). However, if the exponent can be linearized, it is possible to define an equivalent (spectrum weighted) photon energy E_{eff} and absorption coefficient $\mu_{\text{eff}}(z)$, representing the entire spectrum. Contrarily, if the transmission T of the low energy spectral components becomes small, $T(E_{\text{min}}) \ll 1$, the spectral distribution of the transmitted radiation will significantly differ from the input, i.e., the beam will be hardened. Under these conditions, the spatial distribution $\mu(\underline{r})$ obtained from backprojection algorithms will no longer be correct. Consider, for example, an object in the reconstruction volume that has very low transmission apart from the high energy tail of the spectrum. In medical radiology, this may be a metal implant in the body. The reconstructed density in the 'shaded' vicinity of the metal will receive 'distorted' signal from a large number of projections. While FBP now assumes that all detector measurements are equally accurate, this will not be the case for projection angles where the integration path traverses the metal. As a result, strong streak artifacts will be observed [8]. In fact, in radiological practice, metal objects often give rise to artifacts for a number of combined effects: beam-hardening, scatter, noise from low photon counts and the exponential edge gradient effect [8, 9]. To avoid reconstruction artifacts from beam hardening, one can try to narrow the spectrum down, for example, by a suitable choice of filters. This is sometimes called prehardening of the source. If this is not possible, artifacts can be suppressed by special reconstruction schemes, for example, by the metal deletion technique proposed in [8]. It is numerically more involved, but certainly also goes further, to simultaneously track a sufficient number of spectral components in ART, and thereby to overcome beam hardening by modeling the image formation more completely. Finally, we note that the nuisance of a broad spectrum and beam hardening can become a virtue, if the transmitted energy can be recorded in an energy resolved manner. The number of spectral components distinguished then gives the number of chemical or material components to be distinguished based on the reconstructed absorption coefficients, varying for each material component in characteristic and known ways with photon energy. This approach was introduced 40 years ago as the so called dual energy CT in [3] but has only now been translated to broader clinical applications.

Exponential edge gradient effect: In Section 4.4.2, we already addressed the error that arises from taking the exponential mean over a voxel for absorption CT. If the function f appears in the exponent of the integration kernel, it is necessary to take the logarithm of the detector intensity before filtering and backprojection. This operation is not exact if the object function varies over the size of a pixel, since the intensity corresponds to the exponential mean, not the arithmetic mean of f over the pixel. This leads to a systematic overestimation, which depends on the variance of f within a pixel. If strong gradients are present at the subpixel scale (i.e., a pronounced but unresolved substructure) such as resulting from metal edges, artifacts due to the so called exponential edge gradient effect appear; these were first described in [21].

4.5.2 Algebraic reconstruction techniques (ART) and Helgason–Ludwig consistency conditions

Correcting artifacts in FBP is unsatisfactory, since the 'corrections' are likely to introduce additional errors. Even well controlled operations such as median filtering, which we can certainly not do without, will have subtle effects, for example, regarding correlations in the noise. Therefore, an entirely different approach should be considered, which is better suited to adapt the reconstruction to the given experimental conditions. We must take the actual data for what they are, rather than what we would like to believe them to be. The image reconstruction must follow the same conditions and constraints (geometry, interaction of beam and sample, spectral distribution) under which the image was formed.

ART: So far, we have treated the RT and its inverse as operators defined over functional spaces with suitable properties, i.e., in the framework of analytical theory. Discretization required for numerical processing was only considered as a secondary step of numerical implementation. Since measurements are always based on discrete detection points, one can directly formulate a tomographic problem algebraically, with each detector pixel providing one equation and each object voxel representing one unknown. In fact, since the RT is a linear operation, tomography in the discrete setting is essentially a linear algebra problem for a very large system matrix R, as long as the inconsistencies described above can be neglected. Let $\underline{f} = (f_1, \dots, f_{N_v})^T$ denote the object function with a total of N_v sampling points (voxels) in the FOV. All values are arranged in a column vector $\underline{f} \in \mathbb{R}^{N_v}$. Likewise, we arrange all available data from the projections in a column vector, $\underline{p} = (p_1, \dots, p_{D\times N_\theta})^T$ with $\underline{p} \in \mathbb{R}^{D\times N_\theta}$, representing the inhomogeneous part of the set of linear equations; D denotes the total number of measurements in one projection, i.e., the number of detector pixels or scan points evaluated, and N_θ the number of projections. The tomographic problem then amounts to solving

$$\underline{p} = R\underline{f} \tag{4.157}$$

i.e., to the inversion of the so called system (or design) matrix R; R is a $(D \cdot N_\theta) \times N_v$ matrix, accounting for the projection geometry including pixel weights. In other words, R encodes which voxels projects with what weight to a given detector element for a given projection. The corresponding experimental configuration can be adopted by intelligent design, hence the name design matrix. Importantly, many experimental effects such as cone beam geometry, beam hardening or inhomogeneous sampling can be directly implemented in R. Examples for inhomogeneous sampling are different detection pixel size or different angular increments between projections. However, the following issues need attention, see the detailed discussion in [12]: (i) The size of R is prohibitively large for direct inversion, in particular since the structure of the matrix is not well known, even if we know the matrix to be sparse. (ii) It can be shown that R has many small eigenvalues, i.e., is almost singular. The inversion problem is, therefore, ill posed, calling for regularization techniques, as addressed in Chapter 7. (iii) In practice, we will, therefore, have to over determine the matrix with $(D \cdot N_\theta) > N_v$. Further, it will be highly unlikely that an exact solution exists, owing to systematic inconsistencies and experimental errors. Rather, we seek to minimize the least square deviation

$$\chi^2 = |R\underline{f} - \underline{p}|^2 \tag{4.158}$$

with respect to the measurements. This minimization is carried out by an iterative projection algorithm known as Kaczmarz's method [22], illustrated graphically in Figure 4.26 (a). Gordon et al. [15] formulated Kaczmarz's method for the tomographic problem, consisting of the following steps:

1. Choose an initial guess for the object $\underline{f}^{(0)}$.
2. Calculate a forward projection according to $\underline{p}^{(0)} = R\underline{f}^{(0)}$.
3. For a given projection, correct the estimation by adjusting all pixels according to their weights (contribution to the respective line integrals), i.e., find the new update for the object vector according to

$$\underline{f}^{(k+1)} = \underline{f}^{(k)} + \frac{\left(p_j - \underline{r}_j \cdot \underline{f}^{(k)}\right)}{\left|\underline{r}_j\right|^2} \underline{r}_j \,, \tag{4.159}$$

   as illustrated in Figure 4.26 (b). $\underline{r}_j = (r_{j1}, \ldots, r_{jN_v})^T$ denotes the transposed jth row of R.
4. Perform step 3 for all projections j sequentially (random or ordered sequence). Instead of performing the update for each projection, one can also store its value,

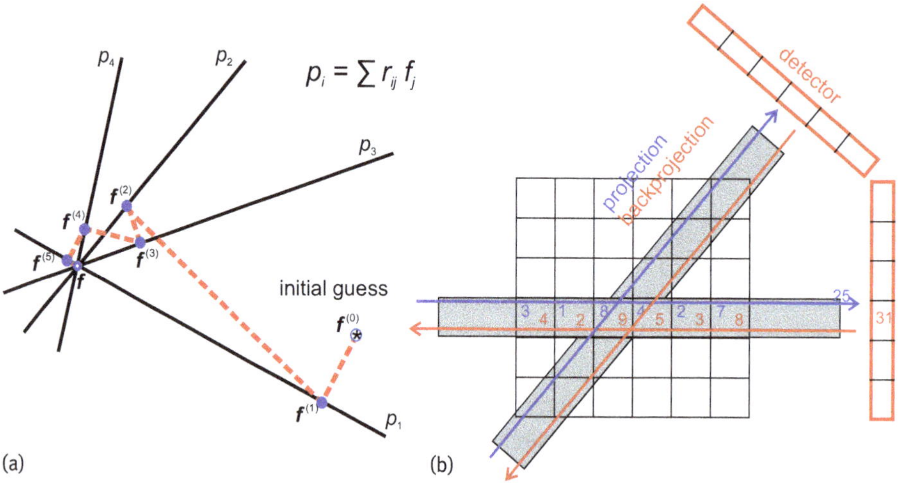

Fig. 4.26: (a) Illustration of Kaczmarz's method to iteratively solve a set of linear equations by consecutive projections. (b) Schematic of the update step, adjusting all voxel values contributing to a detector pixel to a new value by a multiplicative factor or additive term. Since the discrepancy cannot be attributed to a given voxel, all voxels are changed, according to weights r_{ij} determined by the volume of the voxel contributing to the integral over its total volume.

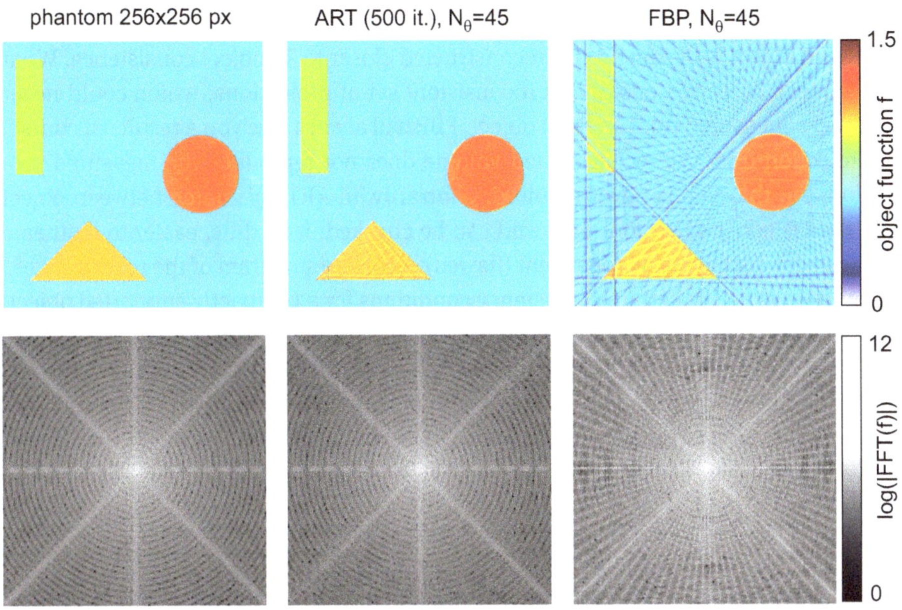

Fig. 4.27: The advantage of ART over FBP is illustrated for a 2d reconstruction of a simple phantom from $N_\theta = 45$ projections, i.e., from undersampled data given the size of the object (256×256 pixels). The undersampling gives pronounced artifacts for FBP. Contrarily, the fact that different projections are not independent from each other in ART, but take into account a consistent coupling, the ART reconstruction is of higher quality. Image quality can also be inspected in Fourier space (*lower row*), where a zoomed image of the low spatial frequencies is shown. From [42].

and then update each voxel only after a complete cycle by the average of the updated values. This has been shown mathematically to converge to the solution minimizing equation (4.158) for inconsistent or noisy data [33].

5. Stop when the changes due to update falls below a threshold, chosen in view of the errors in the data.

Instead of the additive correction, one can also choose a multiplicative variant, the multiplicative ART, by changing equation (4.159) in step 3 to

$$\underline{f}^{(k+1)} = \underline{f}^{(k)} \frac{p_j}{\underline{r}_j \cdot \underline{f}^{(k)}} . \tag{4.160}$$

An example illustrating the advantages of ART in comparison to FBP is shown in Figure 4.27. A further variant of iterative algebraic reconstruction algorithms is denoted by SIRT (simultaneous iterative reconstruction technique), which alternates forward and backprojection [2].

We close by a short note on the tomographic consistency. One advantage of ART is that an object is in some sense reconstructed closer to 3d object consistence. What do we mean by this? Consider an inconsistent set of projections, which could never have been measured from any real object. FBP will always give you a result. Of course, re-projection from the reconstructed volume does not reproduce the measured projection, but this step is never required. Contrarily, in ART, mismatch between object and data is constantly monitored and can be checked. It is, thus, easier to delineate between consistent and inconsistent (flawed) data. The sinogram of the reconstructed object always fulfills certain consistency conditions for a compactly supported object. To this end, consider the sinogram $g(\theta, s)$ of an object supported by the unit circle $-1 \leq s \leq 1$. The Helgason–Ludwig consistency conditions [19, 31] state that the kth moment of the projection $g(\theta, s)$ varies with θ, according to a superposition of sinusoidal curves. Following [27], they can be written as

$$a_k(\theta) := \int_{s=-1}^{s=1} ds\, s^k\, g(\theta, s) = \sum_{l=-\infty}^{\infty} a_{kl} \exp(il\theta)\,, \tag{4.161}$$

where the coefficients a_{kl} must fulfill

$$a_{kl} = 0 \quad \text{for} \quad |l| > k \quad \vee \quad k + |l| : \text{odd}\,. \tag{4.162}$$

In practice, these conditions can help us to inspect the consistency of the data. Furthermore, the constraint that a sinogram has to obey these conditions, can be used as a prior condition for reconstruction, as we will also discuss in Chapter 6.

References

[1] http://www.nobelprize.org/nobel_prizes/medicine/laureates/1979/, last access 31/05/2017.

[2] https://tomroelandts.com/articles/the-sirt-algorithm/.

[3] R. E. Alvarez and A. Macovski. Energy-selective reconstructions in X-ray computerised tomography. *Physics in Medicine and Biology*, 21(5):733, 1976.

[4] A. Averbuch, R. R. Coifman, D. L. Donoho, M. Israeli, and Y. Shkolnisky. A Framework for Discrete Integral Transformations I – The Pseudopolar Fourier Transform. *SIAM Journal on Scientific Computing*, 30(2):764–784, 2008.

[5] M. Bartels. Phasenkontrast-Mikrotomographie an einer Laborröntgenquelle. Master's thesis, Universität Göttingen, 2010.

[6] M. Bartels. *Cone-beam x-ray phase contrast tomography of biological samples: Optimization of contrast, resolution and field of view.* PhD thesis, Universität Göttingen, 2013.

[7] M. Bartels, V. H. Hernandez, M. Krenkel, T. Moser, and T. Salditt. Phase contrast tomography of the mouse cochlea at microfocus x-ray sources. *Appl. Phys. Lett.*, 103(8):083703, 2013.

[8] F. E. Boas and D. Fleischmann. Evaluation of Two Iterative Techniques for Reducing Metal Artifacts in Computed Tomography. *Radiology*, 259(3):894–902, 2011.

[9] F. E. Boas and D. Fleischmann. CT artifacts: Causes and reduction techniques. *Imaging in Medicine*, 4:229–240, 2012.

[10] T. Bortfeld and U. Oelfke. Fast and exact 2D image reconstruction by means of Chebyshev decomposition and backprojection. *Phys. Med. Biol.*, 44:1105–1120, 1999.

[11] A. V. Bronnikov. Cone-beam reconstruction by backprojection and filtering. *J. Opt. Soc. Am. A*, 17(11):1993–2000, 2000.

[12] T. Buzug. *Computed Tomography: From Photon Statistics to Modern Cone-Beam CT.* Springer-Verlag, 2008.

[13] M. D. de Jonge and S. Vogt. Hard X-ray fluorescence tomography – an emerging tool for structural visualization. *Current Opinion in Structural Biology*, 20(5):606–614, 2010.

[14] L. A. Feldkamp, L. C. Davis, and J. W. Kress. Practical cone-beam algorithm. *J. Opt. Soc. Am. A*, 1(6):612–619, 1984.

[15] R. Gordon, R. Bender, and G. T. Herman. Algebraic Reconstruction Techniques (ART) for three-dimensional electron microscopy and x-ray photography. *Journal of Theoretical Biology*, 29(3):471–481, 1970.

[16] P. Grangeat. Mathematical framework of cone beam 3D reconstruction via the first derivative of the Radon transform. In G. T. Herman, A. K. Louis, and F. Natterer, editors, *Mathematical Methods in Tomography*, pages 66–97. Springer, 1991.

[17] P. Grangeat. *Tomography.* Wiley/ISTE, 2009.

[18] M. Guizar-Sicairos, J. J. Boon, K. Mader, A. Diaz, A. Menzel, and O. Bunk. Quantitative interior x-ray nanotomography by a hybrid imaging technique. *Optica*, 2(3):259–266, 2015.

[19] S. Helgason. The Radon transform on Euclidean spaces, compact two-point homogeneous spaces and Grassmann manifolds. *Acta Mathematica*, 113(1):153–180, 1965.

[20] G. N. Hounsfield. Computerized transverse axial scanning (tomography): Part 1. Description of system. *The British Journal of Radiology*, 46(552):1016–1022, 1973. PMID: 4757352.

[21] P. M. Joseph and R. D. Spital. The exponential edge-gradient effect in X-ray computed tomography. *Physics in Medicine and Biology*, 26(3):473, 1981.

[22] S. Kaczmarz. Angenäherte Auflösung von Systemen linearer Gleichungen. *Bull. Acad. Polon. Sci. Lett.*, A35:355–357, 1937.

[23] A. C. Kak and M. Slaney. *Principles of computerized tomographic imaging.* IEEE Press, New York, 1988.

[24] M. Krenkel. Quantitative Phasenkontrast Tomographie. Master's thesis, Universität Göttingen, 2012.

[25] M. Krenkel. *Cone-beam x-ray phase contrast tomography for the observation of single cells in whole organs.* PhD thesis, Universität Göttingen, 2015.

[26] M. Krenkel, M. Töpperwien, C. Dullin, F. Alves, and T. Salditt. Propagation-based phase-contrast tomography for high-resolution lung imaging with laboratory sources. *AIP Advances*, 6(3):035007, 2016.

[27] H. Kudo and T. Saito. Sinogram recovery with the method of convex projections for limited-data reconstruction in computed tomography. *J. Opt. Soc. Am. A*, 8(7):1148–1160, 1991.

[28] A. Kyrieleis, V. Titarenko, M. Ibison, T. Connolley, and P. J. Withers. Region-of-interest tomography using filtered backprojection: assessing the practical limits. *Journal of microscopy*, 241(1):69–82, 2011.

[29] M. Liebi, M. Georgiadis, A. Menzel, P. Schneider, J. Kohlbrecher, O. Bunk, and M. Guizar-Sicairos. Nanostructure surveys of macroscopic specimens by small-angle scattering tensor tomography. *Nature*, 527:349–352, 2015.

[30] L. M. Lohse. On the Extraction of Planar Integrals from X-ray Projections. Bachelor's thesis, Georg-August Universität Göttingen, 2015.

[31] D. Ludwig. The Radon transform on Euclidean space. *Commun. Pure Appl. Math.*, 19(1):49–81, 1966.

[32] B. Münch, P. Trtik, F. Marone, and M. Stampanoni. Stripe and ring artifact removal with combined wavelet-fourier filtering. *Opt. Express*, 17(10):8567–8591, 2009.

[33] F. Natterer. *The mathematics of computerized tomography*. Classics in Applied Mathematics. Society for Industrial and Applied Mathematics, Philadelphia, PA, USA, 2001.

[34] F. Natterer and F. Wübbeling. *Mathematical methods in image reconstruction*. SIAM, 2001.

[35] S. S. Orlov. Theory of three dimensional reconstruction. I: Conditions for a complete set of projections. *Kristallografiya*, 20:511–515, 1975.

[36] D. M. Paganin. *Coherent X-Ray Optics*. New York: Oxford University Press, 2006.

[37] W. J. Palenstijn, K. J. Batenburg, and J. Sijbers. The ASTRA tomography toolbox. In *13th International Conference on Computational and Mathematical Methods in Science and Engineering. CMMSE*, 2013.

[38] F. Pfeiffer, C. David, O. Bunk, T. Donath, M. Bech, G. Le Duc, A. Bravin, and P. Cloetens. Region-of-interest tomography for grating-based x-ray differential phase-contrast imaging. *Phys. Rev. Lett.*, 101(16):168101, 2008.

[39] W. H. Press, S. A. Teukolsky, W. T. Vetterling, and B. P. Flannery. *Numerical Recipes*. Cambridge University Press, 3rd edition, 2007.

[40] J. Radon. Über die Bestimmung von Funktionen durch ihre Integralwerte längs gewisser Mannigfaltigkeiten. *Akad. Wiss.*, 69:262–277, 1917.

[41] A. G. Ramm and A. I. Katsevich. *The Radon Transform and Local Tomography*. CRC Press, 1996.

[42] A. Ruhlandt, M. Krenkel, M. Bartels, and T. Salditt. Three-dimensional phase retrieval in propagation-based phase-contrast imaging. *Phys. Rev. A*, 89:033847, 2014.

[43] F. Schaff, M. Bech, P. Zaslansky, C. Jud, M. Liebi, M. Guizar-Sicairos, and F. Pfeiffer. Six-dimensional real and reciprocal space small-angle x-ray scattering tomography. *Nature*, 527:353–356, 2015.

[44] D. C. Solmon. The X-ray transform. *Journal of Mathematical Analysis and Applications*, 56(1):61–83, 1976.

[45] M. Töpperwien. X-ray phase contrast tomography of neuronal tissue. Master's thesis, Georg-August Universität Göttingen, 2014.

[46] H. K. Tuy. An inversion formula for cone-beam reconstruction. *SIAM Journal on Applied Mathematics*, 43(3):546–552, 1983.

[47] M. Vassholz. 3d Radon Transform based Tomography for Anisotropic X-Ray Optics. Master's thesis, Georg-August Universität Göttingen, 2015.

[48] M. Vassholz, B. Koberstein-Schwarz, A. Ruhlandt, M. Krenkel, and T. Salditt. New x-ray tomography method based on the 3d radon transform compatible with anisotropic sources. *Phys. Rev. Lett.*, 116:088101, 2016.

[49] P. J. Verveer, J. Swoger, F. Pampaloni, K. Greger, M. Marcello, and E. H. K. Stelzer. High-resolution three-dimensional imaging of large specimens with light sheet-based microscopy. *Nature Methods*, 4:311–313, 2007.

[50] Y. Xu. A new approach to the reconstruction of images from Radon projections. *Adv. Appl. Math.*, 36:338–420, 2006.

[51] J. Yang, H. Yu, M. Jiang, and G. Wang. High-order total variation minimization for interior tomography. *Inverse Problems*, 26(3):035013, 2010.

[52] H. Yu and G. Wang. Compressed sensing based interior tomography. *Physics in Medicine and Biology*, 54(9):2791, 2009.

Symbols and abbreviations used in Chapter 4

ART	algebraic reconstruction techniques
δ_{2d}	Dirac delta function (2d)
$\bar{f}$	projection image
FBP	filtered backprojection
FOV	field of view
FST	Fourier slice theorem
$G_n^\lambda(s)$	Gegenbauer polynomials
Γ	source orbit
$\mathcal{H}$	Hilbert transform
I^α	Riesz potential/operator
k_x, k_y	spatial frequencies
M	geometric magnification factor
MRI	magnetic resonance imaging
$n = 1 - \delta + i\beta$	refractive index of x-rays
$\underline{n}_\theta$	unit vector indicating direction of Radon transform
$\underline{n}_\theta^\perp$	unit vector perpendicular to $\underline{n}_\theta$
N_p	number of projections
N_r	number of resolution elements in the reconstruction volume
$P(\theta, x)$	$2d$ projection integral in polar coordinates
$\tilde{P}(\theta, k)$	Fourier transform of $2d$ projection integral in polar coordinates
PSD	power spectral density
R	system (or design) matrix
RT	Radon transformation
$\mathcal{R}$	Radon transform
$\mathcal{R}^{-1}$	inverse Radon transform
$\mathcal{R}_{\underline{n}_\theta}$	Radon transform operator
$\mathcal{R}^\#_{\underline{n}_\theta}$	dual operator to $\mathcal{R}_{\underline{n}_\theta}$
ROI	region of interest
$\mathrm{sgn}(t)$	sign function
2dRT	two-dimensional Radon transformation
3dRT	three-dimensional Radon transformation
$T_n(s)$	Chebyshev polynomials of the first kind
$\hat{\theta}$	unit vector indicating projection direction (parallel beam geometry) or central ray direction (cone beam geometry)
$U_n(s)$	Chebyshev polynomials of the second kind
x, y, z	Cartesian coordinates in laboratory system
x', y', z'	Cartesian coordinates in sample/object system

5 Radiobiology, radiotherapy, and radiation protection

X-ray and gamma ray photons provide unique insights into a patient's body. However, they can also cause considerable harm to living cells and tissues. The same holds for α, β and other types of particle radiation. The main target for radiation causing biological effects is the carrier of the genetic information, deoxyribonucleic acid (DNA), which is located in the nucleus of mammalian cells. Radiation induced DNA damage can kill cells, induce mutations and cause carcinogenesis. Many pioneers of radiology who applied x-rays without caring about radiation protection suffered from serious radiation induced health problems. Awareness for the enormous risks of ionizing radiation has evolved over time. The use of atomic weapons in Hiroshima and Nagasaki, and nuclear power plant accidents such as in Chernobyl or Fukushima, have sadly brought the topic to the attention of the general public.

Since the human body is constantly exposed to radiation of cosmic and terrestrial origin, a variety of DNA repair mechanisms to cope with radiation damage have evolved. Prior to any radiological examination, the risks of ionizing radiation must be carefully assessed and balanced against its possible benefits. Compared to the early days of radiology, there has been enormous progress in dose reduction, but there is still room for further improvement. Already a single CT scan can be associated with radiation doses exceeding the annual natural exposition, since it requires many projections. The medical use of x-rays has become the most important source of man made radiation exposure. The ability of ionizing radiation to kill cells can also be a virtue; in radiotherapy of cancer, it is used to destroy malignant tumors. This requires precise knowledge about the biological response of different tissues to ionizing radiation, as well as the ability to tailor a dose distribution inside the patient's body that closely matches the tumor, while healthy normal tissue is spared. This chapter addresses some fundamental physical principles required for a quantitative description of these processes.

5.1 Interactions of ionizing particles with matter

5.1.1 Introductory remarks

Typical photon energies used in x-ray diffraction and imaging are on the order of about 10 keV, in medical diagnostics of about 100 keV and in radiotherapy of few MeV (cf., Figure 3.1). In contrast, only few eV are usually required for ionization of the low Z elements of biological tissues. If we neglect these small binding energies, the energy transfer from high energy photons to matter (e.g., a biological specimen or a patient's body) is a two step process:

DOI 10.1515/9783110426694-005

1. The primary photon interactions with matter described in Chapter 3 – photoelectric effect, Compton effect and pair production – trigger secondary charged particles.
2. These are much more likely to interact with matter (per unit path length) since they are subject to the Coulomb force and lose their kinetic energy in numerous collision events, foremost with atomic electrons of the absorber.

In addition, the emission of secondary charged particles can produce photons again in the form of bremsstrahlung, fluorescence and annihilation radiation. This cascade of events is called the **coupled transport of photons and electrons/positrons**. Examples of photon and electron trajectories for typical photon energies in medical applications are shown in Figure 5.1. At energies used in diagnostic radiology, the range of secondary electrons is below 1 mm, while it can reach several cm in the MeV range used in radiotherapy. An extreme example is shown in Figure 5.2, where a single high energy primary photon triggers an entire 'shower' of secondary electrons, bremsstrahlung photons and so called δ (or knock on) electrons, which can all excite or ionize biomolecules or water, the most abundant molecular component of most tissues. Since the energy of high energy photons is transferred to an absorbing medium almost exclusively by secondary charged particles, photon radiation is classified as indirectly ionizing radiation. Charged particle radiation, in contrast, is called directly ionizing radiation.

It can be assumed that the effects of radiation to a certain volume element dV of an irradiated absorbing medium, for example, the fraction of cells killed if the medium is living tissue, are determined by the amount of energy $d\epsilon$ deposited per mass $dm = \rho\, dV$. This leads to the definition of the **absorbed dose**, which is the central quantity in this chapter:

$$D = \frac{d\bar{\epsilon}}{dm} = \frac{1}{\rho}\frac{d\bar{\epsilon}}{dV} \tag{5.1}$$

While the energy deposited per unit mass $d\epsilon/dm$ in a small volume is subject to statistical fluctuations due to the discrete character of photons and charged particles, the absorbed dose is a nonstochastic quantity and is defined as the corresponding expectation value, as denoted by the bar. The unit of absorbed dose is 1 J/kg = 1 Gray (Gy), in honor of L. H. GRAY, who established the field of radiobiology. An older unit is the radiation absorbed dose (rad) where 1 rad = 0.01 Gy.

Only if the range of secondary electrons is negligibly small, can the distribution of absorbed dose $D(\underline{r})$ be assumed to be proportional to the primary photon fluence following the Beer–Lambert law (equation 3.5). In the MeV range used in radiotherapy, however, this is no longer possible, since secondary electrons travel macroscopic distances of several cm from the site of the primary interaction, as can be seen in Figure 5.1 and 5.2, and lose their energy along this track. Therefore, methods to calculate the dose distribution $D(\underline{r})$ resulting from high energy x-rays must include not only the primary interactions described in Section 3.2, which determine the linear attenuation

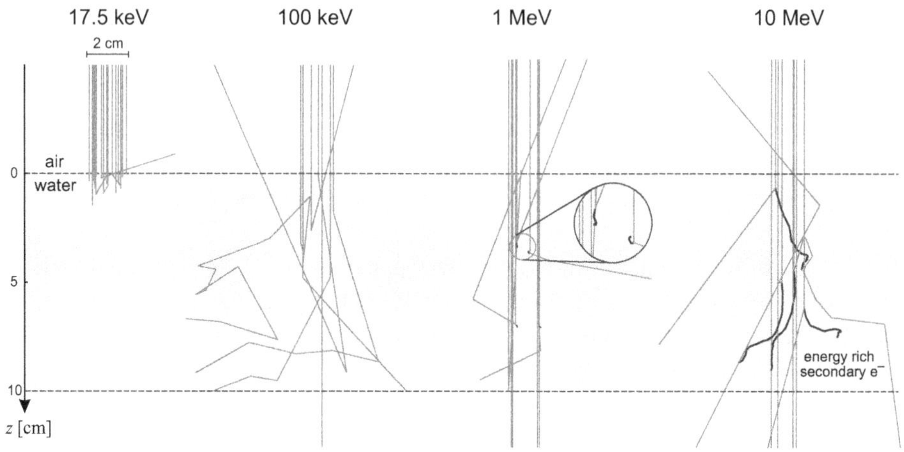

Fig. 5.1: Photons of different energies impinging on a 10 cm thick water layer. Photon trajectories are shown by *thin gray*, secondary electron trajectories by *thick black lines*. At the lowest energy (K_α-line of Mo as used in mammography), the majority of photons is absorbed within few mm by the photoelectric effect. The Compton effect is dominant at 100 keV typical for medical CT. In the energy range of few MeV as commonly used in radiotherapy, the Compton electrons have sufficient kinetic energy to travel several cm. Hence, attenuation of the primary photon beam and energy deposition in the water layer are spatially decoupled for high photon energies (Monte Carlo simulation using `EGSnrc`).

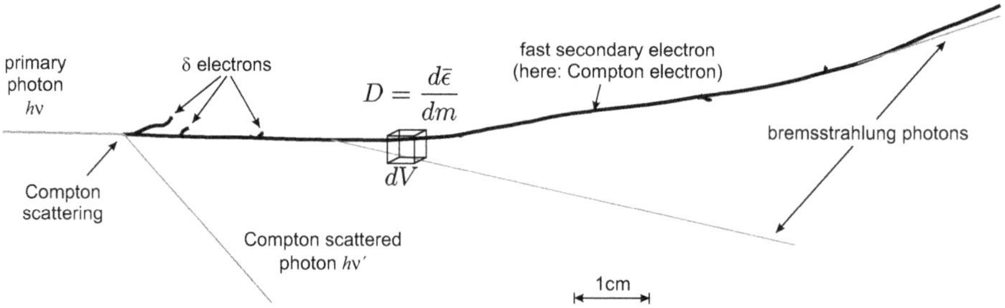

Fig. 5.2: A single primary photon (incident from the left, $E = 50$ MeV, in water) can trigger a cascade or 'shower' of electrons (and positrons) and photons of lower energy. Since quantities like free path lengths, scattering angles and energies of scattered photons and released secondary electrons, etc., are statistically distributed, two photons with the same initial conditions can trigger very different cascades. The most accurate method for calculation of dose distributions resulting from a large number of primary photons or electrons are, therefore, Monte Carlo (MC) techniques, where very large numbers of cascades are simulated to obtain the dose in each volume element (voxel) of a given geometry with sufficient statistical accuracy (cf., Section 5.2.2).

coefficient μ, but also the subsequent interaction of secondary charged particles and matter, as well as the photons resulting from the Compton effect, electron–positron annihilation, bremsstrahlung and fluorescence. This variety of processes in the coupled transport of photons and charged particles (Figure 5.3, left) and their stochastic

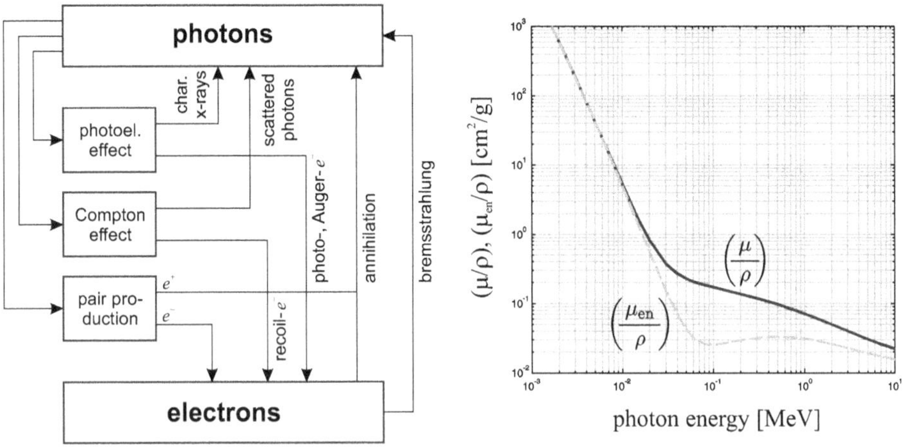

Fig. 5.3: *Left:* Diagram of the different pathways of energy transfer in the coupled transport of photons and electrons. Based on [27]. *Right:* The mass attenuation coefficient (μ/ρ) and the mass energy absorption coefficient (μ_{en}/ρ) of water, as a function of photon energy.

character makes the computation of $D(\underline{r})$ for a known primary photon field a rather complex task, especially in highly inhomogeneous structures like the human body.

In the following, we outline the description of the energy transfer from primary photons to secondary charged particles (Section 5.1.2) and, subsequently, from charged particles to the absorbing medium (Section 5.1.3). This provides the foundations for the computation and measurement of dose distributions (Section 5.2). We consider not only electrons, but also heavy charged particles (protons, α particles and atomic ions beyond helium), which have important applications in radiotherapy.[1] Unlike electrons, these travel on (almost) straight trajectories through matter due to their much larger mass, which allows considerable simplifications.

Like in Section 3.2, also some of the calculations presented in this section require advanced quantum theory. Again, we focus on the most relevant results and provide references for the interested reader. The main textbook references in preparing this section were [6, 27, 28, 37, 40]. In addition, the reports of the International Commission on Radiation Units and Measurement (ICRU) were extremely helpful. We used the online databases provided by the National Institute of Standards and Technology[2] to obtain photon interaction coefficients, as well as electron and proton stopping powers and ranges. The software package `EGSnrc`[3] and the MATLAB script `EGS_WINDOW.m`[4]

1 We do not consider neutrons here, although they also have their place in radiobiology and radiotherapy.

2 https://www.nist.gov/pml/productsservices/physical-reference-data

3 http://www.nrc-cnrc.gc.ca/eng/solutions/advisory/egsnrc_index.html

4 http://medphysfiles.com/index.php

written by M. Bakhtiari were used to produce the figures containing Monte Carlo generated particle tracks.

5.1.2 Energy transfer from photons to charged secondary particles

As described in Chapter 3, the attenuation of a parallel and monoenergetic photon beam in a homogeneous material is given by the Beer–Lambert law

$$\Phi(d) = \Phi_0 \exp\left[-\left(\frac{\mu}{\rho}\right)\rho d\right] , \tag{5.2}$$

where $\Phi = dN/dA$ denotes the fluence (photons per unit area) of unscattered photons. The material specific total linear attenuation coefficient μ results from the contributions of the photoelectric effect, Thomson and Compton scattering and possibly pair production (neglecting photonuclear interactions):

$$\left(\frac{\mu}{\rho}\right) = \frac{N_A}{A}\sigma_a = \frac{N_A}{A}\left(\sigma_a^{\text{p.e.}} + \sigma_a^{\text{Th}} + \sigma_a^{\text{C}} + \sigma_a^{\text{pair}}\right) . \tag{5.3}$$

We follow the usual convention in radiation physics and dosimetry and use the mass attenuation coefficient (μ/ρ). As discussed in Chapter 3, the (total) cross sections σ_a^i depend on the photon energy and the atomic properties of the material. So far, these equations describe how the incident photon field is attenuated, which is fundamental for radiography and computed tomography. Now, we go a step further and consider how the energy of the primary photon field is transferred to charged secondary particles. Photon energy can be converted into kinetic energy of photoelectrons, Compton electrons and to electron–positron pairs (provided that the photon energy exceeds $2m_ec^2 = 1.022$ MeV). By Thomson scattering, on the contrary, no energy is transferred to electrons. For monoenergetic photons of energy $E = h\nu$, the mass energy transfer coefficient is defined as

$$\left(\frac{\mu_{\text{tr}}}{\rho}\right) = \frac{N_A}{A}\left(f_{\text{p.e.}}\sigma_a^{\text{p.e.}} + f_{\text{C}}\sigma_a^{\text{C}} + f_{\text{pair}}\sigma_a^{\text{pair}}\right) . \tag{5.4}$$

The factors f_i denote the average fraction of the energy $E = h\nu$ of the primary photon, which is transferred to a secondary electron as kinetic energy. Based on the results of Section 3.2, one obtains

$$f_{\text{p.e.}} = 1 - \frac{X}{E}, \qquad f_{\text{C}} = 1 - \frac{\langle h\nu' \rangle + X}{E}, \qquad f_{\text{pair}} = \begin{cases} 0 & E < 2m_ec^2 \\ 1 - \frac{2m_ec^2}{E} & E \geq 2m_ec^2 \end{cases} , \tag{5.5}$$

where X denotes the average energy per absorbed photon, which is re-emitted by fluorescence as characteristic x-rays, $\langle h\nu' \rangle$ the average energy of a Compton scattered photon and $2m_ec^2$ the energy of two annihilation radiation photons. For details, see e.g., [6, 63].

Table 5.1: Values of the fraction g of the energy transferred to secondary electrons that is converted back into photons by production of bremsstrahlung. Data from [37], p. 606. Since $E < 2m_ec^2 = 1022$ keV, pair production is not possible, thus no energy is converted into annihilation radiation.

photon energy $E = h\nu$ [MeV]	g_{air}	g_{water}
0.01	$0.11 \cdot 10^{-3}$	$0.09 \cdot 10^{-3}$
0.1	$0.24 \cdot 10^{-3}$	$0.21 \cdot 10^{-3}$
1	$2.49 \cdot 10^{-3}$	$2.22 \cdot 10^{-3}$

By subsequent interactions of the electron resulting in bremsstrahlung, fluorescence or annihilation of positron–electron pairs, a fraction g of its kinetic energy is converted back into photons again. This is accounted for by definition of the mass energy absorption coefficient

$$\left(\frac{\mu_{en}}{\rho}\right) = \left(\frac{\mu_{tr}}{\rho}\right)(1-g) \ . \tag{5.6}$$

However, as shown in Tab. 5.1, g is usually negligibly small in the case of low Z materials and photon energies used in diagnostic imaging. For homogeneous mixtures and compounds, (μ_{tr}/ρ) and (μ_{en}/ρ) are obtained by [6]

$$\left(\frac{\mu_{tr}}{\rho}\right)_{mix} = \sum_i w_i \left(\frac{\mu_{tr}}{\rho}\right)_i \tag{5.7}$$

and

$$\left(\frac{\mu_{en}}{\rho}\right)_{mix} = \left(\frac{\mu_{tr}}{\rho}\right)_{mix}\left(1 - \sum_i w_i g_i\right) , \tag{5.8}$$

where w_i denotes the percentage by weight of atomic component i. Figure 5.3 (right) shows (μ/ρ) and (μ_{en}/ρ) for the case of water (our favorite tissue substitute) over a wide range of photon energies. Below 10 keV, both quantities are almost identical, since the photoelectric effect is the prevalent interaction and the entire photon energy is converted into kinetic energy of photoelectrons (cf., Figure 3.3 and 3.20). At higher energies, however, they differ considerably, especially around 100 keV where Compton scattering dominates. The energy transferred from uncharged (or indirectly ionizing) to charged (or directly ionizing) particles per unit mass, in our case from photons to electrons (and positrons), is called **kerma** (kinetic energy released per unit mass)

$$K = \frac{dE_{tr}}{dm} = \frac{1}{\rho}\frac{dE_{tr}}{dV} \ . \tag{5.9}$$

Consider the situation sketched in Figure 5.4 (left). A parallel beam of monoenergetic photons with energy $E = h\nu$ impinges on a volume element $dV = dA\,dx$. Using the (mass) energy transfer coefficient, the energy transferred to secondary electrons in dV is $dE_{tr} = \mu_{tr}\Phi E\,dA\,dx$. Since $dm = \rho\,dA\,dx$, this directly relates kerma to the photon

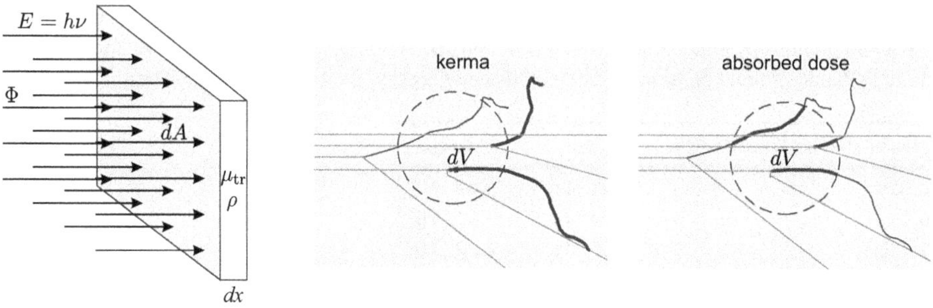

Fig. 5.4: *Left:* Sketch illustrating the parameters relating the kerma K to the energy transfer coefficient μ_{tr}. *Right:* Difference between kerma and dose. Kerma includes the entire kinetic energy of charged particles that are released inside dV, including energy, which is subsequently transferred to the surrounding medium outside dV (as indicated by the *thick lines* crossing the volume boundaries). Dose denotes the entire energy deposited inside dV, including contributions from charged particles set in motion inside as well as outside dV.

fluence

$$K = \frac{\mu_{tr} \Phi E \, dA \, dx}{\rho \, dA \, dx} = \left(\frac{\mu_{tr}}{\rho}\right) \Phi E = \left(\frac{\mu_{tr}}{\rho}\right) \Psi \,. \tag{5.10}$$

$\Psi = \Phi E$ denotes the energy fluence. For a continuous photon spectrum with energy fluence distribution $\Psi_E(E)$, one has to take into account the energy dependence of μ_{tr} and integrate over all energies

$$K = \int\limits_0^{\max(E)} \left(\frac{\mu_{tr}(E)}{\rho}\right) \Psi_E(E) \, dE \,. \tag{5.11}$$

The kerma from some volume element dV centered about r contributes to the absorbed dose in all surrounding volume elements which are within the range of the charged particles released in dV. Kerma is an important conceptual quantity in dosimetry and can directly be related to the primary radiation field but is very difficult to measure. Absorbed dose, on the contrary, can be measured, but is difficult to relate to the primary radiation field. This difference is caused by the energy transport away from the site of the primary interactions by charged secondary particles and photons of lower energy, as sketched in Figure 5.4 (right). In analogy to kerma, **terma** (t̲otal e̲nergy r̲eleased per unit ma̲ss) is defined as the entire amount of energy that is lost from the primary photon field per unit mass. It is thus proportional to the total linear attenuation coefficient μ

$$T = \left(\frac{\mu}{\rho}\right) \Psi \,. \tag{5.12}$$

If all energy was deposited locally, terma would yield the dose. However, this can only be a very coarse approximation, applicable only if the range of secondary charged particles and mean free path length of scattered photons are negligibly small. We will need kerma and terma in Section 5.2 for dose calculation and measurement.

5.1.3 Energy loss of charged particles in matter

Overview

Now consider a charged particle that traverses an absorbing material. It could be either a primary particle emitted from some external source (e.g., by radioactive decay) or a secondary electron triggered by a primary photon, as described in the previous section. The mean energy loss of a charged particle per unit path length dl along its track due to Coulomb interactions with bound electrons and nuclei of the traversed material (these are denoted as 'atomic' electrons and nuclei in the following) is called the **linear stopping power**

$$S(E) = -\left\langle \frac{dE}{dl} \right\rangle . \tag{5.13}$$

The stopping power varies with the particle energy and, thus, changes along the particle track. The brackets indicate that S is an average quantity due to the statistical character of the interaction processes involved – the energy loss of individual particles and their precise trajectory can differ considerably; $S(E)$ is usually given in units of MeV/cm. A charged particle can transfer energy (and momentum) to another charged particle in the absorbing material, which is termed a collision. The 'collision partner' can be an atomic electron or an atomic nucleus of the absorber. In addition, bremsstrahlung can be generated. Therefore, following the latest recommendations of the ICRU [21], three components of the linear stopping power are distinguished:

$$\left(\frac{S}{\rho}\right) = \left(\frac{S_{\mathrm{el}}}{\rho}\right) + \left(\frac{S_{\mathrm{nuc}}}{\rho}\right) + \left(\frac{S_{\mathrm{rad}}}{\rho}\right) . \tag{5.14}$$

As in Section 5.1.2, we have divided by the mass density ρ, which yields the **mass stopping power** (S/ρ). Plots of the stopping power components of protons and electrons in water are given in Figure 5.5. In analogy to the linear attenuation coefficient for the case of photons, the stopping power is an extremely important quantity for charged particles. We start with a brief summary and rough estimates for each component:

- **Electronic stopping power S_{el}:** This component is due to interactions with atomic electrons, leading to excitations (in which orbital electrons are raised to higher energy levels) and ionizations (in which orbital electrons are ejected) of the atoms and molecules of the absorbing medium, including breaking of chemical bonds. Sometimes S_{el} is also called the collisional stopping power and denoted S_{col}. A first estimate of the energy dependence of S_{el} can be made as follows (p. 264 in [36]). Assume that an incoming, charged particle with nonrelativistic velocity ($\beta = \frac{v}{c} \ll 1$) interacts with an atomic electron of the absorbing material via Coulomb interaction over a characteristic interaction time Δt. The momentum transfer is proportional to the particle charge z and the duration of the interaction, $\Delta p \propto z\Delta t$, and thus inversely proportional to the particle velocity, $\Delta p \propto v^{-1}$. If the atomic electron is initially approximately at rest, its kinetic energy after Δt is $E = \Delta p^2/2m \propto z^2 v^{-2} = z^2(\beta c)^{-2}$. Therefore, in this first approximation, the

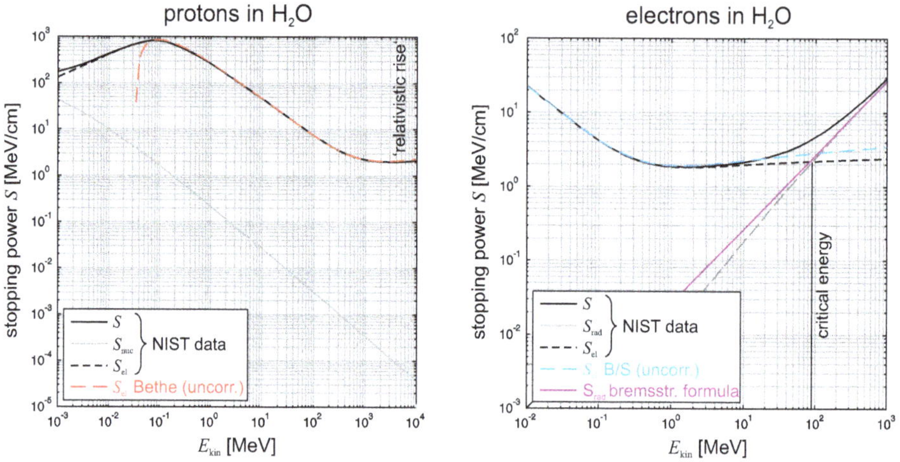

Fig. 5.5: Stopping power of (*left*) protons and (*right*) electrons in water. The total stopping power S of protons (and of other charged heavy particles) is almost entirely determined by the energy transfer to atomic electrons of the absorbing material, the electronic stopping power S_{el}. Above $E_{kin} \approx 0.1$ MeV, S_{el} can be calculated by the Bethe formula (5.44). Interactions with atomic nuclei (nuclear stopping power S_{nuc}) give a relevant contribution to S only at small particle energies. The radiative stopping power S_{rad}, which describes the energy loss by generation of bremsstrahlung, is negligible for heavy particles. The case is different for electrons. Here, S_{rad} calculated by equation (5.49) is much higher due to the small electron mass $m_e \ll m_p$ and dominates S_{el} calculated by equation (5.45) above a critical energy of ~ 100 MeV in water. NIST data are taken from the ESTAR and PSTAR databases and contain further corrections beyond the scope of this book.

energy transferred to the atomic electron and thus S_{el} are $\propto z^2\beta^{-2}$. Contrarily, for highly relativistic particles with $\beta \approx 1$, the particle velocity hardly changes if kinetic energy is lost (Figure 5.7), thus $S_{el} \approx$ const. In addition, we expect S_{el} to be proportional to the electron density $\rho_e = \rho N_A Z/A$ of the absorber. Collecting these results yields

$$\left(\frac{S_{el}}{\rho}\right) \propto \frac{Z}{A}\frac{z^2}{\beta^2} \tag{5.15}$$

as a first estimate for the electronic stopping power. The slower a particle, the higher the energy loss during the subsequent step Δl, until it finally comes to rest. This already gives a qualitative explanation of the so called Bragg peak, i.e., the dose maximum at the end of a charged particle's track. It turns out that S_{el} is by far the most relevant mechanism of energy loss of heavy ($m_0 \gg m_e$) charged particles (Figure 5.5). Its full description is given by the **Bethe formula**. Due to its fundamental importance, its derivation is sketched further below (p. 182).

– **Nuclear stopping power S_{nuc}:** This term comprises energy loss due to elastic Coulomb interactions with atomic nuclei, to which some recoil energy is imparted, while the total kinetic energy remains constant. By definition, S_{nuc} does *not* include inelastic nuclear interactions by nuclear forces, which change the structure of the target nucleus. In comparison to S_{el}, the nuclear stopping power S_{nuc} is a relevant contribution to the total stopping power only in the case of slow heavy particles, but is negligible at higher energies and for electrons (Figure 5.5), see, e.g., Section 2.1.4 in [40] for a more detailed discussion.

– **Radiative stopping power S_{rad}:** Directional changes of a charged particle in the Coulomb field of atomic electrons and nuclei are equivalent to (negative) acceleration, which leads to the generation of bremsstrahlung (see also Section 3.3.1). Consider the energy radiated by the point charge ze accelerated in the Coulomb field of a nucleus in the absorber (target) with atomic number Z. The radiated energy per unit time is proportional to the square of the acceleration [28],

$$\frac{dE_{rad}}{dt} \propto \left|\frac{d\vec{v}}{dt}\right|^2 , \tag{5.16}$$

with the acceleration given by the Coulomb force $\frac{dv}{dt} \propto zZe^2/m$, resulting in

$$\left(\frac{S_{rad}}{\rho}\right) \propto \left(\frac{ze}{m}\right)^2 Z^2 . \tag{5.17}$$

Hence, S_{rad} becomes more relevant for high Z materials. Precise calculations are given, e.g., in [28, 40]. Using the rest mass of electrons and protons, we can estimate

$$\frac{S_{rad}(\text{protons})}{S_{rad}(\text{electrons})} = \left(\frac{m_e}{m_p}\right)^2 \approx \left(\frac{1}{1836}\right)^2 \approx 0.3 \cdot 10^{-6} , \tag{5.18}$$

showing the radiative stopping power of heavy charged particles (protons, α-particles, and heavy nuclei) to be more than six orders of magnitude smaller than that of electrons. Thus, S_{rad} is only a relevant contribution to S in the case of electrons (Figure 5.5, right), while it can be neglected for radiotherapy protons or heavier ions.

After this general introduction to (mass) stopping power components, we briefly examine the relationship between stopping power and range. Subsequently, we consider the most relevant stopping power component, the electronic (or collisional) stopping power in more detail and sketch the derivation of the important Bethe formula for heavy particles. Finally, we give some more details on the stopping power components of electrons. For further details, see e.g., [1, 19, 20, 40, 61].

Range and CSDA approximation

If the functional form of $S(E)$ is known, the average track length of a charged particle that enters a homogeneous absorber with initial energy E_0 and finally comes to rest is given by the integral

$$R_{CSDA} = -\int_{E_0}^{0} \frac{1}{S(E)}\, dE . \tag{5.19}$$

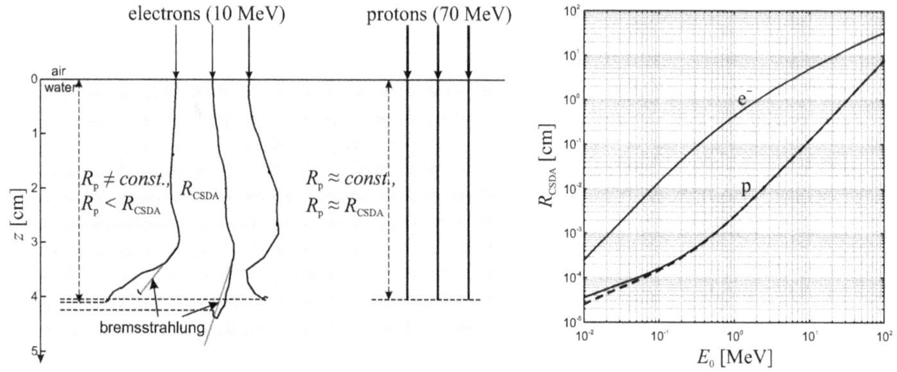

Fig. 5.6: Schematic illustration of the differences between the tracks of MeV electrons and protons in an absorbing water layer. Both particle types lose their initial kinetic energy E_0 in a large number of interaction events and finally come to rest. Due to their small mass, electrons undergo many statistically distributed deflections from their original direction, resulting in a contorted path. Their projected range R_p is considerably smaller than the total track length calculated by the CSDA approximation. Protons, in contrast, travel on an (almost) straight track due to their much higher mass; their projected range is practically equivalent to the total track length. The R_{CSDA} ranges as a function of energy are shown for both types of particles on the *right*. For protons, also the projected range is shown (*dashed line*). For energies in the MeV range as used in proton therapy, R_p and R_{CSDA} are practically the same.

Since this consideration ignores the discrete nature of the statistical collisions, it is called the **continuous slowing down approximation** (CSDA). Due to statistically distributed impact parameters (Figure 5.9), the range varies for the same initial energy E_0. The range distribution can be assumed to be approximately Gaussian[5] about a mean range $\overline{R}$. This is known as **range straggling**. Importantly, the projected range (or average penetration depth) R_p given by the projection of the start and end point of the particle trajectory in an absorber onto the surface normal is always smaller than R_{CSDA} (Figure 5.6). However, there is an important difference between electrons, on the one hand, and heavy charged particles like protons, on the other hand, due to their large mass difference, which should be remembered throughout this chapter:

- Electrons (either primary electrons or secondary electrons set in motion by photons) can easily be deflected from their original direction in collisions with atomic electrons.
- Protons and heavier ions with kinetic energies in the MeV range hardly change their direction in collisions with electrons and thus travel on an almost straight path ('straight ahead approximation' [20]), except for the very final part of their track, where they have almost come to rest (not shown for simplicity in Figure 5.6), and R_{CSDA} is a very good approximation for R_p. Strong deflections only occur due

5 The actual distribution differs somewhat from a Gaussian, see [20] and references therein.

to elastic interactions with atomic nuclei of the absorbing material, but this is a rather rare event (as known, e.g., from Rutherford's scattering experiments with α particles and gold foil), and the corresponding nuclear stopping power S_{nuc} can usually be neglected. Inelastic interactions with the nuclei of the absorber, on the contrary, play a certain role, but we do not consider them here due to length restrictions.

Kinematics of two-body scattering

Before the derivation of the Bethe formula can be sketched, some fundamental considerations on the interaction of charged particles with matter are required. To this end, we treat the interaction of a fast charged particle with an atomic electron or nucleus as a two-body scattering process [40]. Let us examine the kinematics of such a process, i.e., the consequences that arise from the conservation laws for energy and momentum for two major classes of charged particles: on the one hand, electrons (and positrons) with rest mass m_e, on the other hand, protons with much larger rest mass $m_p \approx 1836\, m_e$. The enormous mass difference leads to a quite different behavior when fast electrons and protons penetrate into a layer of an absorbing material. We start by recalling some formulas of relativistic kinematics. Let c denote the vacuum speed of light and v the velocity of a particle with rest mass m_0. Using the common notation

$$\beta = \frac{v}{c} \quad \text{and} \quad \gamma = \frac{1}{\sqrt{1 - v^2/c^2}} = \frac{1}{\sqrt{1-\beta^2}}, \tag{5.20}$$

the relativistic mass m, momentum $\underline{p}$ and total energy E of the particle are

$$m = \gamma m_0 , \tag{5.21}$$

$$\underline{p} = \gamma m_0 \underline{v} = \gamma m_0 c \underline{\beta} , \tag{5.22}$$

$$E = \gamma m_0 c^2 = E_{\text{kin}} + m_0 c^2 = \sqrt{|\underline{p}|^2 c^2 + m_0^2 c^4} . \tag{5.23}$$

Firstly, let us examine to what extent relativistic effects play a role at the typical energies of charged particles in medical diagnostics and therapy. As a rule of thumb, relativistic effects can be neglected if $\beta < 0.1$, which leads to an error in γ (and thus $m, \underline{p}, E$) of about 0.5%. We need to express β in terms of the kinetic energy E_{kin}. Some rearrangements of the above formulas yield

$$\beta = \frac{\sqrt{\mathcal{E}(\mathcal{E}+2)}}{\mathcal{E}+1} \quad \text{where} \quad \mathcal{E} = \frac{E_{\text{kin}}}{m_0 c^2}, \tag{5.24}$$

as plotted in Figure 5.7 for electrons and protons. For 100 keV electrons typically used in radiological diagnostics $\beta \approx 0.5$, and for MeV electrons in a medical linear accelerator used in radiotherapy $\beta \longrightarrow 1$. Hence, accelerated electrons in typical medical applications require a relativistic description.

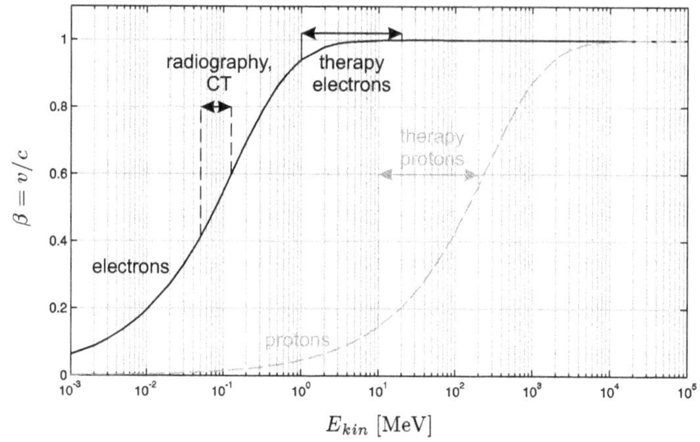

Fig. 5.7: Particle speed of electrons and protons $\beta = v/c$ as a function of particle energy. For free electrons generated in medical diagnostic imaging, a typical value is $\beta \approx \frac{1}{2}$, similar to the case of proton beams in radiotherapy, while the electrons generated in radiotherapy reach the super relativistic limit $\beta \to 1$ (conventional clinical linear accelerators).

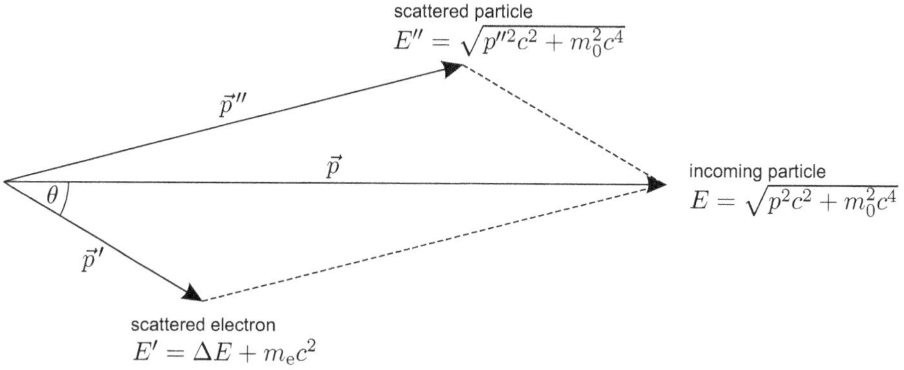

Fig. 5.8: Energy and momenstum conservation in two-body scattering of a charged massive particle with an electron initially at rest. Adapted from [40].

The most relevant interaction of accelerated charged particles with an absorbing material is the energy transfer to atomic electrons by collisions, described by the electronic stopping power S_{el}. Therefore, we consider the kinematics of two-body scattering of an incident particle with rest mass m_0 and a resting target electron with mass m_e. This process is sketched in Figure 5.8. Conservation of momentum requires

$$\underline{p}'' = \underline{p} - \underline{p}' \quad \Rightarrow \quad p''^2 = p^2 + p'^2 - 2pp' \cos\theta \,, \tag{5.25}$$

and conservation of energy yields

$$\sqrt{p^2c^2 + m_0^2c^4} + m_e c^2 = \sqrt{p''^2c^2 + m_0^2c^4} + \Delta E + m_e c^2 \,. \tag{5.26}$$

Using these conservation laws, it is possible to obtain the energy ΔE transferred to the target particle (i.e., its kinetic energy after the interaction process) as a function of initial momentum p and angle θ

$$\Delta E = \frac{2m_e p^2 c^4 \cos^2\theta}{\left(m_e c^2 + \sqrt{p^2c^2 + m_0^2c^4}\right)^2 - p^2c^2\cos\theta} . \tag{5.27}$$

The energy transfer is maximized in the case of a central or 'head on' collision with $\theta = 0$

$$\Delta E_{\text{max}} = \frac{p^2c^2}{\frac{1}{2}\left(\frac{m_0}{m_e} m_0 + m_e\right)c^2 + \sqrt{p^2c^2 + m_0^2c^4}} . \tag{5.28}$$

Inserting $p = \gamma m_0 c\beta$ and $E = \gamma m_0 c^2 = \sqrt{p^2c^2 + m_0^2c^4}$ gives

$$\Delta E_{\text{max}} = 2m_e c^2 \beta^2 \gamma^2 \left[1 + \left(\frac{m_e}{m_0}\right)^2 + 2\gamma\frac{m_e}{m_0}\right]^{-1} . \tag{5.29}$$

The maximum energy transfer depends on the ratio of the rest masses of the electron and the incoming particle $\frac{m_e}{m_0}$, as well as on the ratio of the initial particle velocity and the speed of light $\beta = \frac{v}{c}$, which is contained in γ. We distinguish four cases:

	nonrelativistic case $\beta \ll 1,\ \gamma \approx 1$	**extreme relativistic case** $\beta \longrightarrow 1,\ \gamma \gg 1$
incoming proton $m_0 = m_p$	$\Delta E_{\text{max}} \approx 2m_e v^2$	$\Delta E_{\text{max}} = E$
incoming electron $m_0 = m_e$	$\Delta E_{\text{max}} = E$	$\Delta E_{\text{max}} = E$

In the case of nonrelativistic protons, the relative energy loss in a single interaction with a resting electron is limited by

$$\frac{\Delta E_{\text{max}}}{E} \approx \frac{2m_e v^2}{\frac{1}{2}m_p v^2} = 4\frac{m_e}{m_p} \approx \frac{1}{500} . \tag{5.30}$$

Since collisions are usually noncentral, $\Delta E/E$ is typically even much lower. Therefore, a nonrelativistic proton loses only a tiny fraction of its total kinetic energy in the collision with a single electron due to its much greater mass $m_p \gg m_e$. Typical energy values are on the order of few tens of eV [61]. The corresponding change in momentum is also very small; the proton is hardly deflected from its initial direction and travels approximately on a straight path. This has already been sketched in Figure 5.6. The same applies to heavier particles such as α particles or carbon ions. As outlined in Section 5.2.1, this quasi continuous interaction allows for an analytical description of the energy loss and dose deposition in matter for this type of primary radiation. Modifications are necessary for $\beta \longrightarrow 1$ and thus $\Delta E_{\text{max}}/E \longrightarrow 1$, but this is currently not relevant for medical applications. The situation is quite different for electrons, either, e.g., a primary electron from a linear accelerator or a photo or Compton electron set in motion by primary photon radiation. An electron can transfer its entire kinetic energy to another electron that is initially at rest,[6]

$$\frac{\Delta E}{E} \leq \frac{\Delta E_{\max}}{E} = 1 . \quad (5.31)$$

Thus, electrons can lose a large portion of their initial kinetic energy in a single interaction and undergo considerable directional changes along their trajectory due to their small mass. Since all these events are statistically distributed, the paths of several electrons with the same initial conditions can look very different (cf., Figs. 5.6 and 5.14).

Heavy ions: the Bethe formula

The correct description of the energy loss of charged particles traversing matter is fundamental in various fields of science and has attracted the interest of several very well-known physicists, as reviewed in [1, 23, 66]. In the following, we sketch the simplest case, the classical derivation of the energy loss of a heavy, charged particle due to inelastic collisions with the electrons of an absorbing material (such as water or human tissue), leading to numerous excitations and ionizations in matter. Consider Figure 5.9, left: A heavy, charged particle (mass $m_0 \gg m_e$, charge $q = ze$) with initial speed $v \ll c$ (nonrelativistic case) passes a single electron (mass m_e, charge $-e$) at a minimum distance b (impact parameter). We make the following assumptions [1, 23]:

1. The trajectory of the heavy particle can be approximated by a rectilinear path, the projected range is very close to the total track length (Figure 5.6).

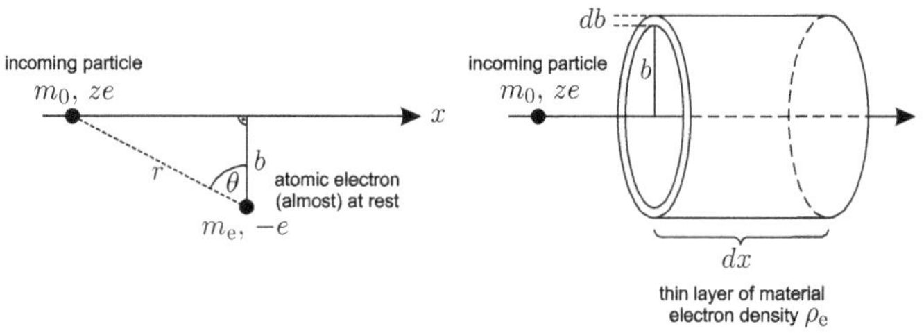

Fig. 5.9: Sketch illustrating the energy loss of a heavy, charged particle (mass $m_0 \gg m_e$, charge ze) which passes a resting electron (mass m_e, charge $-e$) at a minimum distance (impact parameter) b (*left*). Since $m \gg m_e$, the incident particle travels on an (almost) straight path. The total energy loss is obtained by integration over cylindrical shells of the absorbing layer (*right*).

6 By convention, since one cannot distinguish the two scattering particles in electron–electron scattering, the electron with higher kinetic energy after the scattering process is considered to be the primary electron, and the maximum energy transfer is, therefore, limited to $\frac{1}{2}E$.

2. The radiative stopping power S_{rad} due to bremsstrahlung production is negligible compared to the electronic stopping power S_{el} (Figure 5.5, left).
3. The electrons of the absorbing material are initially at rest (or at least have a speed much smaller than the speed v of the particle) and do not move considerably during the interaction process.

Note that assumptions 1 and 2 do *not* hold if the passage of electrons through matter is considered. The incoming particle and a single electron of the absorbing material interact by the Coulomb force (in SI units)

$$F_{\text{C}} = \frac{1}{4\pi\epsilon_0} \frac{ze^2}{r^2} . \tag{5.32}$$

The total momentum transferred to the electron is given by the integral

$$\Delta p = \int_{-\infty}^{+\infty} F_{\text{C}}(t)\, dt . \tag{5.33}$$

For symmetry reasons, only the component of the Coulomb force perpendicular to the direction of the particle needs to be considered

$$F_{\text{C}}^{\perp} = \frac{1}{4\pi\epsilon_0} \frac{ze^2}{r^2} \cos\theta = \frac{1}{4\pi\epsilon_0} \frac{ze^2}{b^2} \cos^3\theta . \tag{5.34}$$

Using the substitution

$$t = \frac{x}{v} = \frac{b}{v} \tan\theta \qquad \Rightarrow \qquad dt = \frac{b}{v\cos^2\theta} d\theta , \tag{5.35}$$

the integration over time can be replaced by an integration over the angle θ, yielding

$$\Delta p = \int_{-\infty}^{+\infty} F_{\text{C}}^{\perp}\, dt = \int_{-\frac{\pi}{2}}^{+\frac{\pi}{2}} \frac{1}{4\pi\epsilon_0} \frac{ze^2}{b^2} \cos^3\theta \frac{b}{v\cos^2\theta}\, d\theta = \frac{1}{4\pi\epsilon_0} \frac{ze^2}{bv} \int_{-\frac{\pi}{2}}^{+\frac{\pi}{2}} \cos\theta\, d\theta = \frac{1}{4\pi\epsilon_0} \frac{2ze^2}{bv} . \tag{5.36}$$

The corresponding energy transfer (in the nonrelativistic limit) is

$$\Delta E = \frac{\Delta p^2}{2m_{\text{e}}} = \left(\frac{1}{4\pi\epsilon_0}\right)^2 \frac{2z^2e^4}{m_{\text{e}}b^2v^2} . \tag{5.37}$$

Note that the mass of the target particle enters this formula in the denominator. Therefore, if we considered an atomic nucleus instead of an electron, the energy transfer would be more than three orders of magnitude smaller. Now, let the particle traverse a thin layer of thickness dx of a homogeneous material (Figure 5.9, right). The particle passes $\rho_{\text{e}} 2\pi b\, db\, dx$ electrons at a distance between b and $b + db$ (i.e., the electrons in the cylindrical shell), where $\rho_e = \rho N_{\text{A}} \frac{Z}{A}$ denotes the electron density. The energy transferred to these electrons is

$$dE = -\left(\frac{1}{4\pi\epsilon_0}\right)^2 \frac{2z^2e^4}{m_{\text{e}}b^2v^2} \rho_{\text{e}} 2\pi b\, db\, dx = -\left(\frac{1}{4\pi\epsilon_0}\right)^2 \frac{4\pi\rho_{\text{e}}z^2e^4}{m_{\text{e}}v^2} \frac{db}{b} dx . \tag{5.38}$$

The minus sign indicates the energy loss of the incoming particle. The energy loss per unit path length, i.e., the electronic stopping power, is obtained by integration over b

$$S_{\mathrm{el}} = -\frac{dE}{dx} = \left(\frac{1}{4\pi\epsilon_0}\right)^2 \frac{4\pi\rho_e z^2 e^4}{m_e v^2} \int_{b_{\min}}^{b_{\max}} \frac{db}{b} = \left(\frac{1}{4\pi\epsilon_0}\right)^2 \frac{4\pi\rho_e z^2 e^4}{m_e v^2} \ln\left(\frac{b_{\max}}{b_{\min}}\right) . \quad (5.39)$$

The remaining difficulty is the correct choice of the integration limits $b_{\min}$ and $b_{\max}$, which is made as follows. The lower limit $b_{\min}$ corresponds to the maximum energy transfer. In the previous section, we derived $\Delta E_{\max} \approx 2m_e v^2$ for a nonrelativistic heavy particle. Inserting this result into equation (5.37) and solving for b yields the lower limit

$$b_{\min} = \frac{1}{4\pi\epsilon_0} \frac{ze^2}{m_e v^2} . \quad (5.40)$$

For $b_{\max}$, the binding effects of the atomic electrons are the relevant energy range. Assuming that the lowest possible energy transfer to an atomic electron is denoted by $\bar{I}$ (as discussed below) and inserting this for ΔE in equation (5.37) yields

$$b_{\max} = \frac{1}{4\pi\epsilon_0} \frac{ze^2}{\sqrt{\frac{m_e}{2} v^2 \bar{I}}} . \quad (5.41)$$

Plugging $b_{\min}$ and $b_{\max}$ into equation (5.39) yields the energy loss formula

$$S_{\mathrm{el}} = \left(\frac{1}{4\pi\epsilon_0}\right)^2 \frac{2\pi\rho_e z^2 e^4}{m_e v^2} \ln\left(\frac{2m_e v^2}{\bar{I}}\right) , \quad (5.42)$$

as first derived by Bohr, who also provided a more rigorous justification of the initial assumptions [11]. Later, it was generalized using quantum mechanics and extended to relativistic particle velocities by Bethe [8, 9], resulting in

$$S_{\mathrm{el}} = \left(\frac{1}{4\pi\epsilon_0}\right)^2 \frac{2\pi\rho_e z^2 e^4}{m_e c^2 \beta^2} \left[\ln\left(\frac{2m_e c^2 \beta^2 \gamma^2 \Delta E_{\max}}{\bar{I}^2}\right) - 2\beta^2\right] , \quad (5.43)$$

where $\Delta E_{\max}$ is given by equation (5.29). For the nonrelativistic limit $\beta \to 0$, $\gamma \to 1$ and $\Delta E_{\max} = 2m_e v^2$ this again yields equation (5.42). For $m_0 \gg m_e$ and γ not too large, inserting $\Delta E_{\max} \approx 2m_e c^2 \beta^2 \gamma^2$ from equation (5.29) yields

$$S_{\mathrm{el}} = \left(\frac{1}{4\pi\epsilon_0}\right)^2 \frac{4\pi\rho_e z^2 e^4}{m_e c^2 \beta^2} \left[\ln\left(\frac{2m_e c^2 \beta^2 \gamma^2}{\bar{I}}\right) - \beta^2\right] , \quad (5.44)$$

which is the form commonly known as the **Bethe formula.**[7] The principal material-dependent quantity $\bar{I}$ that determines the electronic stopping power is called the

7 Equation (5.44) is often also called the Bethe–Bloch equation, since Bloch contributed an approximation for the mean ionization potential, $\bar{I} \approx (10\,\mathrm{eV})\,Z$, and a correction term that accounts for deviations from the first Born approximation.

mean excitation energy. It depends on the specific electronic structure of the atom, molecule or solid of interest and is difficult to calculate, but can be obtained experimentally by a number of methods. The bar indicates the average over different electron states. For hydrogen, one obtains $\bar{I} = 19.2$ eV, which is larger than its ground state binding energy of 13.6 eV, since the ejected electrons have an average kinetic energy greater than zero [27]. Further values are, e.g., $\bar{I} = 75$ eV for water and $\bar{I} = 823$ eV for lead [19].[8] The accuracy of the Bethe formula is improved by several correction terms, which are beyond the scope of this book:

- The shell correction is required when the particle velocity v is no longer much larger than the velocity of bound atomic electrons, i.e., at low particle energies.
- Further correction terms (Barkas correction and Bloch correction) are due to a treatment beyond the first Born approximation used by Bethe and also become relevant at low energies.
- The density effect correction accounts for a reduction of the electronic stopping power due to polarization effects, especially at high particle energies. This effect is particularly strong in dense media, hence its name.

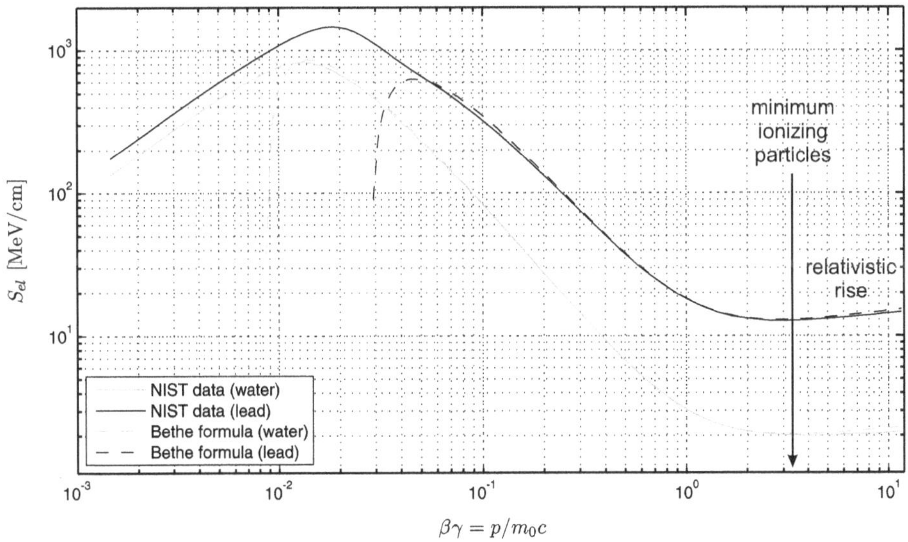

Fig. 5.10: Plot of proton stopping power S_{el} due to collisions with atomic electrons of the traversed material (here: water and lead) as calculated by the Bethe formula (5.44), as a function of $\beta\gamma = \frac{p}{m_p c}$. For sufficiently high $\beta\gamma$, the data are very well described by the Bethe formula. The small discrepancy beyond the minimum at $\beta\gamma \approx 3$ is due to neglection of the density effect. Particles with momentum $\beta\gamma \approx 3$ are 'called minimum ionizing particles'.

8 A table with values of $\bar{I}$ is available at http://physics.nist.gov/PhysRefData/XrayMassCoef/tab1.html.

For protons in water, a plot of the 'pure' Bethe formula (5.44), other stopping power components and NIST data where several corrections are incorporated is given at the beginning of this section in Figure 5.5 (left). The Bethe formula fails at proton kinetic energies below ~ 0.5 MeV, while agreement is very good in an intermediate energy range of ~ 1 to 1000 MeV, relevant, e.g., in proton therapy (Section 5.2.1). A second plot, S_{el} as a function of momentum $\beta\gamma = \frac{p}{m_{\mathrm{p}}c}$ (5.22) is shown in Figure 5.10. Agreement of the Bethe formula with NIST data is very good for sufficiently high energies. For different absorber materials (here: H_2O and Pb), the curves approach a broad minimum at $\beta\gamma \approx 3$ corresponding to a proton energy $\approx 3m_{\mathrm{p}}c^2 \approx 3$ GeV (the exact value of the minimum depends slightly on the material), followed by the so called 'relativistic rise' due to divergence of the factor γ^2.

For a more detailed discussion of the energy loss of heavy charged particles, see, e.g., [1, 20, 40] and references therein. Further effects relevant for heavy particles with $z > 1$ (e.g., α particles and carbon ions) are changes in the effective charge state upon slowing down.

Electrons and positrons

If the incident primary particles are electrons, e.g., from a medical linear accelerator (typical energies between 3 and 20 MeV) or positrons, several assumptions that are made in the derivation of the Bethe formula do *not* hold as a consequence of their small mass:

- Radiative energy loss (bremsstrahlung emission) is more significant (Figure 5.5).
- In a single collision, an electron (or positron) can be strongly deflected, and large energy transfers (in terms of the electron rest energy) are possible. An individual electron's trajectory resembles a "drunken man's walk" (Figure 5.6).
- Relativistic effects become relevant at much lower energies (Figure 5.7).

For electron–electron scattering (MØLLER scattering, –) and electron–positron scattering (BHABHA scattering, +), one finds the following equivalent of the Bethe formula (5.44) [19, 66]:

$$S_{\mathrm{el}}^{\pm} = \left(\frac{1}{4\pi\epsilon_0}\right)^2 \frac{2\pi\rho_e e^4}{m_{\mathrm{e}}c^2\beta^2}\left[\ln\left(\frac{\mathcal{E}^2(\mathcal{E}+2)}{2(\bar{I}/m_{\mathrm{e}}c^2)^2}\right) + f^{\pm}(\mathcal{E}) - \delta\right] \tag{5.45}$$

where

$$\mathcal{E} = \frac{E_{\mathrm{kin}}}{m_{\mathrm{e}}c^2}, \tag{5.46}$$

$$f^{-}(\mathcal{E}) = \left(1-\beta^2\right)\left[1 + \frac{\mathcal{E}^2}{8} - (2\mathcal{E}+1)\ln 2\right], \tag{5.47}$$

$$f^{+}(\mathcal{E}) = 2\ln 2 - \frac{\beta^2}{12}\left[23 + \frac{14}{\mathcal{E}+2} + \frac{10}{(\mathcal{E}+2)^2} + \frac{4}{(\mathcal{E}+2)^3}\right]. \tag{5.48}$$

As in the case of the Bethe formula, the mean excitation energy $\bar{I}$ is the principal material parameter, and a correction term δ due to polarization effects is required for $\beta \longrightarrow 1$. Differences between electrons and positrons arise since Møller scattering occurs between two identical particles. The radiative stopping power is due to two components: repulsive Coulomb interaction with the electron shell and screened attractive Coulomb interaction with atomic nuclei. For electrons with sufficiently high kinetic energy E, it can be approximated by

$$S_{\text{rad}}^{\pm} \approx 4\alpha r_0^2 \rho_e Z^2 \ln\left(\frac{183}{Z^{1/3}}\right) E\,, \tag{5.49}$$

where $\alpha = e^2/(4\pi\epsilon_0 \hbar c) \approx 1/137$ denotes the fine structure constant and $r_0 = e^2/(4\pi\epsilon_0 m_e c^2)$ the classical electron radius [40]. Equations (5.45) and (5.49) are plotted in Figure 5.5 (right), along with NIST data that take into account additional effects and corrections. The radiative stopping power of positrons is different, since they are attracted by atomic orbital electrons and repelled by the nuclei [19]. An approximative formula for the ratio of radiative and electronic stopping power for electrons of energy E in a material with atomic number Z is [18]

$$\frac{S_{\text{rad}}}{S_{\text{el}}}(\text{electrons}) = kZE \tag{5.50}$$

where E is given in units of MeV and $k \approx 1.25 \cdot 10^{-3}\,\text{MeV}^{-1}$. For example, in an x-ray tube with a tungsten anode ($Z = 74$) and a voltage of 100 kV, less than 1% of the kinetic energy of the electrons is converted into bremsstrahlung, while the collisional energy loss generates the enormous heat that makes cooling mechanisms and, e.g., rotating anodes necessary. In water, S_{rad} becomes the dominant energy loss mechanism of electrons at $E \approx 100\,\text{MeV}$ (Figure 5.5, right), while in lead already at $E \approx 10\,\text{MeV}$, which has to be taken into account in the design of proper shielding in radiotherapy [36].

5.2 Dose calculation and measurement

In the following, we apply the results of the previous section to illustrate some key principles for the calculation and measurement of absorbed dose distributions $D(\underline{r})$. All examples are derived from radiotherapy, where the generation of (possibly very complex) dose distributions inside a patient's body are an integral part. For an in depth treatment and many further interesting topics, e.g., dose calculation in nuclear medicine or Fermi–Eyges theory for analytical calculation of dose distributions from primary electron beams, we refer to the given textbook references.

5.2.1 Analytical example: proton Bragg peak

Proton therapy is an ideal example to apply the physics discussed above, regarding the energy loss of heavy ionizing particles in matter. In the following, we briefly review a method to obtain a fully analytic description of the so called Bragg curve $D(z)$, which describes the deposited dose as a function of depth z in the tissue, starting from a phenomenological range–energy relationship (BORTFELD et al., [12, 14]). With particle energies typically ≤ 200 MeV, the following approximations can be made:

- A nonrelativistic treatment is justified, since the energies are well below $m_p c^2 = 938.27$ MeV [51] (cf., Figure 5.7).
- As for the derivation of the Bethe formula, protons can be assumed to travel on a straight path. Lateral deflections and straggling effects are negligible in a first approximation. The radiative stopping power (bremsstrahlung production) can be neglected.
- The range of secondary electrons is small compared to the length scale of interest (~ 1 mm). The maximum energy transfer from a nonrelativistic proton to a single electron is limited by equation (5.30), which yields about 0.4 MeV corresponding to $R_{\text{CSDA}} = 0.2$ cm (Figure 5.6). The mean energy transfer is much lower, on the order of tens of eV only [37, 61]. Therefore, it is a justified approximation that the energy loss to electrons in a depth z determines the absorbed dose $D(z)$.

With the approximations given above, equation (5.15) or (5.44) yield $S(E) \propto E^{-1}$ as a crude first approximation for the total stopping power of nonrelativistic protons with $E \gg \bar{I}$. Using the CSDA approximation (5.19), and since R_{CSDA} and the projected range are approximately the same due to the (almost) straight path of protons, this yields

$$R_p \approx R_{\text{CSDA}} = \int_{E_0}^{0} \frac{1}{S(E)}\, dE \propto \int_{0}^{E_0} E\, dE \propto E_0^2 \,. \tag{5.51}$$

Indeed, empirically one finds that R_p of a heavy charged particle in a given material varies with its initial kinetic energy E_0 approximately according to a power law[9]

$$R_p = A E_0^p \tag{5.52}$$

with an exponent p close to 2. The parameters A and p depend on the properties of the particle (mass, charge) and of the traversed material (density, atomic composition).

9 An energy–range relation of this type follows from the ansatz (with $\gamma, q = $ const.)

$$m\frac{dv}{dt} = -\gamma v^{-q}$$

for the damping of the particle motion, which can be considered as a modified Langevin equation [67, 68].

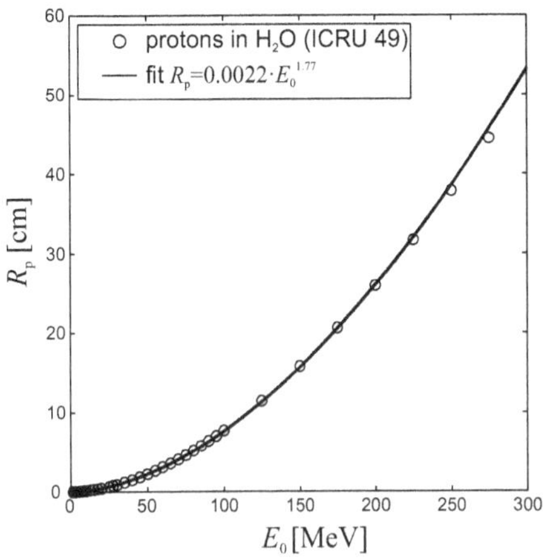

Fig. 5.11: Plot of experimental values of the mean proton range R_p in water as a function of the initial kinetic energy E_0 (*circles*). The data points are well approximated by an analytical fit $R_\text{p} = AE_0^p$ with $A = 0.0022\,\text{cm} \cdot \text{MeV}^{-p}$ and $p = 1.77$. From the convex shape, one can derive an analytical expression that gives a good approximation of experimental proton Bragg peak data. Based on [12].

For the case of low energetic α particles, one finds $p = 1.5$ (Geiger's rule). By fitting measured energy–range values of protons in water with equation (5.52), one finds $A \simeq 0.0022\,\text{cm} \cdot \text{MeV}^{-p}$ and $p \simeq 1.77$ for energies $E \leq 200\,\text{MeV}$ used in proton therapy (Figure 5.11). The residual distance $R_\text{p} - z$ and corresponding energy $E(z)$ are obtained by substituting $R_\text{p} \longrightarrow R_\text{p} - z$ and $E_0 \longrightarrow E(z)$, which yields

$$E(z) = \frac{1}{A^{1/p}} \left(R_\text{p} - z\right)^{1/p} \qquad \Rightarrow \qquad -\frac{dE(z)}{dz} = \frac{1}{pA^{1/p}} \left(R_\text{p} - z\right)^{1/p-1} . \tag{5.53}$$

The energy fluence can be written as the product of the particle fluence $\Phi(z)$ (protons per cross sectional area of the beam) and the residual proton energy $E(z)$ in depth z:

$$\Psi(z) = \Phi(z)E(z) \tag{5.54}$$

The proton terma (total energy released in matter per unit mass) in depth z can be written as

$$T(z) = -\frac{1}{\rho}\frac{d\Psi}{dz} = -\frac{1}{\rho}\Bigg(\underbrace{\Phi(z)\frac{dE(z)}{dz}}_{\substack{\text{energy transfer}\\ \text{protons} \to \text{electrons}}} + \underbrace{\frac{d\Phi(z)}{dz}E(z)}_{\substack{\text{proton loss by}\\ \text{nuclear interactions}}} \Bigg) . \tag{5.55}$$

The first term in equation (5.55) accounts for the loss of kinetic energy of the protons, which is equivalent to the electronic stopping power, since radiative and nuclear stopping power are orders of magnitude lower and can, thus, be neglected (Figure 5.5). Since the range of secondary electrons released by ionization is negligible, this first term fully contributes to the absorbed dose $\hat{D}(z)$ (the hat symbol in $\hat{D}(z)$ is used to denote the 'ideal' dose distribution without range straggling, while for the 'real' dose $D(z)$ will be used further below).

The second term in equation (5.55) describes energy transfer to the medium due to inelastic nuclear interactions of protons with atomic nuclei, which reduces the proton fluence Φ. A variety of secondary particles can be produced (n, α, ...). Per 1 cm of traversed tissue, about 1% of the protons are subject to these inelastic nuclear interactions in the energy range used in proton therapy, which makes this effect very important for clinical applications [37, 47]. As shown in [12], it can be incorporated into the calculation rather easily, but we neglect it in this didactic example for the sake of convenience and assume a constant proton fluence $\Phi(z) = \Phi_0$ up to the depth $z = R_p$, where all protons come to rest. With these approximations, the 'ideal' absorbed dose is given by

$$\hat{D}(z) = -\frac{1}{\rho}\left(\Phi(z)\frac{dE(z)}{dz}\right) \tag{5.56}$$

Inserting equation (5.53) into equation (5.56) yields

$$\hat{D}(z) = \begin{cases} \Phi_0 \dfrac{(R_p - z)^{1/p-1}}{\rho p A^{1/p}} & z \leq R_p\,, \\ 0 & z > R_p\,. \end{cases} \tag{5.57}$$

This is plotted in Figure 5.12 (dashed line). Upon slowing down, the absorbed dose $\hat{D}(z)$ increases with depth z, and then drops to zero beyond the range R_p. The function diverges for $z \longrightarrow R_p$, since the exponent of the first term $\frac{1}{p} - 1 < 0$. So far, it has been assumed that all protons behave in exactly the same way. For a realistic description, the statistical nature of the interaction processes leading to range straggling effects,

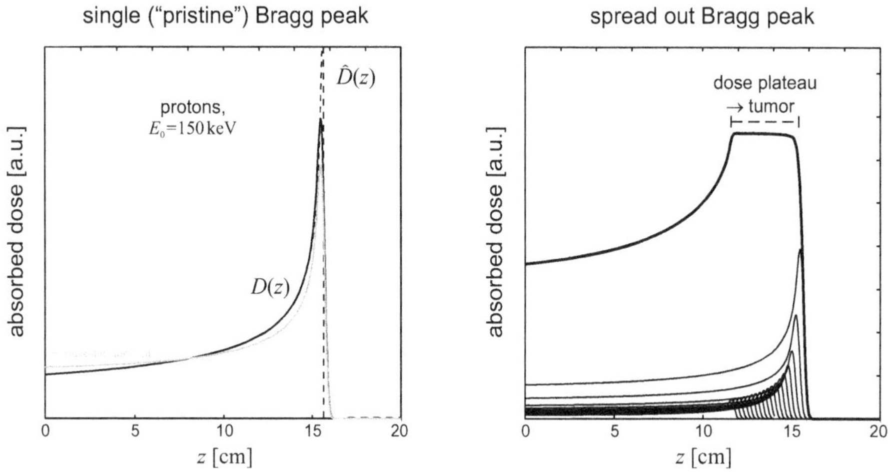

Fig. 5.12: *Left:* Bragg curves of monoenergetic protons (E_0 = 150 MeV) without (*dashed line*) and with range straggling (*solid line*) computed by equation (5.57) and equation (5.58), respectively. Adapted from [12]. The *gray curve* indicates how the Bragg curve changes if inelastic nuclear interactions are taken into account. *Right*: Spread out Bragg peak created by superposition of several pristine proton Bragg peaks of different energies based on the methods presented in [30]. The plateau region can be used for homogeneous dose delivery to a tumor. However, note that this benefit comes with the cost of a considerably higher entrance dose.

as well as a slight angular divergence of the proton beam caused by scattering from the atomic nuclei of the absorbing material, must be taken into account. Empirically, this can be incorporated by the convolution of equation (5.57) with a Gaussian with a standard deviation σ,

$$D(z) = \hat{D}(z) * \frac{1}{\sqrt{2\pi}\sigma} \exp\left(-\frac{z^2}{2\sigma^2}\right) = \frac{1}{\sqrt{2\pi}\sigma} \int_{-\infty}^{R_p} \hat{D}(z') \exp\left(-\frac{(z-z')^2}{2\sigma^2}\right) dz' . \quad (5.58)$$

This leads to a rather complex, but closed analytic expression for the 'real' proton Bragg curve $D(z)$, including some tabulated special functions. The Bragg peak now has a finite maximum and is in very good agreement with experimental data [12]. Here, we have carried out the convolution numerically ($\sigma = 0.16$ cm). The resulting Bragg curve is plotted in Figure 5.12 (solid line). A finite spectrum of initial energies further contributes to the broadening of realistic Bragg peaks. In radiotherapy with protons, the superposition of several quasi monoenergetic beams is employed to create a so called spread out Bragg peak with an almost homogeneous dose plateau in a certain depth interval (Figure 5.12, right).

5.2.2 Algorithms for external photon beams

In the case of high energy photon beams with energies of several MeV, which is by far the most common type of radiation used in radiotherapy, the complex statistical nature of the coupled transport of photons and electrons/positrons and the range of secondary electrons of several centimeters preclude a sufficiently accurate analytic calculation of dose distributions. Therefore, powerful algorithms have been developed. In addition to being able to calculate dose distributions with sufficient accuracy, algorithms must be fast, since treatment planning is an iterative optimization process where dose calculations are repeated many times for each patient. Here, we give a brief overview on the basic ideas of dose calculation algorithms for external photon beams. Our summary is based on the reviews [3, 44, 55], as well as the textbooks [47, 48], which are recommended for further details.

Correction based models

The oldest group of methods for dose calculation is purely empirical and relies on data measured at various square and rectangular field sizes in a water phantom, i.e., in a homogeneous medium. A set of percentage depth dose curves (PDDs) and lateral dose profiles, as well as tissue:phantom ratios (TPR) and further similar quantities is measured (Figure 5.13). Dose distributions expected in different geometries and materials are then determined by means of interpolation or extrapolation (for details, see, e.g., [33], Chapters 9 and 10). This works sufficiently well for homogeneous media and large field sizes, but is too inaccurate for the human body with its tissue inho-

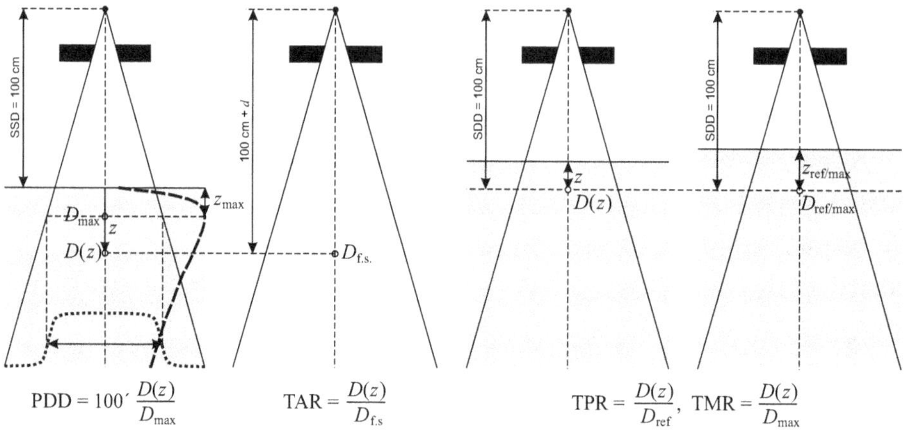

Fig. 5.13: Correction based dose algorithms use percentage depth dose curves (*bold dashed curve*) and lateral dose profiles (*bold dotted curve*), as well as several dose ratios (tissue:air ratio TAR, tissue:phantom ratio TPR and tissue:maximum ratio TMR) and other similar quantities measured at fixed sample surface distance (SSD) or sample detector distance (SDD) and different square and rectangular field sizes to approximate the dose distribution for different conditions (different SSD, more complex field shapes, ...) by interpolation or extrapolation. TAR is defined as the quotient of the dose measured in a certain water depth z at a fixed SSD and the dose measured in air (free space f.s.) at the same position if the water phantom is removed. TPR and TMR, in contrast, use a fixed SDD and two dose values measured in water, which makes them more useful for isocentric treatment techniques and high photon energies.

mogeneities and modern treatment techniques utilizing small field segments. Hence, these methods have been superseded by the more advanced dose calculation algorithms explained in the following. However, they are still in use for independent plausibility checks.

Monte Carlo (MC) methods

The principle of MC calculations of radiation transport can be summarized as follows (see, e.g., [5, 59, 70] for a review). The fundamental interactions of photons and charged particles with matter are stochastic by nature. However, the corresponding probability distributions of distances between two interactions, scattering angles, etc., are very well understood and readily available as a function of energy and material composition. In addition, if also the used radiation source (i.e., probability distributions of initial particle energy, position and direction) and the material composition in some geometry are accurately modeled, it is possible to simulate a very large number of realistic 'particle histories' by random sampling from these probability distributions.

In several figures in this chapter, resulting particle tracks are shown for the simple case of a water phantom, but the method can easily be extended to more com-

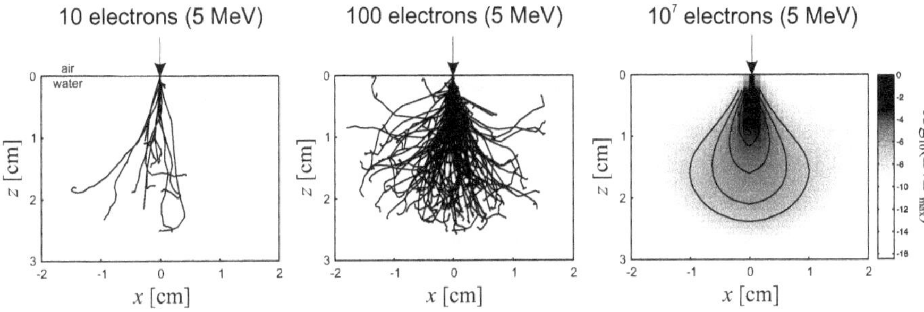

Fig. 5.14: MC simulation of a narrow electron beam impinging onto a water surface. There are 10 electron tracks, 100 electron tracks and normalized dose distribution (with isodose curves) from 10^7 electrons. Due to lateral scatter, the projected range in the *z* direction is considerably lower than the total track length. An analytical method for electron dose calculations that yields similar results (but with some approximations) is the so called Fermi–Eyges theory (e.g., [47, 48]).

plex geometries. By 'bookkeeping' of the events in each voxel, it is possible to obtain numerical estimates of all radiation field characteristics and dosimetric quantities, in particular those that are inaccessible by experiments. Therefore, MC calculations have become an indispensable tool in radiation physics and dosimetry. By scoring the energy deposited in each voxel of the geometry (as shown for the simple example of a narrow electron beam in Figure 5.14), precise numerical dose calculations become possible, both for problems related to dosimetry (e.g., an ionization chamber inserted into some phantom) and also based on an actual CT scan of a patient, provided the Hounsfield units (HU) are correctly mapped to electron density.

In principle, the precision of the MC method is only limited by the accuracy of the interaction data and the model of the radiation source and the geometry. However, there is always a tradeoff between statistical accuracy, voxel size and the computational time required. For N particle histories, the standard deviation σ_D of the dose D in a cubic voxel with volume $V = \ell^3$ scales as [29]

$$\sigma_D \propto \frac{\sqrt{D}}{\ell\sqrt{N}} \,. \tag{5.59}$$

The first MC simulations of coupled photon/electron transport already date back to around 1960 [59], yet they remained impractical for many applications. A number of so called variance reduction techniques were developed to speed up calculations. Today, due to the ever increasing availability of computational power, computing time is becoming less and less critical, and the MC method has entered the domain of routine treatment planning [58].

Since MC simulations correctly capture the 'true' physics of coupled electron/photon transport, they are considered the 'gold standard' for dose computations and are used to benchmark other methods that may be faster but employ more approximations. Some widely used Monte Carlo software packages in medical physics are EGS

(Electron Gamma Shower)[10], GEANT4 (*Geometry And Tracking*)[11] PENELOPE (*Penetration and Energy Loss of Positrons and Electrons*)[12] and MCNP (*Monte Carlo N-Particle*)[13]. Also codes for simulation of protons are available, e.g., TOPAS (*Tool for Particle Simulation*)[14].

So far, in this chapter there have already been several figures showing photon and electron tracks generated by the code DOSRZnrc included in the EGS package in cylindrical geometry (r, z). In the same way, Figure 5.15 shows exemplary dose distributions $D(r, z)$ in a water phantom (radius 10 cm, height 20 cm) resulting from some standard photon and electron fields with typical energies used in radiotherapy. All modeled fields result from a point source with source-to-surface distance SSD = 100 cm and have a diameter of 10 cm at the water surface, a 'voxel' size of $0.2 \times 0.2\,\text{cm}^2$ (slabs in the z direction, shells in the r direction) was used. We use these to discuss some general characteristics of photon and electron fields used in radiotherapy:

In case of photons (Figure 5.15, top panel), the spectra of a 250 kV x-ray tube, a ^{60}Co source and two linear accelerators included in EGS were used. An important characteristic of a radiation field is its percentage depth dose curve

$$\text{PDD}(z) = 100\% \cdot \frac{D(z)}{D_{\text{max}}} \tag{5.60}$$

measured along the central axis in a water phantom. While the PDD curve resulting from the 250 kV x-ray tube spectrum has its maximum very close to $z = 0$, it moves into greater depth for the spectra with higher maximum energy. This so called **buildup effect** is due to the fact that the secondary electrons become more forward directed with increasing energy, and their maximum range becomes larger than the voxel size used for photon energies exceeding ~ 0.5 MeV and reaches several centimeters in the case of the two linear accelerator (Linac) beams (cf., Figure 5.6). In greater depth, the PDD curves decay exponentially due to attenuation of the primary photon fluence. In addition, also beam divergence adds a factor $\propto (\text{SSD} + z)^{-2}$ to the PDD curve. In the radial direction r, the dose profiles do not immediately drop to zero outside the geometric field borders, but exhibit a region of finite extent where the dose continuously goes to zero. This is called the *penumbra* and is due to photons scattered and secondary electrons deflected in lateral direction. In a real beam, also the finite extent of the source (instead of a point source) and transmission through the collimator contribute to the penumbra.

In contrast to photons, the PDD curves of monoenergetic electron beams (Figure 5.15, bottom panel) show a very broad maximum, before they practically go to

10 http://www.nrc-cnrc.gc.ca/eng/solutions/advisory/egsnrc_index.html
11 http://www.geant4.org/geant4/
12 https://www.oecd-nea.org/tools/abstract/detail/nea-1525
13 https://mcnp.lanl.gov/
14 http://www.topasmc.org/

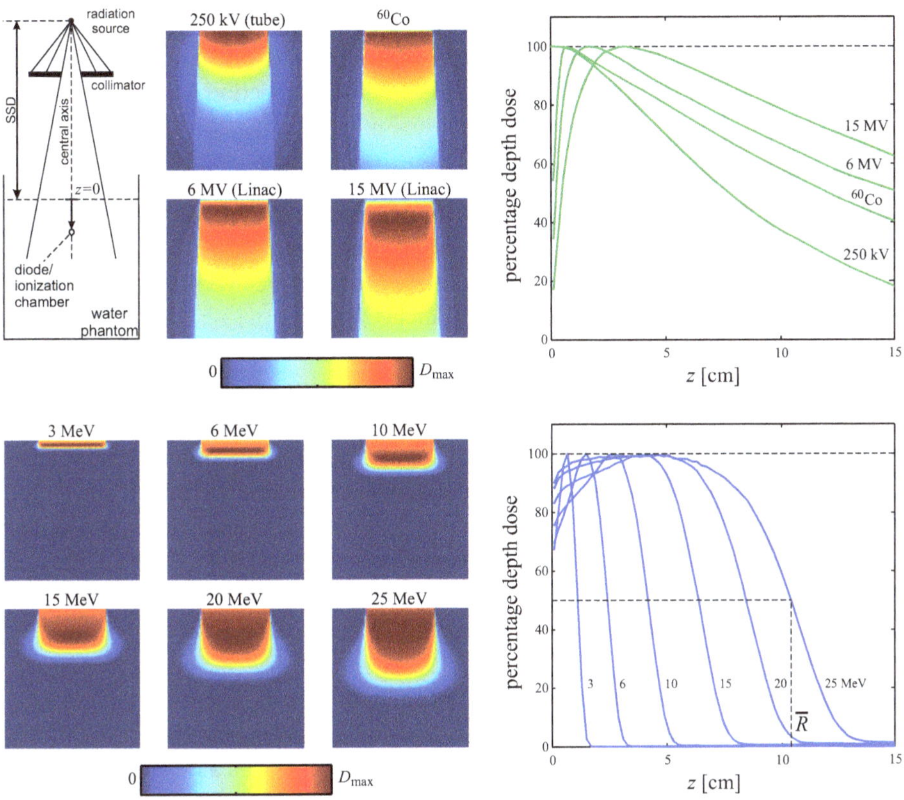

Fig. 5.15: MC simulations of dose distributions in water (cylindrical phantom, radius 10 cm, height 20 cm) resulting from several radially symmetric photon beams with different energy spectra (*top panel*) and several monoenergetic electron beams with energies from 3 to 25 MeV (*bottom panel*) and the corresponding percentage depth dose curves. For all fields, the field diameter at the water surface $z = 0$ is 10 cm, and the point source is located at a distance of 100 cm from the water surface. The results are based on between 10^7 (high energy electrons) and 10^9 (low energy photons) particle histories, which took about 8 to 12 h per energy on a standard desktop PC.

zero at a depth z corresponding to the R_{CSDA} range (Figure 5.6). As a rule of thumb for typical energies used in radiotherapy, the mean range $\bar{R}$ (indicated for 25 MeV in Figure 5.15) of electrons of initial energy E in water or soft tissue is [37]

$$\bar{R}\,[\text{cm}] \approx \frac{E\,[\text{MeV}]}{2.33} \qquad \text{for} \qquad 1 < E_0 < 50\,\text{MeV}\,. \tag{5.61}$$

A small dose 'tail' extending to larger z is due to bremsstrahlung. In the lateral direction, the dose distributions bulge out with increasing depth due to the scattering of electrons, as is also visible in the electron tracks of Figure 5.14.

Due to their different characteristics, photons and electrons are used for different purposes in radiotherapy. Electrons are ideal for irradiation of superficial regions that do not extend beyond a depth of few cm. Photons are used for deeply seated tumors and allow a superposition of beams from several angles to generate very conformal dose distributions by a variety of techniques explained in Section 5.4.

Kernel based convolution/superposition

In this class of algorithms developed in the 1980s and afterwards, a dose distribution $D(\underline{r})$ is considered as the superposition of 'elementary' dose distributions. These are called **energy deposition kernels**. Different geometries can be distinguished, the most common ones are point and pencil beam kernels.

A point kernel $h_{\text{point}}(\underline{r})$ represents the normalized dose distribution resulting from an infinitely thin photon beam that interacts in a single point of an infinite homogeneous medium. In analogy to image formation, it can be understood as a PSF that converts the total energy released in matter in a singular point to the extended dose distribution in its surroundings [2]. It is not accessible in an experiment, since photons cannot be forced to interact in a single point only. Also analytical calculations from first principles are precluded by the complexity and stochastic nature of the processes involved. However, point kernels are easily generated by Monte Carlo calculations. An example is shown in Figure 5.16. Dose calculation by point kernels involves the following steps [47]:

1. The energy fluence $\Psi(\underline{r})$ of the photons emitted from the radiation source, usually a linear accelerator (linac) must be modeled. This comprises both unscattered bremsstrahlung photons originating from the target, as well as scattered photons from beam modifying devices such as collimators or wedges.
2. Using the Beer–Lambert law, the energy fluence in the absorbing medium is obtained, yielding the total energy released per unit mass by equation (5.12), $T(\underline{r}) = \frac{\mu}{\rho}\Psi(\underline{r})$.
3. The absorbed dose distribution $D(\underline{r})$ is obtained by superposition of all point kernels weighted by $T(\underline{r})$. If the kernel is spatially invariant (which is the case only for a parallel photon beam and in an homogeneous and infinite absorber), this superposition can be written as a convolution integral

$$D(\underline{r}) = \iiint_V T(\underline{s})\, h_{\text{point}}(\underline{r} - \underline{s})\, d^3 s = (T * h_{\text{point}})(\underline{r})\,. \tag{5.62}$$

Pencil beam kernels $h_{\text{pencil}}(\underline{r})$ are the normalized dose distributions resulting from a narrow monochromatic beam impinging onto a semi-infinite homogeneous medium (Figure 5.16). They can be computed by MC methods either directly or as a superposition of point kernels. The dose distribution in three dimensions is obtained by convolution of the energy fluence in the entrance plane ($z = 0$) with the pencil beam kernel

$$D(\underline{r}) = \iint_A T(u, v, 0)\, h_{\text{pencil}}(x - u, y - v, z)\, du\, dv = (T * h_{\text{pencil}})(\underline{r}) \tag{5.63}$$

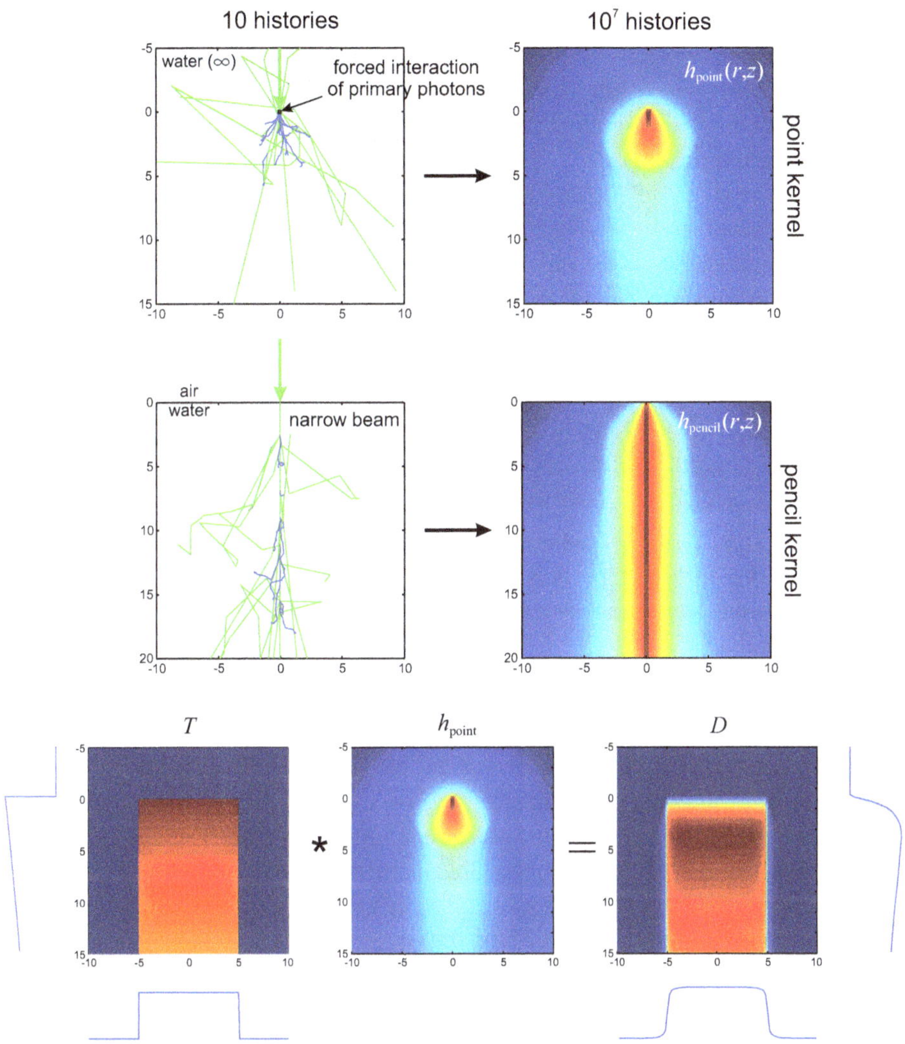

Fig. 5.16: Energy deposition kernels can be obtained by MC simulation and are employed for dose computation by convolution/superposition methods. *Top:* Point kernels are obtained in an infinite water geometry where photons (here 10 MeV) from a narrow beam are forced to interact in a small volume element. *Center*: Pencil beam kernels are obtained from a narrow photon beam impinging on a semi-infinite water layer. *Bottom*: Convolution of the total energy released per unit mass $T(r, z) = (\mu/\rho)\, h\nu\Phi_0 \exp(-\mu z)$ with the point kernel $h_{point}(r, z)$ yields the resulting dose distribution, which displays the buildup effect and is smeared out at the field edges. Note that the energy deposition kernels are plotted on logarithmic scale, but T and D on linear scale.

The important point is that each energy deposition kernel for a given photon energy needs to be calculated only once. Therefore, the calculation speed compared to full MC methods is increased enormously. In addition, note that the convolution theorem allows for a numerical evaluation of equation (5.62) and equation (5.63) using the FFT algorithm (cf., Section 2.7).

So far, a monochromatic and parallel photon beam has been assumed. For application in radiotherapy, the formalism has to be extended to incorporate continuous photon spectra and beam divergence (the corresponding modifications of equation (5.62) and equation (5.63) are outlined, e.g., in [2, 47]), as well as tissue inhomogeneities and the finite extent of a patient's body. Since convolution implies a spatially invariant kernel, the term *superposition* is more suitable if noninvariant kernels are used. A number of techniques have been developed to apply inhomogeneity corrections, either only in beam direction (*pencil beam convolution*) or also in lateral direction (*collapsed cone convolution* (CCC), analytical anisotropic algorithm (AAA)). However, it has been demonstrated that this can still be associated with considerable errors in dose calculation close to the interface of two media with very different densities (e.g., at the lung surface), especially in the case of pencil beams [39]. For accurate skin dose, also the electrons generated by scattering processes in the linac head must be considered [47].

Deterministic models of radiation transport

In the last years, a different approach to radiation dose calculation, which is based on transport theory (as applied, e.g., in astrophysics or particle physics), has gained interest. The main idea is that photon and electron radiation fields are described *exactly* by two coupled (integro differential) linear Boltzmann transport equations (LBTE), which can be solved by numerical techniques and some approximations. In the second step, the absorbed dose $D(\underline{r})$ is calculated from the electron fluence obtained. A meaningful discussion requires a considerable amount of definitions and is beyond the scope of the one semester course this book is based on. For further reading, Chapter 4 in [47] is recommended. First results (Varian AcurosXB algorithm) indicate that this class of algorithms can yield an accuracy that is close to Monte Carlo methods, but is considerably faster [31, 69].

5.2.3 Principles of ionization dosimetry

For the measurement of absorbed dose, several effects can be exploited depending on the absorbing material, for example, generation of heat, ionization, luminescence, chemical reactions and biological effects (e.g., DNA damage or chromosome aberrations). The corresponding theory and experimental aspects, including the technical design of detectors, would easily fill entire books on its own. Here, we give a brief

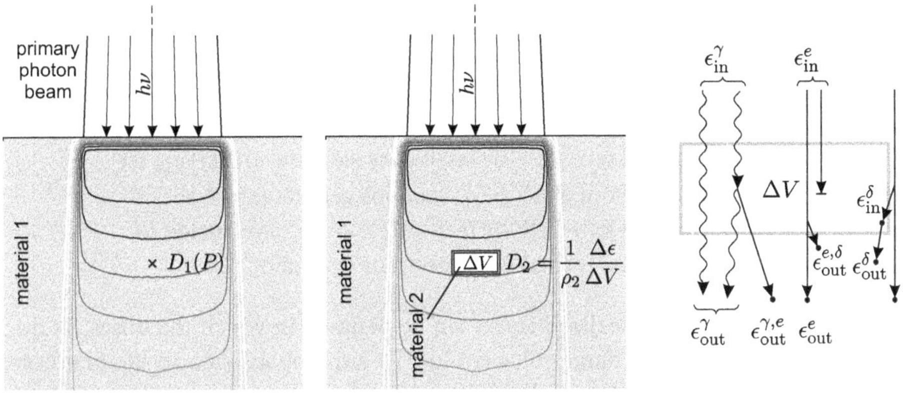

Fig. 5.17: *Left*: Illustration of the fundamental problem of dosimetry. The absorbed dose D_1 at point P in some irradiated material 1, e.g., in a water phantom irradiated by an external photon beam, is to be determined. To this end, some probe like an air filled ionization chamber is placed at P, yielding the dose D_2 to the probe material 2 averaged over the detecting volume ΔV. How and under which conditions can D_1 be determined from D_2? *Right*: Sketch of the different ways of energy transport into and out of ΔV.

summary of some fundamental dosimetric principles that illustrate the application of some concepts of the coupled transport of photons and electrons we have discussed so far in this chapter. Reference textbooks on dosimetry are [6, 34, 37], and further helpful comments for the preparation of this section were found in [75].

Consider the example sketched in Figure 5.17 (left). The goal is to measure the dose D_1 generated by an external photon beam in some point P of an absorbing material 1. To this end, a probe is placed in P, and the dose D_2 averaged over the active volume ΔV of the probe composed of material 2 is measured. Materials 1 and 2 differ in atomic composition and density, thus the components of the mass attenuation coefficient μ/ρ of photons and mass stopping power S/ρ of charged particles are usually different in materials 1 and 2 (plus the wall of the probe surrounding ΔV). Hence, the fundamental problem of dosimetry is: *How is the absorbed dose in a probe inserted in a medium related to that in the medium itself?* (p. 232 in [6]). In addition, placing the probe in P may disturb the primary photon and electron fluences in P. How does this have to be taken into account, e.g., in the construction of an ionization chamber?

We address these questions for the case of external photon irradiation. A useful relation for the dose in a volume ΔV is obtained by considering all possibilities of radiation transport into and out of the volume as sketched in Figure 5.17 (right) [37, 48]:

$$D = \frac{1}{\rho}\frac{1}{\Delta V}\left(\epsilon^{\gamma}_{\mathrm{in}} - \epsilon^{\gamma}_{\mathrm{out}} - \epsilon^{\gamma,\mathrm{e}}_{\mathrm{out}} + \epsilon^{\mathrm{e}}_{\mathrm{in}} - \epsilon^{\mathrm{e}}_{\mathrm{out}} - \epsilon^{\mathrm{e},\delta}_{\mathrm{out}} + \epsilon^{\delta}_{\mathrm{in}} - \epsilon^{\delta}_{\mathrm{out}}\right) = \frac{1}{\rho}\frac{\Delta\epsilon}{\Delta V}\,. \tag{5.64}$$

The different terms are (the bar over each ϵ indicating the expectation value has been dropped):

$\epsilon^{\gamma}_{\text{in}}$: energy transported into ΔV by photons
$\epsilon^{\gamma}_{\text{out}}$: energy transported out of ΔV by photons
$\epsilon^{\gamma,\text{e}}_{\text{out}}$: energy transported out of ΔV by secondary electrons generated in ΔV
$\epsilon^{\text{e}}_{\text{in}}$: energy transported into ΔV by secondary electrons generated outside ΔV
$\epsilon^{\text{e}}_{\text{out}}$: energy transported out of ΔV by secondary electrons traversing ΔV
$\epsilon^{\text{e},\delta}_{\text{out}}$: energy transported out of ΔV by δ electrons generated in ΔV
$\epsilon^{\delta}_{\text{in}}$: energy transported into ΔV by δ electrons generated outside ΔV
$\epsilon^{\delta}_{\text{out}}$: energy transported out of ΔV by δ electrons traversing ΔV

From this relation, two idealized cases can be defined, in which the doses D_1 and D_2 (as measured by the charge collected in ΔV) are proportional. In the first case, **charged particle equilibrium** (CPE), all electron terms cancel, and the net energy deposition in ΔV is related solely to photon interactions. In the other case, **Bragg–Gray conditions** (BGC), only secondary electrons determine the absorbed dose in ΔV. These conditions are of fundamental importance in dosimetry, and determine how ionization chambers need to be constructed.

Charged particle equilibrium (CPE)

In the case of primary photon radiation, one also speaks of electronic equilibrium, since pair production is usually negligible and electrons are the only charged particles of relevance. In this case of CPE, all electron terms in (5.64) cancel, and the *net* energy deposition in ΔV is equivalent to the difference in energies transported into and out of ΔV as photons:

$$D = \frac{1}{\rho}\frac{1}{\Delta V}\left(\epsilon^{\gamma}_{\text{in}} - \epsilon^{\gamma}_{\text{out}}\right) \tag{5.65}$$

Now we can use some results of Section 5.1.2: The difference in photon energy must be the energy transferred from photons to secondary electrons in ΔV (the reduction of the primary photon fluence), the kerma K, reduced by the fraction g, which is re-irradiated by bremsstrahlung production. Thus, in the case of CPE, the dose is given by

$$D \overset{\text{CPE}}{=} K(1-g) = K_{\text{col}} = \left(\frac{\mu_{\text{en}}}{\rho}\right)\Psi \tag{5.66}$$

where the collision kerma K_{col} is directly proportional to the energy fluence $\Psi = \Phi E$ of the primary photons. Therefore, in the case of CPE and if the primary photon fluence is not perturbed by the probe ($\Psi = \text{const.}$), equation (5.66) yields

$$\frac{D_1}{D_2} = \frac{\left(\frac{\mu_{\text{en}}}{\rho}\right)_1}{\left(\frac{\mu_{\text{en}}}{\rho}\right)_2} . \tag{5.67}$$

The doses in two materials generated by identical photon fields can be converted into each other using the ratio of their mass energy absorption coefficients. In the case of polyenergetic photon beams with spectral distribution $\Psi_E(E)$, one has to use average values

$$\left(\frac{\bar{\mu}_{en}}{\rho}\right) = \frac{\int_0^{\max(E)} \left(\frac{\mu_{en}(E)}{\rho}\right) \Psi_E(E)\, dE}{\int_0^{\max(E)} \Psi_E(E)\, dE} \quad \text{where} \quad \int_0^{\max(E)} \Psi_E(E)\, dE = \Psi\,. \tag{5.68}$$

Now we must explore under which circumstances CPE can exist. Let a broad parallel photon beam impinge onto a homogeneous medium. We neglect the δ electron terms and ask for existence of secondary electron equilibrium

$$\epsilon_{in}^{e} - \epsilon_{out}^{e} - \epsilon_{out}^{\gamma,e} = 0\,. \tag{5.69}$$

In lateral direction (i.e., perpendicular to the photon beam, Figure 5.18 left), each secondary electron leaving region 1 will, on average, be replaced by another one with the same energy entering region 1, since it lies within a homogeneous photon beam and is sufficiently far away from the borders. This obviously does not hold for region 2, where an overweight of electrons entering from the direction towards the beam center exists.

In longitudinal direction (i.e., in the direction of the photon beam), CPE can exist beyond a depth that corresponds to the (projected) secondary electron range R, provided that $\mu = 0$. This is sketched in Figure 5.18 (center) for a simplified case where

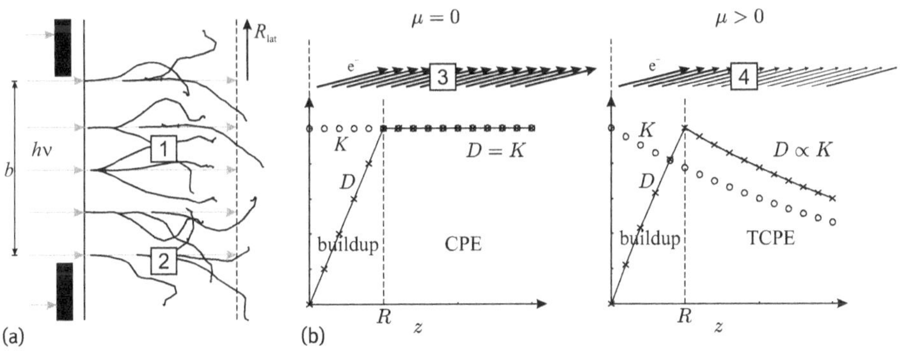

Fig. 5.18: Illustration of the concept of CPE for the case of an external photon beam. (a) In lateral direction, CPE is given in a volume ΔV if it is located within the beam of width b and further away from the outer beam margin than the maximum lateral electron range R_{lat}, i.e., when increasing the beam width no longer increases the dose in ΔV. This is true for volume 1, but not for volume 2. (b) In beam direction, true CPE is given in a depth z larger than the secondary electron range $\geq R$ in beam direction if photon attenuation is negligible ($\mu = 0$). In this case, dose is equal to (collision) kerma. However, if $R \ll \mu^{-1}$ in the absorbing material, $D \approx K$ is still a valid approximation (transient charged particle equilibrium TCPE).

all secondary electrons move in forward direction (the arrows are tilted due to space limitations), have the same initial energy and range and lose their energy with constant rate and without bremsstrahlung emission along their track. Without photon attenuation, the photon fluence Φ and hence the kerma K are constant. The dose $D(z)$ (which is proportional to the sum of the thicknesses of the arrows crossing z) reaches its maximum at a depth corresponding to R (which is known as the buildup effect) and is constant and equal to the kerma K beyond this depth.

Since $\mu > 0$ in any real experiment, true CPE cannot exist. The number of secondary electrons released and thus K decrease exponentially with d, as indicated by the thickness of the arrows in Figure 5.18 (right). However, at a depth beyond R, dose and kerma are at least proportional to each other, and their quotient is close to unity if $R \ll \mu^{-1}$. This situation is called transient charged particle equilibrium (TCPE)

$$D \overset{\text{TCPE}}{=} \beta K(1-g) \tag{5.70}$$

where β is called the energy transport factor. For ^{60}Co radiation (1.17 and 1.33 MeV), $\beta(1-g) \simeq 1.002$ for water, air and graphite, and the value approaches unity for lower energies [35]. Hence, equation (5.66) is still a very good approximation.

The conditions for (T)CPE are not fulfilled, e.g., in the buildup region, at the edges of a photon field, in strongly divergent beams and close to inhomogeneities of the absorber, e.g., at the interface between soft tissue and bone or lung. In addition, inhomogeneous electric or magnetic fields must not influence the secondary electron tracks. Dose measurements relying on the assumption of CPE and equation (5.67) in such regions can yield incorrect results. At the interface between two materials that differ in atomic composition and density, the mass energy absorption coefficient (μ_{en}/ρ) and, thus, the collision kerma $K_{\text{col}} = (\mu_{\text{en}}/\rho)\,\Psi$ usually displays a discontinuity, which perturbs (T)CPE within the secondary electron range R and would affect the result of a dose measurement that relies on the assumption of CPE and equation (5.67). This has to be taken into account in the design of ionization chambers intended for measurement under conditions of CPE (e.g. [6, 37, 48]).

Bragg–Gray conditions (BCG):

In the other set of idealized conditions, photon interactions inside ΔV are negligible, and only the secondary electrons entering and leaving ΔV determine the net energy deposition (δ electrons are neglected again),

$$D = \frac{1}{\rho}\frac{1}{\Delta V}\left(\epsilon^{\text{e}}_{\text{in}} - \epsilon^{\text{e}}_{\text{out}}\right) . \tag{5.71}$$

The rate of energy loss per unit path length of electrons to their local environment is the electronic stopping power S_{el} (equation (5.15) and (5.45)). For a single electron that traverses a distance Δx with almost constant energy E, the energy loss is simply $S_{\text{el}}(E)\Delta x$. For a fluence $\Phi = N/A$ of monoenergetic electrons, the resulting dose in

$\Delta V = A\Delta x$ is

$$D = \frac{1}{\rho}\frac{S_{el}\Phi A\,\Delta x}{A\,\Delta x} = \left(\frac{S_{el}}{\rho}\right)\Phi\,. \tag{5.72}$$

Since secondary electrons have a continuous fluence distribution $\Phi_E(E)$, the dose is given by

$$D = \left(\frac{\bar{S}_{el}}{\rho}\right)\Phi \tag{5.73}$$

with the energy averaged mass electronic stopping power and total fluence

$$\left(\frac{\bar{S}_{el}}{\rho}\right) = \frac{1}{\Phi}\int_0^{\max(E)}\left(\frac{S_{el}(E)}{\rho}\right)\Phi_E(E)\,dE \quad \text{and} \quad \Phi = \int_0^{\max(E)}\Phi_E(E)\,dE\,. \tag{5.74}$$

In analogy to equation (5.67), the doses generated by an identical electron fluence Φ in two different materials are thus related by their mass electronic stopping powers according to the Bragg–Gray relation

$$\frac{D_1}{D_2} = \frac{\left(\frac{\bar{S}_{el}}{\rho}\right)_1}{\left(\frac{\bar{S}_{el}}{\rho}\right)_2}\,. \tag{5.75}$$

In contrast to detectors designed for CPE, which are photon based, detectors designed for BGC can also be used to measure dose from primary electron radiation, since equation (5.75) applies both to secondary electrons generated by a primary photon field and to primary electron beams. The BGC required for the application of these equations can be formulated as follows [35, 37]:

1. The fluence of secondary electrons and their spectral and angular distribution are not changed by the probe.
2. The energy deposited in ΔV by secondary electrons released by photons inside ΔV is negligible compared to the total energy deposited inside ΔV.
3. The spectral distribution of all electrons inside ΔV is constant over ΔV.

The two first conditions imply that the extent of the detector must be small compared to the mean range of secondary electrons. Therefore, plane parallel ionization chambers with thin walls and small extent in beam direction but considerably larger lateral extent are often used. Condition 3 implies that the (energy averaged) mass electronic stopping power $(\bar{S}_{el}/\rho)$ is constant inside ΔV. Hence, BGC require that the energy of secondary electrons is not too small; a typical lower limit given in the literature is about 600 keV. In addition, a thin graphite layer with $(\rho d)_{wall} \geq 2\text{mg/cm}^2$ is used to establish δ electron equilibrium [35, 37].

Ionization chamber dosimetry

A typical example from clinical practice is the measurement of dose to water in a water phantom by an air filled ionization chamber, such as in the commissioning and regular checks of linear accelerators. If a charge per mass dQ/dm_a is generated in the air filled cavity of the chamber, the corresponding dose to water is obtained by

$$D_w = \frac{\left(\frac{\bar{S}_e}{\rho}\right)_w}{\left(\frac{\bar{S}_e}{\rho}\right)_a} D_a = \frac{\left(\frac{\bar{S}_e}{\rho}\right)_w}{\left(\frac{\bar{S}_e}{\rho}\right)_a} \left(\frac{\bar{W}}{e}\right) \frac{dQ}{dm_a} \tag{5.76}$$

if a Bragg–Gray detector is used. If the energy averaged mass electronic stopping powers and the mean energy $(\bar{W}/e) = 33.97$ eV required to produce an ion pair in air, as well as the chamber volume and air pressure are known, it is possible to obtain the dose by measuring the generated charge Q. The corresponding currents are very small, usually on the order of 10^{-12} A.

In practice, each ionization chamber is calibrated in a reference laboratory using standardized irradiation conditions with a known radiation quality Q (for example, ^{60}Co). In subsequent measurements by the end user, dose is determined according to

$$D_w(Q) = N_w(Q) \cdot M \tag{5.77}$$

where $N_w(Q)$ denotes the calibration factor determined by the reference laboratory and M the readout signal. If the conditions in the actual measurement differ from the calibration conditions (e.g., a different energy spectrum), this has to be taken into account by a number of correction factors on the right hand side of equation (5.77).

5.3 Radiobiology

For radiotherapy and radiation protection, it is important to know about the effects of ionizing radiation on living organisms. This in itself is an interesting topic of molecular biophysics and cell biology. Relevant biological functions such as cell cycle control, DNA replication and repair are well treated by textbooks on cell biology, e.g., [4, 42]. Our main references for the following brief account of radiobiology are the books by Hall and Kelsey [25, 32].

5.3.1 The cell cycle

Most cells in our body do not exist over our entire lifetime, but are eliminated and replaced with a characteristic period ranging from a few days to several years. Typical turnover times for many cell types are given in [50]. According to our current under-

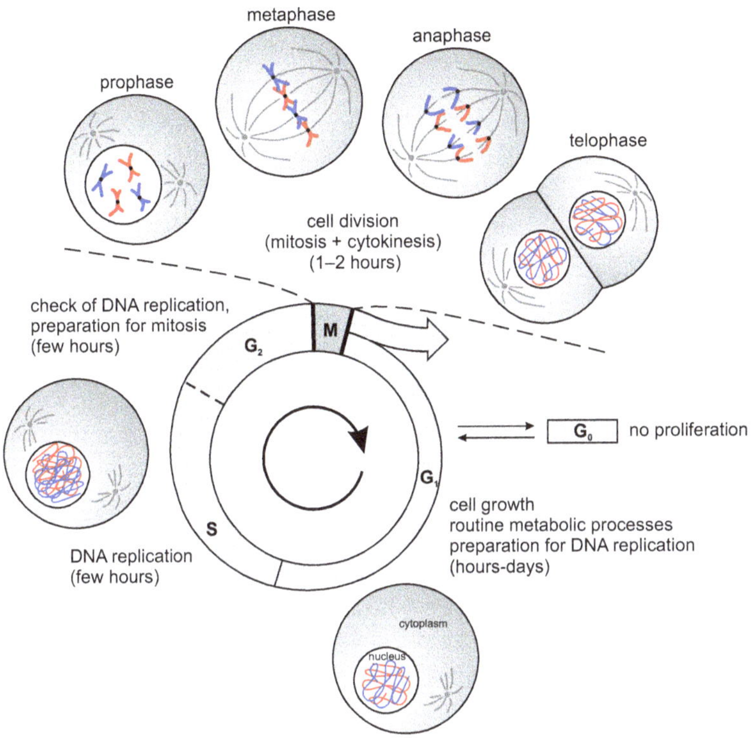

Fig. 5.19: Schematic of the cell cycle. The interphase comprises the G_1, S and G_2 phases, in which the cell grows, replicates its DNA and prepares for division. Subsequently, in the M phase, the cell divides into two genetically identical daughter cells. For different mammalian cell types, the length of the G_1 phase can vary considerably, while the duration of the other phases show only little variation. Differences among cell cycle times are, therefore, primarily due to the length of the G_1 phase. Typical durations for human cells are indicated. Cells that do not show any proliferative activity are said to be in the G_0 phase. In cancerous cells, the molecular mechanisms controlling the cell cycle are insufficient, leading to unrestricted and autonomous proliferation. Radiosensitivity is highest during the M phase, followed by the G_2 phase. Therefore, cells that proliferate quickly (e.g., in the skin, mucosa or bone marrow) are typically more sensitive to radiation, which contributes to some of the side effects of radiotherapy.

standing, each tissue contains a small fraction of stem cells, which are able to undergo an unlimited number of cell divisions and provide replacement for old, worn out cells. Cell division proceeds by the so called **cell cycle**, a tightly orchestrated sequence of molecular processes (Figure 5.19). Five phases are distinguished:

- G_1: After cell division, each daughter cell is concerned with protein synthesis and growth to perform its specific task and prepare for replication of the DNA, the carrier of all genetic information.
- S: DNA replication leads to chromosomes consisting of two sister chromatids with identical base sequence that are held together in a region called the centromere.

- G_2: Proteins that are required to perform the subsequent cell division are synthesized. Highly complex mechanisms check if complete DNA replication has been achieved and repair possible errors.
- *M*: It is only now that the cell is ready for division. The orderly distribution of the sister chromatids to opposite poles of the cell and formation of two nuclei, each one containing the full genetic information, is known as mitosis. It can be further subdivided into prophase, metaphase, anaphase and telophase. Only in the metaphase, the chromosomes are fully condensed and exhibit (in the light microscope) the typical X like shape often used in illustrations. Mitosis is usually followed by cytokinesis, the division of the cell body, resulting in two separate and genetically identical daughter cells.[15]
- G_0: Cells can leave the cell cycle and enter a resting or quiescent phase, in which they simply pursue their specific tasks, and no preparations for further division are being made. In fact, this is the state many cell types are usually found in. Upon certain stimuli, they can re-enter the cell cycle and start to proliferate again.

In normal and healthy tissue, the cell cycle is tightly controlled to maintain the cell number at a constant level adjusted to the requirements of the organism and to the supply with nutrients. The replacement of old cells by new ones is known as cell turnover in most tissues. For example, the intestinal mucosa or the bone marrow contain cells that have a mean lifetime of only a few days. In contrast, e.g., neurons are considered postmitotic and do not divide any further. Very importantly, the radiation sensitivity of cells shows some characteristic variation with the phases of the cell cycle. This has important implications for radiotherapy, which will be discussed in Section 5.4.

5.3.2 Ionizing radiation and biological organisms

The events in a biological organism that follow irradiation with high energy photons or charged particles can be divided into three subsequent stages:

1. The physical stage (time scale about 10^{-16} to 10^{-13} s) comprises the excitation and ionization events in which energy is transferred from the incident radiation to the molecules of the biological organism. In the case of photons, this can occur by the primary interactions described in Section 3.2, in the case of charged particles by collisions with atomic electrons as described in Section 5.1. As noted above, the energy of a single high energy photon with an energy of ~ 10 keV to 10 MeV as used in radiological diagnostics or radiotherapy is several orders of magnitude higher than the typical excitation and ionization energies of biomolecules.

15 In some tissues like skeletal muscle, mitosis can occur without cytokinesis, resulting in a so called syncytium.

2. In the chemical stage (~ 10^{-13} to 10^{-2} s), the absorbed energy leads to changes in the structure of the four classes of biomolecules – nucleic acids (DNA and RNA), proteins, carbohydrates and lipids. The result can be ionization, formation of radicals, disruption of covalent or hydrogen bonds, loss of functional groups, etc. However, since approximately 80% of a cell's content is water, the most frequent event is the radiolysis of water:

$$\begin{aligned} H_2O &\xrightarrow{\text{ion. rad.}} H\cdot + OH\cdot \\ H_2O &\xrightarrow{\text{ion. rad.}} H_2O^+ + e^-_{aq} \end{aligned} \tag{5.78}$$

This is immediately followed by further reactions, like

$$\begin{aligned} H_2O^+ &\longrightarrow H^+ + OH\cdot \\ OH\cdot + OH\cdot &\longrightarrow H_2O_2 \\ O_2 + e^-_{aq} &\longrightarrow O_2^- \\ O_2^- + H_2O &\longrightarrow H_2O_2 + 2OH^- + O_2 \end{aligned} \tag{5.79}$$

which produce so called **reactive oxygen species** (ROS). The hydroxyl radical $OH\cdot$ with an unpaired electron, hydrogen peroxide H_2O_2 and the superoxide anion O_2^- are chemically highly reactive and can cause numerous modifications of biomolecules, including several types of DNA damage discussed below.[16]

In irradiation with photons and electrons, the **indirect action of radiation** via radicals and ROS created from water and molecular oxygen accounts for about 70% of radiation induced cell death, whereas with **direct action of radiation**, the immediate damage of DNA by the primary radiation or secondary electrons dominates in the case of massive charged particles (protons, α-particles, heavy ions). The two different mechanisms are sketched in Figure 5.20.

3. In the biological stage (10^{-2} s to years), the chemical modifications of biomolecules result in a cellular response. This can be an arrest of the cell cycle and DNA repair processes, changes in cellular metabolism or, if the damage is severe, cell death. If a proliferating cell survives despite damage to its DNA, the DNA mutations are passed on to subsequent cell generations. When a certain number of mutations has accumulated, which may last many years, this can finally lead to the transformation into a cancer cell (carcinogenesis). Mutations in germ cells are passed on to the progeny and eventually manifest themselves generations later.

---

16 Remarkably, ROS are also a byproduct of normal aerobic cell metabolism. It is estimated that about 50 000 DNA lesions *per cell* are induced by ROS every day [43]. In addition to the natural level of ionizing radiation, this is another reason why highly efficient DNA repair mechanisms are indispensable.

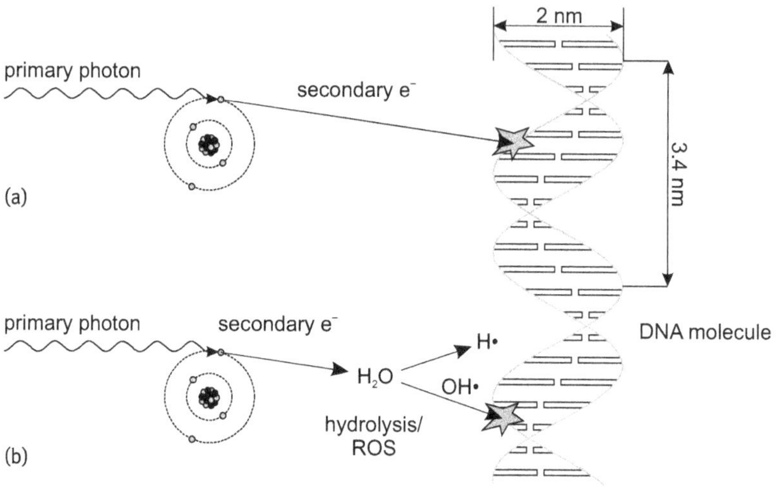

Fig. 5.20: Direct and indirect action of radiation. A secondary electron released by an energy rich photon, for example, can (a) directly cause DNA damage or (b) create ROS, here a hydroxyl radical OH·, which subsequently reacts with DNA and thus indirectly leads to DNA damage.

5.3.3 Radiation damage to DNA

For humans, a total body dose of 4 Gy (that is, *each* mass element of the body receives a dose of about 4 Gy) is lethal with a probability of 50%. Using $\Delta Q = cm\Delta T$ with the specific heat capacity $c = 4.18\,\text{kJ/(kg}\cdot\text{K)}$ of water, one can easily estimate that this amount of energy can lead to a rise in temperature of only $\Delta T \approx 0.001\,°\text{C}$. Thus, it is immediately clear that biological effects of ionizing radiation like carcinogenesis or radiation-induced cell death must be due to damage to specific molecular targets, rather than to the total amount of energy deposited inside a biological system.

Radiation damage to proteins, carbohydrates and lipids, as well as to RNA, does usually not constitute a severe problem (at least with moderate radiation doses), since all molecules belonging to these three groups are present in multiple copies in each cell, and old or damaged molecules are permanently replaced by freshly synthesized ones. Only at very high radiation doses can cell death be caused by the loss of enzymatic activity of too many proteins and severe damage to cellular membranes.

This is fundamentally different for DNA, which carries the genetic information. In human cells, the total genetic code is distributed over 46 DNA molecules called chromosomes; 23 inherited from the mother and 23 from the father, and concentrated in the cell nucleus (Figure 5.21). The base sequence (about $3 \cdot 10^9$ bases in total) encodes for the amino acid sequence and thus the structure of all proteins, the molecules that perform all sorts of tasks in our body. In a nonproliferating cell in the G_0 phase, each DNA sequence that encodes for a certain protein (a gene) is available only twice, as one maternal and one paternal copy.

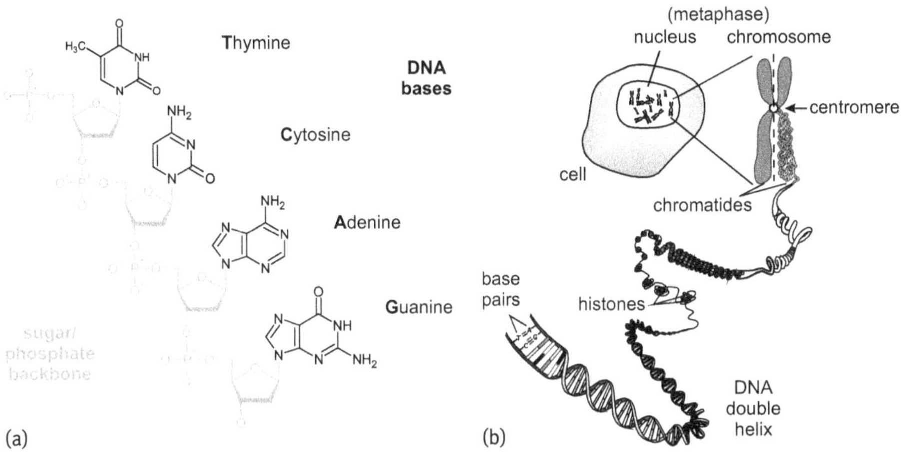

Fig. 5.21: Structure of DNA and its compaction to chromosomes in the eukaryotic cell nucleus. *(a)* A DNA single strand molecule consists of a backbone of deoxyribose molecules linked by phosphate groups. To each furanose ring (consisting of four carbon atoms and one oxygen atom), one of the nucleobases adenine (A), guanine (G), cytosine (C) and thymine (T) is covalently bound. A DNA double strand is formed by the specific pairing of A and T by two and by C and G by three hydrogen bonds, respectively. *(b)* Several levels of compaction are required to fit the entire DNA ($\sim 3 \cdot 10^9$ base pairs for humans, total length $\sim$ 2 m) into a cell nucleus with a typical diameter of $\sim$ 5 µm. Full DNA condensation to chromosomes that are visible in the light microscope only occurs during cell division in the metaphase. Figure source: https://commons.wikimedia.org/wiki/File:Chromosom.svg?uselang=de#file.

Different types of DNA damage can be caused by ionizing radiation (Figure 5.22):
- base or sugar (deoxyribose) damage
- covalent cross links between two DNA strands or DNA and proteins
- single strand breaks (SSBs)
- double strand breaks (DSBs)
- clustered (or bulky) lesions, i.e., combinations of the above types in close proximity

It has been estimated that an x-ray dose of 1 Gy causes about 1000 base damages, 1000 SSBs and 20–40 DSBs in a typical, sufficiently oxygenated mammalian cell [25, 43]. Several DNA repair mechanisms, each one specialized for a certain type of damage, have evolved to cope with the inevitable natural levels of ionizing radiation from cosmic or terrestrial sources (see also Section 5.5.3) as well as with endogenously produced ROS and thus warrant the integrity of the genetic material. In general, DNA repair mechanisms are extremely efficient.[17] For an overview, see, e.g., [46, 62, 74]. In brief, base damages and SSBs can be repaired more reliably and with fewer errors

17 In 2015, the Nobel Prize in Chemistry was awarded for elucidation of these DNA repair mechanisms.

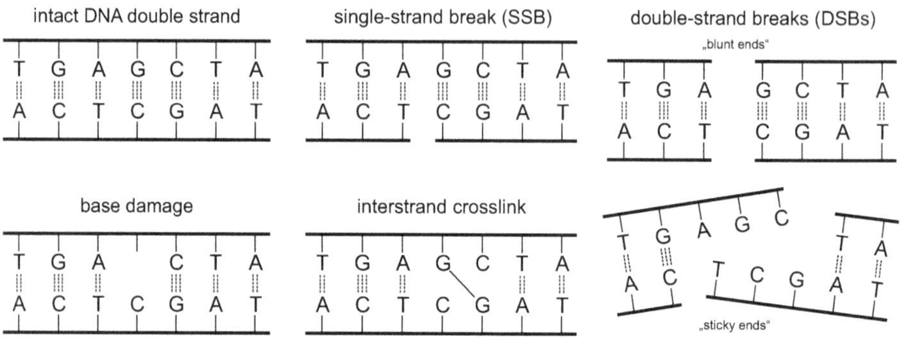

Fig. 5.22: Schematic representation of several types of damage that can be inflicted on DNA by ionizing radiation, either by direct hits or indirectly by ROS. A number of highly efficient repair mechanisms exist. The biological effects of ionizing radiation are attributed to DSBs.

than DSBs, since the intact complementary DNA strand can be used as a template. Therefore, it is generally accepted that DSBs are the principal lesions leading to the following radiation induced biological effects:

- **Genomic instability and carcinogenesis**: If repair mechanisms fail (but the cell survives), radiation damage can modify the DNA base sequence and, thus, cause genetic mutations. A single mutation in a cell does not usually have a great effect. However, since mutations are passed on to all daughter cells, they can accumulate over the lifetime of an organism. With regard to cancer, mutations in DNA regions that encode for proteins involved in cell cycle control or DNA repair are critical. If sufficiently many of these mutations (the exact number can vary from a few up to more than 100, depending on the type of cancer) have accumulated in a single (stem) cell, the result can be unrestricted cell proliferation and invasive growth – the cell has transformed into a malignant cancer cell. For an overview on the 'hallmarks' of cancer cells, see e.g., the famous review paper by Hanahan and Weinberg [26].
- **Cell death**: Ionizing radiation can kill cells. Different pathways are distinguished:
  1. *Mitotic* or *clonogenic* cell death: If DNA repair mechanisms fail, irradiation can produce chromosomal aberrations (Figure 5.23), which lead to errors in the next or one of the subsequent cell divisions after irradiation and thus trigger cell death. This is considered to be the most important form of cell death from irradiation.
  2. *Apoptosis*: This 'programmed cell death' warrants homeostasis of the cell number in normal tissue. It can also be triggered by failure of DNA repair mechanisms; a cell 'senses' that something has seriously gone wrong with its DNA. As a safeguard to protect the entire organism, it undergoes a defined sequence of morphological changes and finally decays into smaller, membrane bound apoptotic bodies, which are eliminated by phagocytosis. At all

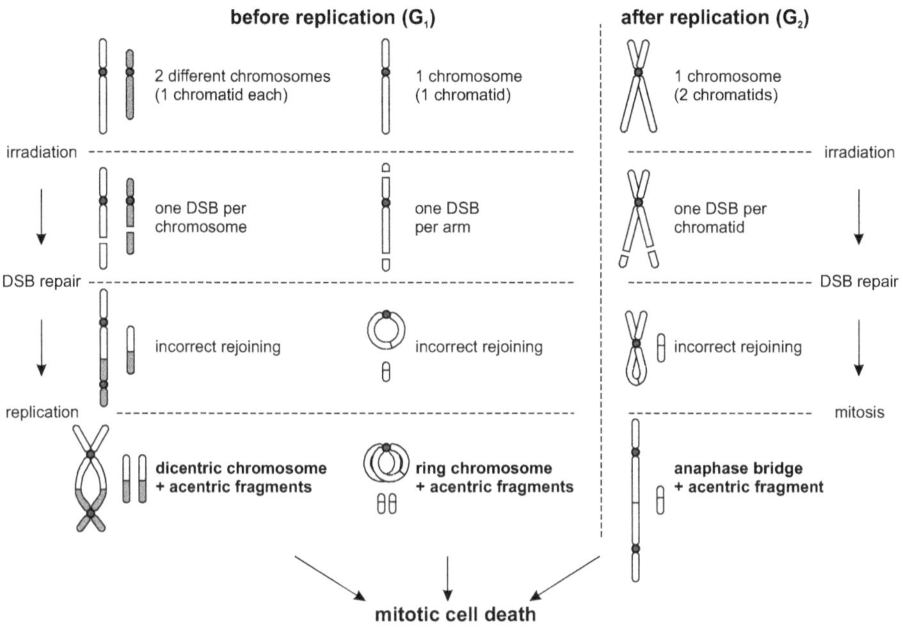

Fig. 5.23: DNA repair mechanisms are, in general, extremely efficient. However, especially DSB repair is sometimes erroneous. Three examples of resulting chromosome aberrations that lead to mitotic cell death are dicentric chromosomes, ring chromosomes and anaphase bridges. Acentric chromosome fragments without a centromere are not distributed evenly between the two daughter cells during mitosis. Based on [25], p. 27.

stages of this process, the cell's content remains separated from the extracellular space. Importantly, one of the main characteristics of cancer cells is their increased resistance to pro-apoptotic signals, which promotes tumor growth [26].

3. *Necrosis*: High radiation doses can cause a massive loss of the function of many proteins and damage membranes, leading to immediate cell death, including loss of cellular integrity. In contrast to apoptosis, cell content leaks into the extracellular space and causes an inflammatory reaction.

In summary, ionizing radiation is a 'double edged sword' [43] and can both cause and cure cancer. On the one hand, it can induce genetic mutations that finally lead to unrestricted cell division and tumor growth. On the other hand, large radiation doses targeted at tumor cells can kill these by inducing massive DNA damage, especially DSBs. The former effect is the reason for measures of radiation protection, while the latter is applied in radiotherapy.

5.3.4 Linear energy transfer and relative biological effectiveness

On the microscopic scale, the energy imparted to a biological system by ionizing radiation is not deposited uniformly. Energy transfer takes place in discrete ionization events, which are associated to the tracks of charged particles. This is illustrated in Figure 5.24: In two volume elements of equal size (e.g., two cell nuclei), the same absorbed dose D (total energy deposited in the volume divided by its mass) is deposited either by several ionizing particle tracks (left), or by a single track of a different particle that produces considerably more ion pairs per unit path length (right). The field of **microdosimetry** is concerned with theoretical and experimental methods to quantify such dose inhomogeneities on the microscopic level [18, 60].

The absorbed dose D alone is, therefore, not sufficient to characterize the effect of different types of ionizing radiation on living organisms. The probability of DSB production increases if ionization events occur in a spatially and temporally correlated pattern, and thus will be quite different for the two cases shown in Figure 5.24. If the mean distance between two subsequent ionizations along a track are close to the diameter of a DNA molecule, DSBs are produced more efficiently, and the radiation has a more pronounced biological effect (i.e., a higher rate of cell killing). This is quantified by the **relative biological effectiveness** (RBE) defined as the ratio of the dose of a reference spectrum (often 250 kV x-rays) and the dose of the test radiation that produces the same biological effect (e.g., the same survival probability of cells),

$$\mathrm{RBE} := \left.\frac{D\ (\text{reference radiation})}{D\ (\text{test radiation})}\right|_{\text{isoeffect}} \tag{5.80}$$

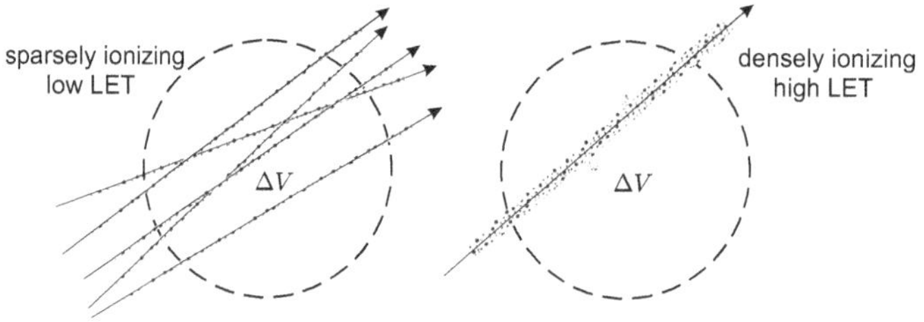

Fig. 5.24: Microscopic energy deposition pattern of two different types of charged particles. Each black dot symbolizes an ionization event. In both volume elements ΔV, the same absorbed dose $D = \Delta\bar{\epsilon}/\rho\Delta V$ is generated. On the *left*, D results from several incident charged particles with low ionization density along their track, e.g., secondary electrons set in motion by primary photon radiation. On the *right*, an identical value D is created by a single charged particle that produces considerably more ionization events per unit path length, e.g., a proton or carbon ion. The probability of the generation of DNA DSBs does not only depend on the value of D, but also on the spatial distribution of ionization events on the microscopic scale, since one DSB requires two ionization events in close proximity and within a sufficiently short period of time.

The higher RBE, the more efficient the test radiation in production of DSBs compared to the reference radiation.

A frequently used quantity to characterize the microscopic pattern of dose deposition is the so called **linear energy transfer** (LET), which is usually given in units of keV/µm. It is defined as the fraction of the electronic stopping power[18] S_{el} due to ionization events in which the kinetic energy transferred to an atomic electron of the medium is below a cutoff value Δ [21]:

$$\mathrm{LET}_{\Delta} = S_{el,\Delta} = S_{el} - \frac{dE_{k.e.<\Delta}}{dl} \tag{5.81}$$

This way, energy rich knock on electrons that lead to energy dissipation away from the primary particle track are excluded. For given Δ, a high LET implies that ionization events resulting from a primary particle are 'densely concentrated' along its track. The choice of Δ depends on the application. If no cutoff is employed, the unrestricted LET

$$\mathrm{LET}_{\infty} = S_{el} \tag{5.82}$$

is nothing else but the electronic stopping power and includes all collisions with atomic electrons regardless of the transferred energy.[19] LET_{Δ} can be interpreted as a measure of how 'dense' the ionization track of a given type of radiation is. In radiobiology, the LET concept is often used to coarsely distinguish between two categories of ionizing radiation:

- **Densely ionizing radiation**: Two subsequent interaction events of a single incident particle are often close enough to induce DNA DSBs. Protons and heavy ions (i.e., α particles and beyond) usually fall into this category, especially towards the end of their track in the region of the Bragg peak. Also neutrons (which interact with atomic nuclei and produce recoil protons and eventually heavier spallation products) belong to this class.
- **Sparsely ionizing radiation**: The typical distance between two interaction events is much larger than the diameter of the DNA molecule. Creation of DSBs usually requires that two different particle tracks happen to cross the same DNA region. This category comprises photons and electrons.

Since it is derived from S_{el}, also LET_{Δ} is an average quantity by definition due to the stochastic nature of interaction events. It depends on a particle's kinetic energy E and increases when a particle slows down towards the end of its track (cf., Figure 5.5).

18 When the term LET is used to characterize photon radiation, it actually refers to the secondary electrons released by the photons. The same applies to the case of neutrons, where LET refers to recoil protons released in collisions with atomic nuclei.

19 The symbol ∞ must not be taken literally, it is used to indicate that no energy transfers are excluded. The maximum energy transfer is, of course, limited.

Therefore, the LET of a given type of radiation (especially if it contains a broad spectrum of initial particle or photon energies) is fully characterized only if its spectral distribution is known [17]. However, values averaged over the track of particles with known initial energy (or initial energy spectrum) are usually used as an approximation. Frequently, LET is simply used synonymously with electronic stopping power according to equation (5.82). Therefore, LET usually involves a considerable amount of approximation, and the use of the term LET without further specification can be misleading. This has evoked some criticism of the LET concept (p. 104–106 in [25]). However, keeping these shortcomings in mind, there are two major applications [17]:

- If some pairs of LET and corresponding RBE for a given type of radiation are known, estimates for LET and RBE at other energies (or spectra) of this type of radiation can be made by interpolation and using RBE vs. LET curves, as shown in Figure 5.25. It has been demonstrated that the maximum RBE for mammalian cells is reached around LET ≈ 100 keV/μm. This 'optimal' LET roughly corresponds to a mean distance between ionization events along a charged particle's track which is on the order of the diameter of a DNA molecule, i.e., the energy is spent most efficiently for DSB production. At still higher LET, energy is deposited inefficiently in ionization events that are too close together, and RBE decreases again.
- In radiation protection (Section 5.5), LET is used to determine quality factors for the risk assessment of different types of radiation.

type/energy of radiation	**⟨$LET_{0.1\,keV}$⟩ [keV/μm]**
50 kV x-ray spectrum	6.3
200 kV x-ray spectrum	1.7
^{60}Co γ radiation	0.22
22 MV x-ray spectrum	0.19
18.6 keV electrons	4.7
2 MeV electrons	0.2
5.3 MeV α particles	43

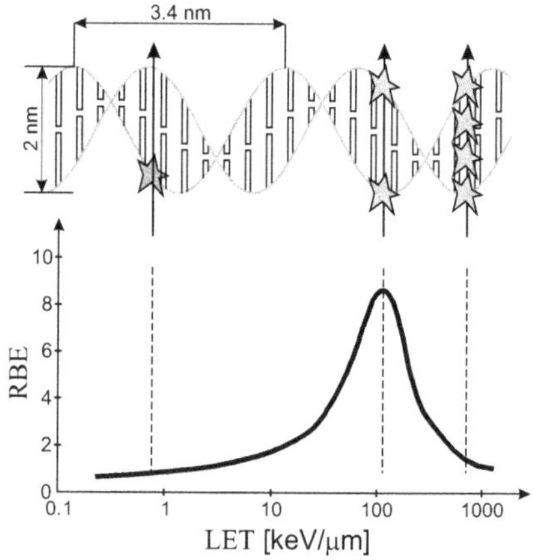

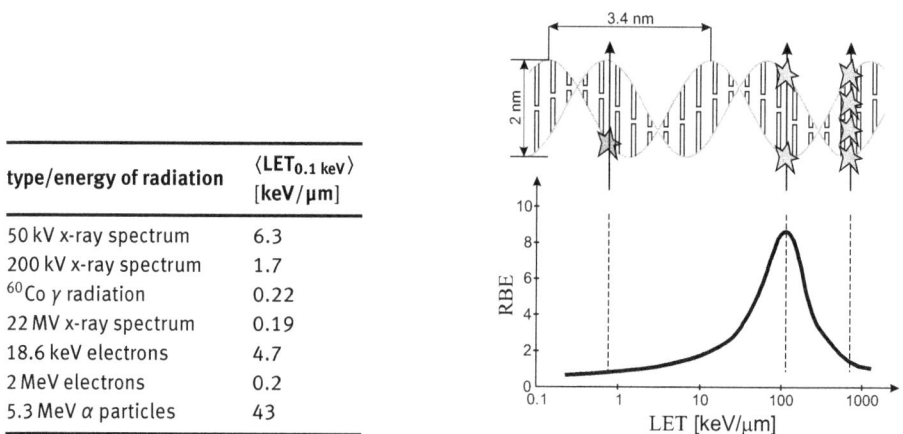

Fig. 5.25: *Left*: Literature values for LET_Δ with $\Delta = 0.1$ keV, averaged along the particle track [17]. For a given type of radiation, LET increases if the kinetic energy of charged particles decreases. In the case of photon radiation, LET refers to the released secondary electrons. *Right*: Plot of (average) LET vs. RBE in cell killing. The curve reaches a peak at about 100 keV/μm, where the energy deposited in the medium is used most efficiently to induce DNA DSBs. Note that the absolute value of RBE is not unique, but depends on the chosen biological effect. Redrawn from [25, 40].

5.3.5 Cell survival after irradiation: linear quadratic model

The effect of a given type of radiation on a certain type of cells can be studied by in vitro experiments using cell cultures in Petri dishes [25]. Empirically, one finds that the survival probability of cells

$$S := \frac{\text{number of surviving cells after irradiation}}{\text{number of cells prior to irradiation}} \tag{5.83}$$

after irradiation with dose D (not exceeding few Gy) within a short time (on the order of few minutes) can be modeled with satisfying accuracy by

$$S(D) = \exp(-\alpha D - \beta D^2) \qquad \Rightarrow \qquad \log S = -\alpha D - \beta D^2 \tag{5.84}$$

Due to the combination of a term linear and a term quadratic in dose D, this is known as the linear quadratic (LQ) model of cell survival after irradiation. The parameters α, β (not to be confused with α or β radiation) are radiation and cell type dependent. Although first used as an empirical fitting formula, it was shown by CHADWICK and LEENHOUTS that equation (5.84) can be derived from the assumption that the central event leading to cell death is an unrepaired DSB, and thus be put on a solid biophysical basis [16].

For a start, consider a cell that initially contains N_0 targets, which can be hit and 'destroyed' by an irradiation event (either directly or indirectly by ROS). Let D denote the absorbed dose and k the probability per unit dose that a target is hit. The decrease in the number of targets dN follows the differential equation

$$dN = -kNdD \qquad \Rightarrow \qquad N(D) = N_0 e^{-kD} ; \tag{5.85}$$

the number of remaining targets decreases exponentially with dose D. The mean number of hits is

$$N_0 - N = N_0 \left[1 - e^{-kD}\right] \tag{5.86}$$

and approaches the total number of targets N_0 for large doses.

Now, assume that the critical event that eventually leads to cell death ('lethal lesion') is a DSB in the DNA. This may occur in two different ways:

(i) A single incident particle breaks two bonds, one in each DNA strand, along its track:

n_{DSB}:	number of critical sites for DSBs
k_{DSB}:	probability of DSB creation per critical site per unit dose
D:	absorbed dose
Δ:	percentage of D that leads to DSBs by mode (i)

In analogy to equation (5.86), the mean number of DSBs per cell caused by mode (i) after receiving the dose D is

$$n_{\text{DSB}} \left[1 - e^{-k_{\text{DSB}}\Delta D}\right] . \tag{5.87}$$

(ii) Two independent SSBs (or 'sublethal lesions') caused along two different particle tracks occur in close spatial proximity:

n_{SSB}:	number of critical bonds per DNA single strand
k_{SSB}:	probability of SSB creation per single critical bond per unit dose
r_{SSB}:	probability that a SSB is restored or fixed by enzymatic repair
$f_{\mathrm{SSB}} = 1 - r_{\mathrm{SSB}}$:	probability that SSB persists
D:	absorbed dose
$1 - \Delta$:	percentage of D that leads to DSBs by mode (ii)

In this case, the mean number of critical SSBs per single strand is

$$f_{\mathrm{SSB}} n_{\mathrm{SSB}} \left[1 - e^{-k_{\mathrm{SSB}}(1-\Delta)D}\right] . \tag{5.88}$$

The mean number of DSBs per cell caused by mode (ii) is proportional to the product of the number of SSBs in each strand

$$\epsilon f_{\mathrm{SSB}}^2 n_{\mathrm{SSB}}^2 \left[1 - e^{-k_{\mathrm{SSB}}(1-\Delta)D}\right]^2 , \tag{5.89}$$

where ϵ denotes the probability that two independent SSBs occur in sufficiently close spatiotemporal correlation to yield a DSB.

Combining modes (i) and (ii), the mean number of DSBs per cell generated by irradiation with dose D is given by the sum of equation (5.87) and equation (5.89). If we assume that also DSBs can be fixed by enzymatic repair, only a fraction $f_{\mathrm{DSB}} < 1$ persists. In addition, if a DSB causes cell death with probability p, the mean number of lethal DSBs per cell is

$$p f_{\mathrm{DSB}} \left(n_{\mathrm{DSB}} \left[1 - e^{-k_{\mathrm{DSB}}\Delta D}\right] + \epsilon f_{\mathrm{SSB}}^2 n_{\mathrm{SSB}}^2 \left[1 - e^{-k_{\mathrm{SSB}}(1-\Delta)D}\right]^2 \right) := \lambda . \tag{5.90}$$

Upon irradiation (with a constant dose rate $\frac{d}{dt}D = \dot{D}$), the creation of a DSB in a cell can be assumed to occur with a known average rate, but independent of the time since the last DSB was created, i.e., subsequent DSBs in a cell are statistically independent. In addition, for very short time intervals, the probability that a DSB is created is proportional to the length of the interval. Therefore, the creation of DSBs can be considered as a Poisson process (see also Section 3.5). With λ defined by equation (5.90), the probability $P(n)$ for n lethal DSBs is given by

$$P(n) = \frac{\lambda^n}{n!} e^{-\lambda} . \tag{5.91}$$

Using equation (5.90), the survival probability is given by the probability of zero lethal DSBs,

$$\begin{aligned} P(0) &= \frac{\lambda^0}{0!} e^{-\lambda} \\ &= \exp\left[-p f_{\mathrm{DSB}} \left(n_{\mathrm{DSB}} \left[1 - e^{-k_{\mathrm{DSB}}\Delta D}\right] - \epsilon f_{\mathrm{SSB}}^2 n_{\mathrm{SSB}}^2 \left[1 - e^{-k_{\mathrm{SSB}}(1-\Delta)D}\right]^2 \right)\right] . \end{aligned} \tag{5.92}$$

The probabilities per unit dose k_{SSB} and k_{DSB} that a certain critical bond or site is affected by a radiation event are very small. As noted above, an x-ray dose of 1 Gy generates about 1000 SSBs and 20–40 DSBs per cell, which are very small numbers compared to the number of possible sites in a DNA molecule containing about $3 \cdot 10^9$ base pairs. Using the expansion $\exp(-x) \simeq 1 - x$ (for $x \ll 1$) yields

$$P(0) = \exp\left[-pf_{DSB}\left(n_{DSB}k_{DSB}\Delta D - \epsilon f_{SSB}^2 n_{SSB}^2 k_{SSB}^2 (1-\Delta)^2 D^2\right)\right] . \tag{5.93}$$

With the abbreviations

$$\alpha = pf_{DSB}n_{DSB}k_{DSB}\Delta \qquad \text{and} \qquad \beta = pf_{DSB}\epsilon f_{SSB}^2 n_{SSB}^2 k_{SSB}^2 (1-\Delta)^2 , \tag{5.94}$$

the survival probability $S = P(0)$ of a large number N of identical cells that have been irradiated with a homogeneous dose D adopts the very simple form

$$S(D) = \exp\left(-\alpha D + \beta D^2\right) \tag{5.95}$$

which is identical to equation (5.84). This description of radiation induced cell death was originally termed a 'molecular theory', since it was the first model explicitly based on the double strand architecture of DNA and the rationale that DNA DSBs are the lethal damages leading to cell death. Today it is widely known as the LQ model. The linear term $-\alpha D$ accounts for DSB creation by mode (i) along a single particle track. At small doses, cell killing is most likely caused by this mode, since it is very unlikely that two independent SSBs are created sufficiently close to each other and persist long enough to form a DSB. The quadratic term $-\beta D^2$ describes cell death due to mode (ii), two statistically independent SSBs that result in a potentially lethal DSB if they occur in close proximity, and becomes the dominant effect in cell killing at higher doses (Figure 5.26).

The parameters α and β can be obtained by fitting equation (5.95) to measured survival curves. They depend on cellular properties (e.g., the enzymatic repair capacities encoded in f_{SSB} and f_{DSB} or the likelihood that apoptosis is initiated if a DSB is detected, which is encoded in p) as well as on the properties of the radiation that generates the dose D. Due to the Δ dependence, the linear term $-\alpha D$ dominates for densely ionizing radiation (high LET: e.g., protons, α particles, neutrons), whereas the quadratic term $-\beta D^2$ is more relevant for sparsely ionizing radiation (low LET: photons, electrons). The dose

$$D = \frac{\alpha}{\beta} \tag{5.96}$$

where both terms contribute equally to cell killing, is called the α/β ratio and is used to classify the response of tissues to irradiation (Section 5.4.4). The LQ model is by far the most prevalent model for cell survival in radiobiology. For typical fraction doses of 1.8 or 2.0 Gy of conventional fractionation schemes in radiotherapy, it adequately describes the experimental cell survival curves. Also extensions incorporating further effects relevant in radiotherapy, like tumor repopulation or chemotherapy, exist (see e.g., [48, 53, 71] and references therein).

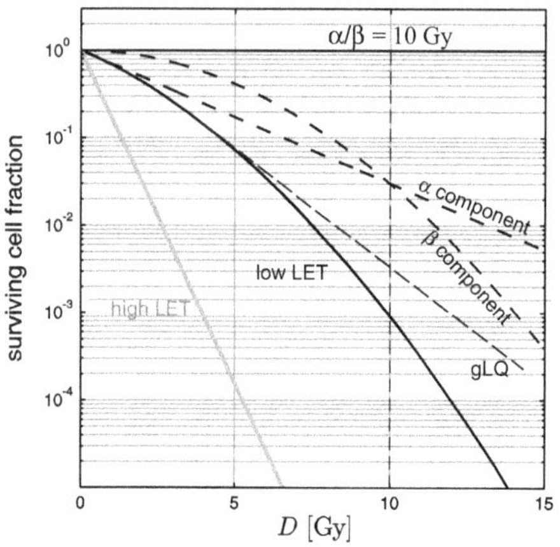

Fig. 5.26: Plot of typical cell survival curves in single dose experiments (i.e., cells receive the dose D within a short period of time) as predicted by the LQ model (equation 5.95) for low LET (photons, electrons) and high LET radiation (e.g., α particles, carbon ions, neutrons). The low LET cell survival curve bends continuously due to the quadratic component, while the high LET curve is almost a straight line. For higher doses corresponding to longer irradiation times, the generalized LQ (gLQ) model (not discussed here) in which the bent curve becomes straight above few Gy, provides a more appropriate description of measured data [71].

For higher doses that reduce the cell number by several orders of magnitude and require longer delivery times, however, experimental data indicate that $\log(S)$ vs. D approximates a straight line ('shoulder curve'), while the cell survival curves given by the LQ model are continuously bending due to the quadratic term $-\beta D^2$ (Figure 5.26). Therefore, the LQ model as described so far tends to overestimate the effect of cell killing for several modern radiotherapy techniques like stereotactic radiosurgery (SRS), stereotactic body radiation therapy (SBRT) or high dose rate (HDR) brachytherapy, where fraction doses considerably larger than 2 Gy are used.[20] Measured cell survival curves are described more consistently over a wide dose range by extending the LQ model to take into account the exact temporal pattern of the production of SSBs and their repair or conversion to DSBs [53, 71].

5.4 Fundamentals of radiotherapy

In this section, we give a very brief overview of some general principles of modern radiotherapy. Many additional topics that would certainly fit here, e.g., the generation of megavoltage electron and x-ray beams by linear accelerators, the production of beams of heavy charged particles, or the algorithms used in inverse treatment planning, have

20 For example, doses of 18 Gy are delivered in a single fraction in SRS to destroy intracranial metastases, or 9 Gy in HDR brachytherapy of prostate carcinomas.

to be skipped due to length restrictions. For an in depth introduction to the physics of radiotherapy, we refer to the textbooks [33, 45, 48, 56].

5.4.1 Motivation

In normal cells, the cell cycle (Section 5.3.1) is tightly regulated by a number of proteins. However, if a critical number of DNA mutations in the corresponding genes has accumulated in a single cell, these control mechanisms can fail. The resulting cancer cells usually proliferate at a much higher rate and have lost the ability to eliminate themselves by apoptosis. They take up nutrients and space that normal cells need. This uncontrolled growth can damage tissues and organs, disseminate over the entire body by forming metastases, and finally lead to death. There are also a few other indications for radiotherapy, but the vast majority of patients receiving radiotherapy suffer from one of the very different kinds of cancer. Therefore, the terms radiation therapy (or radiotherapy) and radiation oncology are used almost interchangeably.

Almost every second adult (in western countries) is confronted with a cancer diagnosis during his or her lifetime [10]. On the one hand, this is a consequence of the enormous increase of life expectancy during the last century – many types of cancer are 'diseases of age', since a longer life span increases the time over which random mutations in normal stem cells can accumulate and eventually lead to carcinogenesis [64]. On the other hand, the incidence of many cancer types is highly correlated with habits like smoking and alcohol abuse or exposition to certain environmental factors (e.g., ultraviolet radiation, some chemicals like asbestos). Behind cardiovascular diseases, cancer ranks second in the causes of death.

Currently, more than 50% of cancer patients receive radiotherapy, making it one of the three cornerstones of cancer treatment along with surgery and chemotherapy. The goal of radiotherapy treatment is to kill all tumor cells (curative treatment) or at least reduce symptoms such as pain (palliative treatment). This must go along with a tolerable level of damage to the surrounding normal tissue, which inevitably receives some dose. In addition, the risk of radiation induced secondary cancer, as discussed in Section 5.5 must be taken into account. If properly applied, the beneficial effects of radiotherapy clearly outweigh the adverse side effects. However, as discussed in Section 5.4.5, the 'therapeutic window', i.e., the dose interval where this can be achieved, is often very narrow. This requires the highest possible accuracy in all treatment steps outlined in the following.

5.4.2 Steps in radiotherapy

The actual radiotherapy treatment usually extends over several weeks and is preceded by a planning phase in which the optimal treatment for given boundary conditions (e.g., type and stage of cancer and available radiation qualities) has to be devised. In addition to purely medical aspects, this involves the following steps where the physics of medical imaging and dose calculation, as well as the principles of radiobiology are essential:

- Imaging of the region of the body involved by tomographic methods (Chapter 4): A CT scan is always required, since it yields density data that are the basis of dose calculations. Additional MRT and/or PET scans often yield useful complementary information. In the CT dataset, the tumor and all critical vital organs (**organs at risk**, OAR) that receive a significant dose are delineated (contouring). An additional safety margin is added to the macroscopically visible tumor to account for microscopic spread, as well as uncertainties due to variation in daily positioning and changes in patient anatomy (e.g., varying volume of the urinary bladder), as well as movements during the irradiation (e.g., breathing motion). This yields the so called **planning target volume** (PTV).
- Choice of the most appropriate treatment technique (Section 5.4.3).
- Consideration of the response of tumor and surrounding normal tissue to high doses of ionizing radiation and choice of total dose and temporal dose delivery pattern (**fractionation**, Section 5.4.4), which yield high tumor control probability (TCP) at an acceptable level of normal tissue complication probability (NTCP, Section 5.4.5).
- Planning of how to generate of a dose distribution $D(\underline{r})$ inside the patient that matches the PTV as closely as possible, while sufficiently sparing the surrounding normal tissue and adjacent OARs. Today, this is achieved with the aid of treatment planning systems (TPS), i.e., software that can model the radiation sources available and calculate the resulting dose distributions in a patient's body by the algorithms outlined in Section 5.2.2, based on CT data.
- Plan verification and quality assurance: At least in the case of more complex techniques, each patient plan is transferred to a dosimetric phantom and agreement of precalculated and actual dose distribution is experimentally checked by dosimetry prior to treatment of the patient. Since the dose interval for sufficient TCP at an acceptable NTCP is usually quite narrow (Section 5.4.5), only small discrepancies (e.g., $< 2\%$) are tolerable. In addition, regular constancy checks are performed to assure that the actual properties of the radiation sources (linear accelerators and brachytherapy sources in a typical radiotherapy department) are still in agreement with the model used by the TPS.

Standards and guidelines exist for different tumor entities, but in principle the treatment planning process needs to be carried out individually for each patient. Due to the technical and dosimetric requirements, the involvement of medical physicists in treatment planning, dosimetric quality control and commissioning and maintenance of the radiation sources and treatment planning software is mandatory.

5.4.3 Radiation modalities and treatment techniques

A first classification of the different techniques used in radiotherapy can be made according to the type of radiation (photons, charged particles, neutrons), as well as the nature of the radiation source and its location relative to the patient:

– **Teletherapy** (or external beam radiation therapy): The prefix *tele* means *far from* or *at a distance*, i.e., the radiation source is well separated from the patient. The workhorse of contemporary radiation therapy is the linear accelerator (linac), in which electrons reach energies of several MeV before they hit a target and generate bremsstrahlung, which shows a pronounced dose buildup effect and is only slightly attenuated by typical patient dimensions (Figure 5.27, blue and pink curves). The distance of the source (where electrons hit the target) and the center of rotation of the linac (where the patient is positioned) is usually 100 cm. By removing the target from the beam, electrons can be directly used for the treatment of superficial regions due to their different depth dose characteristics (Figure 5.27, red curve). With dimensions of few meters, medical linacs can, in principle, be installed in every hospital. They have largely superseded the ^{60}Co machines, which use large amounts of this radioisotope (Figure 5.27, green dashed curve).
  Protons can be advantageous in many cases due to the narrow maximum of their percentage depth dose (PDD) curve (Figure 5.27, black curve), but require consid-

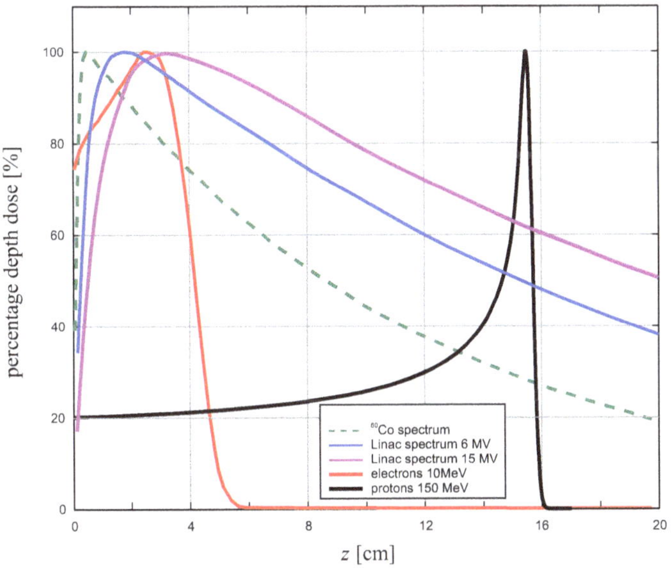

Fig. 5.27: Characteristic percentage depth dose (PDD) curves of different types of radiation. Location and extent of the tumor determine which type of radiation at which energy is suited best. In addition, high LET radiation like protons, carbon ions or neutrons can be advantageous for radiobiological reasons.

Table 5.2: Some radionuclides commonly used in brachytherapy. Only the emitted γ radiation, which accompanies the other types of decay, can pass through the encapsulation and generate a dose distribution to treat the patient. Data from [38].

isotope	$T_{1/2}$	type of decay	E_γ [MeV]
^{60}Co	5.27 a	β^-, γ	1.173, 1.332
^{192}Ir	73.82 d	K, β^+, β^-	0.296–0.612
^{125}I	59.41 d	EC	0.035

erably higher technical effort (e.g., [54]) and are currently only available at a few dozen facilities worldwide. This applies even more to heavy ion therapy (e.g., with carbon ions), which has an even sharper Bragg peak.[21]

- **Brachytherapy** (or internal beam radiation therapy): In contrast, the prefix *brachy* means *short* or *close*. Radioactive sources contained in a protective capsule are placed either very close to or even within the tumor. Radiation can escape from the capsule, but the radioisotope itself remains separated from the organism. The energy of the emitted radiation is usually considerably lower compared to teletherapy, which allows good sparing of healthy tissue. Due to the small distances between source and irradiated tissue, dose gradients are steeper and dose inhomogeneity is higher compared to teletherapy. Some isotopes commonly used in brachytherapy are given in Tab. 5.2. In high dose rate (HDR) brachytherapy, a single source stops at several positions with different dwell times, thus shaping a desired dose distribution $D(\underline{r})$. In low dose rate (LDR) brachytherapy, several sources (so called seeds) are permanently implanted into the region to be irradiated.
- **Unsealed source therapy**: Here, radionuclides are taken up by the organism (either orally or by intravenous injection) and are distributed and metabolized within the body just like their nonradioactive counterparts. They can be coupled to a carrier molecule, yielding a radiopharmaceutical, or be given in pure form as in the case of their most prominent example: The radioisotope ^{131}I is used to treat tumors or functional autonomy of the thyroid gland. Since this is the only organ that stores considerable amounts of iodine, the radioisotope is automatically accumulated here by the human organism. The emitted β^- radiation (electron energy $E_e = 0.606$ MeV) with a range of few mm provides a well localized therapeutic effect; the accompanying γ radiation (photon energy $E_\gamma = 0.364$ MeV) can additionally be used to monitor activity and dose distribution. Another example are the radioisotopes ^{153}Sm and and ^{223}Ra, which accumulate in bones and can be exploited to treat skeletal metastases.

21 However, the number of centers offering proton therapy is steadily increasing. For a current list and further links, see http://www.ptcog.ch/index.php/facilities-in-operation.

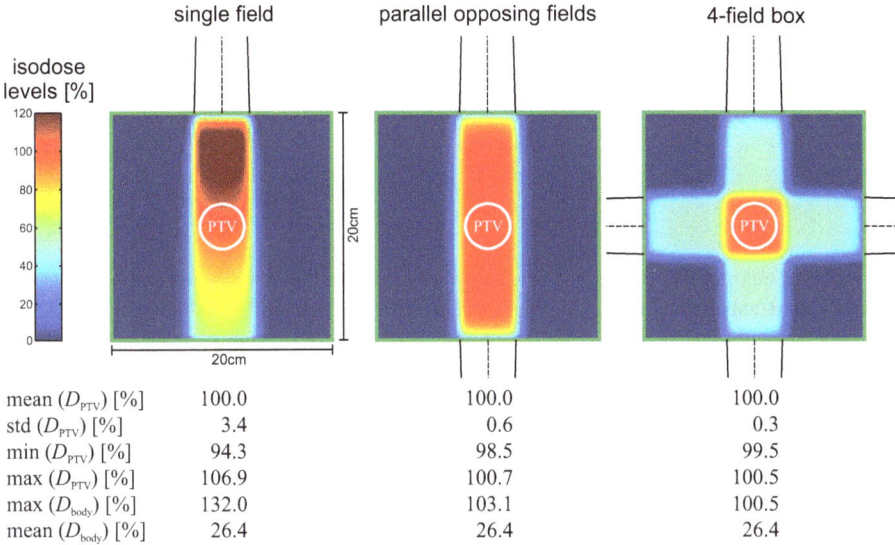

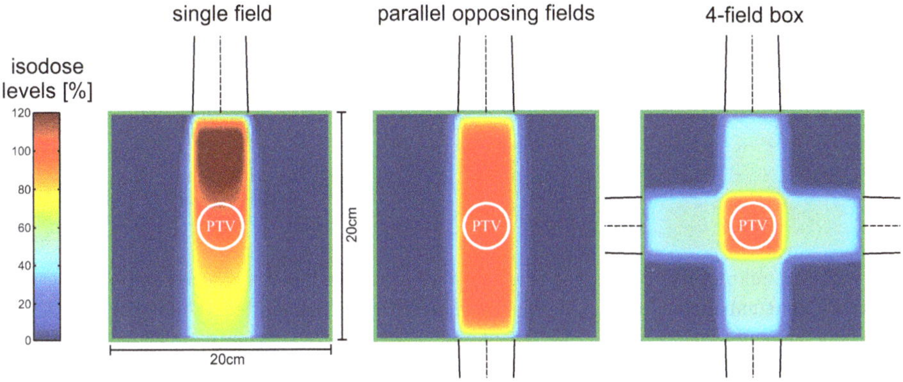

	single field	parallel opposing fields	4-field box
mean (D_{PTV}) [%]	100.0	100.0	100.0
std (D_{PTV}) [%]	3.4	0.6	0.3
min (D_{PTV}) [%]	94.3	98.5	99.5
max (D_{PTV}) [%]	106.9	100.7	100.5
max (D_{body}) [%]	132.0	103.1	100.5
mean (D_{body}) [%]	26.4	26.4	26.4

Fig. 5.28: Illustration of the principle of field superposition in teletherapy. Conformality of the high dose region with the PTV and dose homogeneity inside the PTV (as quantified by the standard deviation of the dose in all voxels belonging to the PTV) can be improved considerably by superposition of several beams from different directions.

Teletherapy with photon radiation from a linear accelerator, which clearly outnumbers all other forms of radiotherapy, usually achieves high conformality (i.e., spatial agreement of the high dose region and the target volume to be irradiated) by superposition of radiation beams from several directions, as illustrated in Figure 5.28. It can be further subdivided into different techniques (the following list is nonexhaustive):

- **Conventional radiotherapy**: Each beam has a simple rectangular shape (Fig. 5.29a). For a long time, further modulation could only be achieved by additional devices like blocks, wedges or compensators. Nowadays, this form of radiotherapy is rather rarely used, e.g., in the first fraction of very urgent cases without time for sophisticated treatment planning.
- **3d conformal radiotherapy** (3d-CRT): The shape of each beam is adapted to the contour of the tumor by a multileaf collimator (MLC) (Figure 5.29b). This yields better spatial agreement of the resulting dose distribution $D(\underline{r})$ with the PTV, which is called *conformality*. A treatment plan is generated by *forward planning* on the basis of a CT dataset. Parameters like beam directions and relative weights are chosen and the resulting dose distribution is calculated using one of the methods given in Section 5.2.2 using dedicated treatment planning software. The parameters are manually refined in an iterative fashion until an acceptable dose distribution is obtained.

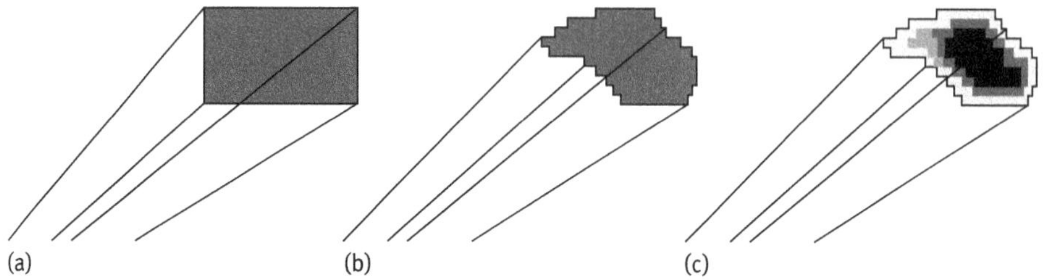

Fig. 5.29: Illustration of the differences between (a) conventional radiotherapy beams (rectangular shape, constant photon fluence), (b) conformal radiotherapy beams (fields shaped by an MLC matching the target volume region, constant photon fluence) and (c) intensity modulated radiotherapy beams (IMRT, conformal fields with additional variation of photon fluence across the field).

- **Intensity modulated radiotherapy** (IMRT): As an additional degree of freedom, the fluence (or intensity) within each beam is modulated using the MLC (Figure 5.29c). This allows to achieve even better dose conformality, also for concave and very complex PTV shapes. Since this involves many degrees of freedom, *inverse planning* is used. Weighted criteria for the desired dose distribution and few starting conditions (like beam number and gantry angles) are defined. The optimal fluences and their realization by the MLC are found by iterative optimization algorithms [13, 72].
- **Volumetric modulated arc therapy** (VMAT): This can be considered as an extension of IMRT. While IMRT uses few (usually 5, 7 or 9) fixed beam directions, VMAT irradiates the target volume continuously during a rotation of the gantry around the patient, while the fluence is simultaneously modulated.
- **Stereotactic radiosurgery** (SRS) and **stereotactic body radiation therapy** (SBRT): These techniques aim to destroy small lesions (e.g., metastases in brain, lung and liver) by high doses given in a single (SRS) or few (SBRT) fractions and include measures for very precise localization and immobilization.

5.4.4 Fractionation

In most cases, the total dose D that is considered sufficient to achieve tumor control (i.e., eradicate all tumor cells) at an acceptable level of damage to normal tissue is not given in a single treatment, but in smaller portions called **fractions**. Here we outline the reason for this. In cell survival experiments (Section 5.3.5), one can determine the LQ parameters α, β of different normal and tumor cell types. It has been found that tissues can roughly be divided into two groups based on their α/β ratio (equation 5.96) [45, 48]:

Table 5.3: Examples of early and late responding tissues and their α/β ratios [25, 57].

early responding organs/tissues	$\frac{\alpha}{\beta}$ [Gy]	**late responding organs/tissues**	$\frac{\alpha}{\beta}$ [Gy]
skin	9–12	spinal cord	1.7–4.9
colon	10–11	lung	2.0–6.3
bronchial carcinoma	50–90	kidney	1.0–2.4
cervical carcinoma	> 13.9	bladder	3.1–7.0

- **Early responding tissues** with $\alpha/\beta \approx 10$ Gy or higher: This group comprises most malignant tumors, as well as quickly proliferating normal tissues such as skin or intestinal mucosa. Typical values are $\alpha = 0.35\,\text{Gy}^{-1}$ and $\beta = 0.035\,\text{Gy}^{-1}$.
- **Late responding tissues** with $\alpha/\beta \approx 3$ Gy: This other group comprises most normal tissues with lower proliferation rates and has typical LQ parameters $\alpha = 0.15\,\text{Gy}^{-1}$, $\beta = 0.05\,\text{Gy}^{-1}$. As an exception, also prostate cancer cells have very low α/β.

Some exemplary α/β values are given in Tab. 5.3. This characteristic difference in response to ionizing radiation has very important implications for radiotherapy (Figure 5.30). Let n denote the number of fractions and d the dose per fraction ($D = nd$). After irradiation with dose d, the probability for cell survival according to the LQ model is

$$S(d) = \exp\left(-\alpha d - \beta d^2\right) . \tag{5.97}$$

If d is small, the linear term $-\alpha d$ in the exponent dominates, and the survival probability will generally be larger for normal tissue cells than for tumor cells due to their

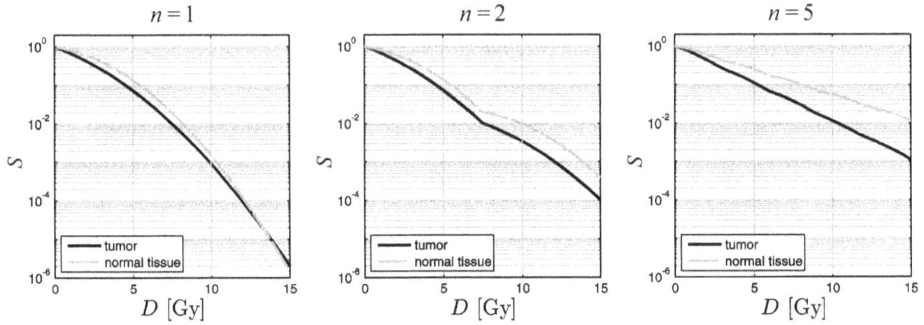

Fig. 5.30: Illustration of the effect of fractionation on the survival of tumor and normal tissue cells with typical LQ parameters (tumor: $\alpha = 0.35\,\text{Gy}^{-1}$, $\beta = 0.035\,\text{Gy}^{-2}$; normal tissue: $\alpha = 0.15\,\text{Gy}^{-1}$, $\beta = 0.05\,\text{Gy}^{-2}$). A total dose $D = 15$ Gy is given in a single, two or five fractions, between which complete repair of SSBs is assumed. The curves indicate that fractionation spares the late responding normal tissue compared to the tumor. However, also the total dose D required to eradicate all tumor cells increases.

smaller value of α. Only if d becomes sufficiently large, does the quadratic term $-\beta d^2$ begin to dominate cell killing; see also Figure 5.26). Since β tends to be larger for normal tissue, the survival curves again cross at some point (in Figure 5.30 (left), this happens at $d \approx 13$ Gy). Thus, if d is chosen small, the chance to survive a fraction is higher for normal cells than for tumor cells. If complete repair of SSBs is always achieved before the next fraction dose d is given, one obtains

$$S(nd) = [S(d)]^n = \exp\left[-nd(\alpha + \beta d)\right] \ , \tag{5.98}$$

i.e., the initial part of the survival curves is repeated in every fraction, and the curves do not cross. This applies to the cases $n = 2$ and $n = 5$ in Figure 5.30. In theory, this effect is maximized if the number of fractions approaches infinity (the survival curves would become straight lines on semilogarithmic scale). In the most common scheme, patients receive one daily fraction of 1.8 or 2.0 Gy, with a break over the weekend. This is a compromise between the sparing of normal tissue and a manageable number of fractions. However, note that the total dose $D = nd$ required for tumor control increases with the number of fractions n. From equation (5.98), one obtains that two fractionation schemes n_1, d_1 and n_2, d_2 have the same cell killing effect if

$$\frac{D_1}{D_2} = \frac{n_1 d_1}{n_2 d_2} = \frac{\frac{\alpha}{\beta} + d_2}{\frac{\alpha}{\beta} + d_1} \ . \tag{5.99}$$

This equation can be used if the fractionation scheme has to be changed for some reason. Every time a tumor receives a fixed fraction dose d, a constant percentage $1 - \exp(-\alpha d - \beta d^2)$ of the tumor cells that have survived up to this point of the treatment is killed according to the LQ model. Roughly speaking, the first fractions of a curative radiotherapy treatment kill the vast majority of tumor cells (which can lead to a quick reduction of acute symptoms, e.g., if the tumor compresses a nerve), while the later fractions are required to ensure that *every* tumor cell is killed and thus minimize the relapse rate.

Very importantly, the effects of fractionation do not only depend on the different shapes of cell survival curves of tumors and normal tissues, but are further modulated by other factors, which are sometimes called the **4 R's of radiobiology** [25]:

- **Repopulation:** During treatment, which typically extends over several weeks, the surviving cells proliferate (which has not been considered in the equations so far). This is advantageous for the regeneration of irradiated normal tissue, but can pose a serious problem in the case of very quickly proliferating tumors and has to be taken into account to achieve high TCP.
- **Redistribution**: The sensitivity of cells to radiation varies with the phases of the cell cycle (Figure 5.19). Normal cells are most sensitive to radiation during the M phase, followed by the G_2 phase. Hence, cells that reduplicate rapidly (interphase short compared to M phase), such as tumor cells or epithelial cells of the intestinal mucosa, are more sensitive to radiation. Fractionation increases the chance that every tumor cell is irradiated during a radiosensitive phase.

- **Repair of sublethal damage:** The above formulas are only valid if complete repair of SSBs is achieved between fractions. For normal cells, an interval of at least 6 h between two fractions is considered sufficient. The repair capacity varies between different tissues and from patient to patient.
- **Reoxygenation:** A DNA radical R· that has been formed by ionizing radiation can react with molecular oxygen O_2 and form an organic peroxide $RO_2\cdot$, which constitutes a type of DNA damage for which no repair mechanisms exist (*oxygen fixation hypothesis*). For fraction doses of 1 to 2 Gy, the *oxygen enhancement ratio*

$$\mathrm{OER} = \left.\frac{D\,[\mathrm{Gy}]\ \text{with}\ \mathrm{O}_2}{D\,[\mathrm{Gy}]\ \text{without}\ \mathrm{O}_2}\right|_{\text{isoeffect}} \tag{5.100}$$

  typically lies between 2.0 and 3.0 in the case of sparsely ionizing radiation (photons and electrons). Hypoxic tumor regions with reduced blood supply (and thus lower partial pressure of O_2) are reoxygenated if cells in the surrounding, better supplied regions are killed in the previous fractions and, thus, become more radiosensitive.

There exist extensions of the LQ model that take into account tumor proliferation, incomplete repair and treatment interruption [45, 48, 53]. In general, it has to be noted that the simplifying models of radiobiology must always be used with caution and can not replace consideration of the actual clinical situation of a patient. However, they provide some useful insights into tumor and normal tissue response to irradiation and can be used as a first guide in radiotherapy.

5.4.5 Tumor control and normal tissue complication probability

Since the interaction of radiation with matter is stochastic by nature, the success of radiotherapy is, to some extent, dependent on chance. Using the cell survival probability S as predicted by the LQ model and some simple statistics, it is possible to construct mechanistic models that yield estimates for the probability of successful treatment or the occurrence of complications. Although clearly a simplification, these models can give some instructive insights into the principles of radiotherapy. This section is based on [47, 48, 73].

The primary goal of a curative treatment is to eradicate all clonogenic tumor cells, i.e., those cells from which the tumor can regenerate even if only a single one survives. Assuming that there are T of these cells and each one is killed with individual probability P_t, the **tumor control probability** (TCP) that none of them survives and the tumor is successfully eradicated is given by the product

$$\mathrm{TCP} = P_1 \cdot \ldots \cdot P_T = \prod_{t=1}^{T} P_t\,. \tag{5.101}$$

Already a single clonogenic cell outside the irradiated area, and hence $P = 0$, reduces TCP to zero. In addition, if the cell number T increases and all other variables are kept constant, TCP decreases, since it becomes more likely that at least one of the T cells survives. If all clonogenic cells have identical LQ parameters and have received the same dose in a fractionated treatment, it is possible to replace P_t by $(1 - S_T)$ according to equation (5.98):

$$\text{TCP} = (1 - S_T)^T \tag{5.102}$$

The subscript is used to distinguish the survival probability of tumor cells S_T from the survival probability of normal cells S_N used below. Writing $\lambda = T \cdot S_T$ for the expected number of surviving tumor cells, inserting equation (5.98) for S and identifying TCP with the probability that no clonogenic cell survives, it follows from Poisson statistics that

$$\begin{aligned} \text{TCP} &= \frac{\lambda^0}{0!} \exp(-\lambda) \\ &= \exp\left(-T \cdot S_T\right) \\ &= \exp\left(-\rho_T V_T \exp[-nd(\alpha_T + \beta_T d)]\right) \end{aligned} \tag{5.103}$$

where ρ_T denotes the density of clonogenic tumor cells and V_T the tumor volume. This way, TCP can be linked to the parameters $\alpha_T, \beta_T, \rho_T, V_T$ characterizing the tumor and n, d characterizing the fractionation scheme. In Figure 5.31 (right), this is plotted for typical values and various tumor sizes. The minimum size of a tumor that is reliably detectable by diagnostic imaging is on the order of 1 cm^3. With typical cell dimensions of 10 µm in each direction, this corresponds to about $\mathcal{O}(10^9)$ cells per cm^3, a common clonogenic cell density often used in the literature is $\rho_T = 10^7\ \text{cm}^{-3}$. The TCP curves exhibit a characteristic sigmoid shape and rise steeply from close to zero to close to unity within a dose interval of a few Gy. The dose required to achieve a certain TCP scales logarithmically with the tumor volume.

The formulation of **normal tissue complication probability** (NTCP) is slightly different. While each clonogenic tumor cell is considered as an independent entity, organs can fulfill their specific tasks only if a number of cells cooperate. An organ is thus modeled as the assembly of N identical functional subunits (FSUs) consisting of k cells each.[22] For simplicity, it is assumed that all k cells of an FSU must be killed to inactivate it. In analogy to equation (5.102), the corresponding probability is

$$P_{\text{FSU}} = (1 - S_N)^k \tag{5.104}$$

provided that all cells of a FSU receive the same dose and are described by the same LQ parameters. The survival probability of an FSU is simply $S_{\text{FSU}} = 1 - P_{\text{FSU}}$. Organs are modeled by two different arrangements of FSUs (Figure 5.31, left). In the spinal

22 Examples for FSUs are alveoli for the lungs, groups of oligodendrocytes for the spinal cord or nephrons for the kidneys.

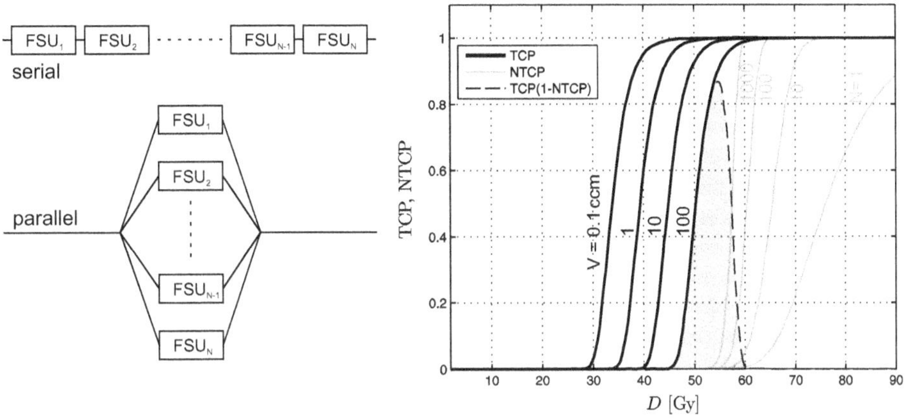

Fig. 5.31: *Left*: Many organs can be modeled either as a serial (e.g., spinal cord, optical nerves) or as a parallel (e.g., lungs, kidneys) arrangement of FSUs. In organs of the serial type, failure of a single FSU leads to failure of the entire organ, while parallel organs can still provide sufficient functionality if a larger number of FSUs fails. *Right*: Tumor control probability (TCP, equation 5.103) for different tumor volumes $V_T = 0.1, 1, 10, 100\,\text{cm}^3$ (using $\rho_T = 10^7\,\text{cm}^{-3}$, $d = 2\,\text{Gy}$, $\alpha = 0.35\,\text{Gy}^{-1}$, $\beta = 0.035\,\text{Gy}^{-2}$) and normal tissue complication probability (NTCP, equation 5.105) for a serial organ if different numbers of FSUs ($N = 1, 10, 100, 1000$) are irradiated ($k = 10000$, $d = 1.2\,\text{Gy}$, $\alpha = 0.15\,\text{Gy}^{-1}$, $\beta = 0.05\,\text{Gy}^{-2}$). The *dashed curve* indicates the probability of tumor control without normal tissue complications (for $V_T = 100\,\text{cm}^3$ and $N = 1000$).

cord, for example, a single lesion that interrupts the signal transduction is sufficient to make the entire structure dysfunctional. This can be described as a serial arrangement of FSUs. The probability of a serial organ composed of N identical FSUs to fail is

$$\begin{aligned} \text{NTCP}_{\text{serial}} &= 1 - S_{\text{FSU}}^N \\ &= 1 - (1 - P_{\text{FSU}})^N \\ &= 1 - \left[1 - (1 - S_N)^k\right]^N \end{aligned} \tag{5.105}$$

for constant dose to all FSUs. This is plotted for some exemplary values in Figure 5.31, right. The larger the number N of irradiated FSUs, the steeper the NTCP curve. In addition, the dashed curve shows the probability TCP(1 − NTCP) for tumor control without normal tissue complications, i.e., the desired treatment outcome, for one combination of parameters. It has a narrow peak (gray area) with a full width at half maximum (FWHM) of a few Gy only. This illustrates that the 'therapeutic window', i.e., the dose interval where successful treatment can be achieved with high probability, can be quite narrow. The steep rise both of TCP and NTCP is also observed in clinical practice and emphasizes the significance of accurate dose calculation and delivery.

Other organs, for example, the lungs or kidneys, are rather described by a parallel arrangement of FSUs, which fulfill their specific task independently. A certain number of at least M out of the total number of N FSUs must fail before a clinically relevant effect can be observed. With equation (5.104) and N independent but identical FSUs, the probability that *exactly* l out of N FSUs fail is given by the binomial distribution (cf., Section 3.5)

$$\binom{N}{l} P_{\mathrm{FSU}}^{l} (1 - P_{\mathrm{FSU}})^{N-l} . \tag{5.106}$$

NTCP for the case of a parallel organ is thus given by the sum of all probabilities where at least M FSUs fail,

$$\mathrm{NTCP}_{\mathrm{parallel}} = \sum_{l=M}^{N} \binom{N}{l} P_{\mathrm{FSU}}^{l} (1 - P_{\mathrm{FSU}})^{N-l} . \tag{5.107}$$

This type of model can be extended to include the more realistic case of inhomogeneous dose distributions, as well as a distribution of LQ parameters α, β to account for tumor heterogeneity and variations between patients [73]. Predicting TCP and NTCP for each patient individually is one of the challenges in the future of radiotherapy [48]. The main implication of the examples presented here is that small deviations from a planned dose can have a considerable effect on the success of radiotherapy due to the sigmoidal shape of TCP and NTCP curves and the resulting narrow therapeutic window. As was mentioned before, actual treatment must be based on clinical data, but the LQ model (Sections 5.3.5 and 5.4.4) and its extensions and the resulting TCP/NTCP curves nicely illustrate some key principles.

5.5 Elements of radiation protection

5.5.1 Deterministic and stochastic effects

It is well established that ionizing radiation has detrimental effects on living organisms, especially by damaging DNA, the carrier of the genetic information (Section 5.3.3). Based on the relationship between dose and effect, the biological effects of radiation can be categorized as follows [22, 25, 52]:

- **Deterministic effects** (or tissue reactions) result from radiation induced cell killing. The loss of a small, randomly distributed fraction of cells in an organ or tissue does not usually affect its function. However, if this fraction increases further, a clinically observable impairment of organ function will occur at some point. The corresponding dose is called the **threshold dose**. For example, a cataract of the eye lens becomes observable after the lens has received an absorbed dose of about 0.25 Gy. Above the threshold dose, the probability of damage to an organ rapidly increases to 100% (as quantified by the sigmoidal shape of TCP/NTCP models), and the degree of severity also increases with dose. In radiotherapy planning, dose limits to organs at risk are used to avoid these deterministic effects. The threshold doses are usually on the order of several tens of Gy for a fractionated treatment (a recent comprehensive summary is given by the QUANTEC study [7]).

- **Stochastic effects**, in contrast, are not due to the loss of cells, but result from radiation induced DNA mutations. As discussed below, it is most likely that there is **no threshold dose** for stochastic effects, which comprise two components:
  - *Radiation carcinogenesis* (or radiation tumorigenesis) is the induction of cancer by DNA mutations in somatic cells of an exposed individual. From epidemiological studies, it is well known that radiation exposure correlates with an increased incidence of most types of cancer.
  - *Heritable diseases* result from DNA mutations in germ cells (sperm cells and oocytes), which are transferred to *all* cells of the subsequent generation. Ionizing radiation does not create specific heritable diseases, but increases the probability of DNA mutations that also occur spontaneously [25].
- Some further effects are listed separately from these two groups. Exposing an unborn child to ionizing radiation can lead to severe developmental abnormalities or even intrauterine death. Ionizing radiation can also cause the induction of diseases other than cancer.

The differences between deterministic and stochastic effects are summarized in Tab. 5.4. In the remainder of this chapter, we focus on the stochastic effects of ionizing radiation. Their probability increases with the total dose an individual has received. However, in contrast to deterministic effects, their severity is dose independent. A malignant cancer resulting from mutations induced – by pure chance – by a few high energy photons is as severe as if it was induced by a much higher radiation dose. For a number of questions, it is very important to quantify the exact functional dependence of low doses of ionizing radiation and stochastic radiation effects. Examples are the use of nuclear power, limits for occupational dose exposure or screening tests like mammography. Even if the individual risk may appear small, a large number of cases of, e.g., carcinogenesis can result if large numbers of healthy individuals are deliberately exposed to ionizing radiation [15].

The so called **linear nonthreshold** (LNT) model assumes strict proportionality between dose and incidence of stochastic effects of radiation (induction of cancer and heritable effects), without a threshold dose (Figure 5.32). As discussed in [15], it is difficult to obtain statistically valid data for very small doses by epidemiological stud-

Table 5.4: Comparison of deterministic and stochastic effects of ionizing radiation [25].

	deterministic effects	**stochastic effects**
mechanism	cell killing	DNA mutations
threshold dose	yes	no
probability	increases with dose	increases with dose
severity	increases with dose	independent of dose
mathematical model	(g)LQ model	LNT model

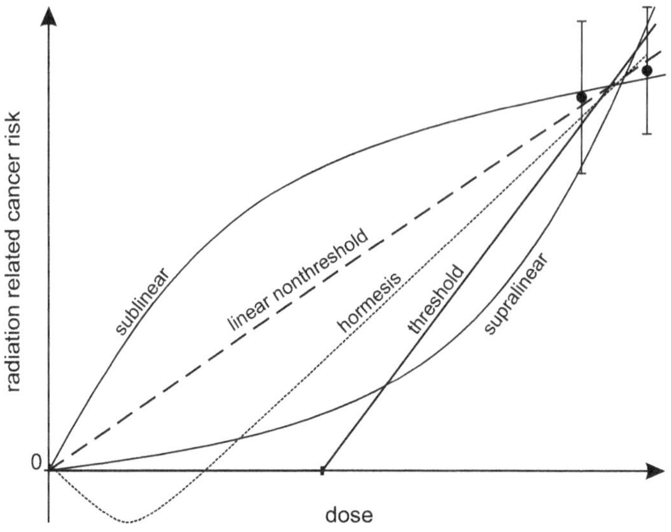

Fig. 5.32: Schematic illustration of different possible extrapolations of the probability of stochastic radiation effects down to very low doses. The LNT model is considered a plausible choice and used as the basis of radiation protection. However, this remains controversial, since other models cannot be ruled out. In the case of radiation hormesis, a beneficial effect of small radiation doses is assumed, e.g., by stimulation of DNA repair mechanisms or the immune system. Adapted from [15].

ies, since the required sample size increases as the inverse square of the dose. Various other types of extrapolation from measured risks down to very small doses as sketched in Figure 5.32 are possible and corresponding models are discussed, see e.g. [41, 65]. Based on biophysical arguments, the LNT model is considered a plausible choice and is, therefore, used as the basis of radiation protection.

5.5.2 Equivalent and effective dose

For the purpose of radiation protection, one is interested a simple system to estimate the total risk of developing *stochastic* effects after exposure of the whole body or some parts of it to low doses of ionizing radiation. This is provided by the concept of the **effective dose**, which is outlined in the following (importantly, note that this is very different from the organ specific dose limits in radiation therapy, which apply to *deterministic* effects at much higher doses).

For a given non-uniform dose of ionizing radiation (e.g. from a diagnostic radiograph or CT scan), the corresponding effective dose can be understood as the theoretical uniform whole-body dose that would lead to the same nominal health risk. Assuming validity of the LNT model, the number of cells in an organ or tissue T that may be affected by radiation induced mutations is proportional to the absorbed dose

Table 5.5: Radiation weighting factors w_R according to current recommendations of the International Commission on Radiation Protection (ICRP) [22]. The values are based on the RBE (Section 5.3.4). Except for the case of neutrons, energy independent values are used as an approximation.

radiation	w_R
photons	1
electrons and muons	1
protons and charged pions	2
neutrons	continuous function of kinetic energy[24]
α particles, heavy ions, fission fragments	20

$\bar{D}_T$ averaged over T. In addition, as discussed in Section 5.3.4, not only the absorbed dose alone, but also the RBE of the radiation determines the probability of radiation induced DNA mutations. RBE depends on the type and energy spectrum of the chosen radiation in a very complex pattern. For the purpose of radiation protection, approximative values, the **radiation weighting factors** w_R (Tab. 5.5) are used to determine the radiation weighted **equivalent dose**

$$H_T = w_R \bar{D}_T \tag{5.108}$$

generated by radiation of type R in an organ or tissue T. The unit of the equivalent dose is $1\,\mathrm{J}\cdot\mathrm{kg}^{-1}$, which is called 1 Sievert (Sv)[23] in demarcation to the absorbed dose $D = \frac{d\bar{\epsilon}}{m}$ measured in Gray (Gy). If an individual is exposed to different types of radiation, the equivalent dose for organ or tissue T is given by the sum

$$H_T = \sum_R w_R \bar{D}_{T,R} \,. \tag{5.109}$$

The central quantity in radiation protection is the **effective dose**, which is given by the weighted sum of the equivalent doses of all organs and tissues in the human body

$$E = \sum_T w_T H_T = \sum_T w_T \left[\sum_R w_R \bar{D}_{T,R} \right] . \tag{5.110}$$

Like the equivalent dose, also the effective dose is given in units of Sv. The weights w_T are called the **tissue weighting factors**. In the following, we briefly summarize

23 After R. Sievert, for his fundamental contributions to medical radiation physics and radiation protection.

24 The radiation weighting factor for neutrons with kinetic energy E is given by the continuous function

$$w_R(E) = \begin{cases} 2.5 + 18.2 \exp\left[-\frac{1}{6}\ln^2(E)\right] & E < 1\,\mathrm{MeV}\,, \\ 5.0 + 17.0 \exp\left[-\frac{1}{6}\ln^2(2E)\right] & 1\,\mathrm{MeV} < E < 50\,\mathrm{MeV}\,, \\ 2.5 + 3.25 \exp\left[-\frac{1}{6}\ln^2(0.04E)\right] & E > 50\,\mathrm{MeV}\,. \end{cases}$$

Table 5.6: The tissue weighting factors w_T used in calculation of the effective dose are based on the nominal risk coefficients R_T (cases per 10.000 persons per Sv), which are weighted by the lethality fraction k_T, the relative life lost l_T and the reduction in life quality as described in the main text. The point 'other solid' comprises all remaining organs and tissues not explicitly listed in the table. Within this group, brain and salivary glands are considered the most relevant and are thus attributed $w_T = 0.01$ each. The ovary tissue weighting factor (∗) is included in $w_T = 0.08$ of the gonads. All data are from ICRP Publication 103 [22].

	nominal risk coeff.	lethality fraction	relative life lost	detriment (absolute)	detriment (relative)	tissue weighting factor
organ/tissue T	R_T	k_T	l_T	D_T	d_T	w_T
gonads (heritable)	20	0.80	1.32	25.4	0.044	0.08
bone marrow	42	0.67	1.63	61.5	0.107	0.12
colon	65	0.48	0.97	47.9	0.083	0.12
lung	114	0.89	0.80	90.3	0.157	0.12
stomach	79	0.83	0.88	67.7	0.118	0.12
breast	112	0.29	1.29	79.8	0.139	0.12
other solid	144	0.49	1.03	113.5	0.198	0.12
bladder	43	0.29	0.71	16.7	0.029	0.04
liver	30	0.95	0.88	26.6	0.046	0.04
oesophagus	15	0.93	0.87	13.1	0.023	0.04
thyroid	33	0.07	1.29	12.7	0.022	0.04
ovary	11	0.57	1.12	9.9	0.017	(∗)
skin	1000	0.002	1.00	4.0	0.007	0.01
bone	7	0.45	1.00	5.1	0.009	0.01
brain	included in 'other solid'					0.01
salivary glands						0.01
$\sum$	1715	–	–	574.3	1.000	1.00

how these are obtained, without going too deep into the details of the full statistical analysis, which is given in [22]. From epidemiological studies, most notably on the atomic bomb survivors of Hiroshima and Nagasaki, the *nominal risk coefficients*

$$R_T = \frac{\text{excess number of cases}}{10.000 \text{ persons} \cdot 1\,\text{Sv}} \tag{5.111}$$

for radiation induced cancers in different organs and tissues of the body, as well as for heritable effects, were obtained (Tab. 5.6). These are read as follows. The excess number of persons who develop, e.g., lung cancer in their further life increases by 114 per 10.000 persons per 1 Sv equivalent dose to the lung (as compared to an unexposed control group), assuming validity of the LNT model. The R_T values are further processed in order to take into account that the consequences of carcinogenesis vary considerably, depending on which organ or tissue is affected (for example, cancer in the liver or the lung is associated with considerably higher lethality than cancer in the breast or skin):

1. For each organ or tissue, the lethality fraction k_T (i.e., the probability that cancer in organ T is lethal) is used to decompose the nominal risk coefficient into lethal and nonlethal cases:

$$R_T = R_{T,\text{lethal}} + R_{T,\text{nonlethal}} = R_T k_T + R_T(1 - k_T) \tag{5.112}$$

2. Adjustment for quality of life. The nonlethal fraction $R_T(1 - k_T)$ of the nominal risk is weighted by a factor

$$q_T = (1 - q_{\min})k_T + q_{\min} \qquad (q_{\min} = 0.1) \tag{5.113}$$

in the interval $[0.1, 1]$ which expresses the reduction of the quality of life in non-lethal cases. It is assumed that this is proportional to the lethality fraction k_T, i.e., that a type of cancer with high lethality also leads to a strong reduction in the quality of life in cases where it is not lethal. The minimum value $q_{\min} = 0.1$ can be interpreted as a means to express that already the diagnosis of cancer in cases that are rarely lethal means a reduction in the quality of life.[25]
3. Adjustment for years of life lost. Each organ or tissue T is weighted by the factor

$$l_T = \frac{\text{average years of life lost due to cancer in organ } T}{\text{average years of life lost due to all types of cancer}}\,, \tag{5.114}$$

which increases the relative weight of organs that often develop cancers that are lethal in younger years, for example, bone marrow (leukemia).

Applying these corrections yields the *radiation detriment*

$$D_T = [R_T k_T + R_T(1 - k_T)q_T]\, l_T \tag{5.115}$$

associated with cancer in organ or tissue T, or heritable effects in the case of the gonads. It takes into account the reduced quality of life in nonlethal cases and increases the weight of cancer types that typically affect young patients. The *relative detriment* is obtained by the normalization

$$d_T = \frac{D_T}{\sum\limits_T D_T} \qquad \text{with} \qquad \sum_T d_T = 1\,. \tag{5.116}$$

Finally, considering the statistical uncertainties in the estimation of the R_T values and in order to keep calculations simple, the d_T values are grouped into four categories, yielding the tissue weighting factors $w_T = 0.12,\ 0.08,\ 0.04$ or 0.01 (Tab. 5.6), which are used to determine the effective dose (equation 5.110) as a measure of the radiation related risk due to carcinogenesis and heritable effects. It is the central quantity in

25 An exception is radiation induced skin cancer, which is almost exclusively of the basalioma type that does practically not metastasize and can be cured very well. Here, the $q_{\min}$ adjustment is omitted.

Table 5.7: Summary of different dosimetric quantities. The absorbed dose is the fundamental physical quantity in this chapter. Equivalent and effective dose are only used in the context of radiation protection and involve several approximations, which are explained in the main text.

quantity	definition	used in
ion dose	charge generated per mass element of air $J = \frac{dQ}{dm_a} = \frac{1}{\rho_a}\frac{dQ}{dV}$ $[J] = 1\,\frac{C}{kg}$	ion chamber dosimetry
absorbed dose	energy deposited per mass element of a medium (usually water) $D = \frac{d\bar{\epsilon}}{dm} = \frac{1}{\rho}\frac{d\bar{\epsilon}}{dV}$ $[D] = 1\,\frac{J}{kg} = 1\,\text{Gy (Gray)}$	dosimetry, radiotherapy
equivalent dose	mean absorbed dose in a certain organ/tissue T multiplied by radiation weighting factor (Tab. 5.5) $H_T = w_R \bar{D}_T = \sum_R w_R \bar{D}_{T,R}.$ $[H_T] = 1\,\frac{J}{kg} = 1\,\text{Sv (Sievert)}$	radiation protection
effective dose	sum of equiv. doses weighted according to risk of stochastic effects (Tab. 5.6) $E = \sum_T w_T H_T = \sum_T w_T \sum_R w_R \bar{D}_{T,R}$ $[E] = 1\,\frac{J}{kg} = 1\,\text{Sv (Sievert)}$	radiation protection

radiation protection, and maximum permissible annual values for different groups of persons (children, general public, radiation exposed personnel) are defined in radiation protection regulations (e.g., *Strahlenschutzverordnung* in Germany).

Based on the detriment values of 574.3 (total) and 25.4 (heritable) per 10.000 persons per Sv (Tab. 5.6), one obtains a risk estimate for developing severe or fatal cancer of about 5.5% per Sv, and a risk estimate for heritable effects of about 0.2% per Sv [22, 25]. Or, expressed differently, an additional effective dose of 2 mSv (which is a typical annual value due to natural or man made sources, see below) causes carcinogenesis in approximately 1 out of 10.000 cases [25]. Again, note that these estimations are based on the validity of the LNT model, as discussed above. Finally, Tab. 5.7 summarizes the different dose concepts used in dosimetry and in radiation protection.

Table 5.8: Mean values of the annual effective dose in Germany [24]. Remarkably, almost 50% is due to medical diagnostics.

	source of radiation	effective dose E [mSv] per year
natural sources	cosmic radiation (sea level)	0.3
	terrestrial sources:	
	– radiation from the ground	0.4
	– radon	1.1
	– ingestion of natural radionuclides	0.3
	$\sum_{\text{natural}}$	$\approx$ **2.1**
man-made sources	medical diagnostics	1.9
	nuclear power plants	< 0.01
	Chernobyl	< 0.01
	Fukushima	no measurable activity
	nuclear weapon tests	< 0.01
	other man-made sources	< 0.02
	$\sum_{\text{man made}}$	$\approx$ **1.9**
total		$\approx$ **4.0**

5.5.3 Natural and man made radiation exposure

The human organism is constantly exposed to ionizing radiation from natural sources. This is one reason why very efficient DNA repair mechanisms have evolved. It is instructive to compare the corresponding natural, unavoidable effective dose (per year) to typical effective doses that can result from man made sources (Tab. 5.8).

The natural radiation exposure contains contributions from cosmic radiation, which primarily consists of protons and atomic nuclei, plus secondary cosmic radiation that is produced when these hit the earth's atmosphere. The second source are primordial radionuclides in the earth's crust. The radioactive isotopes of the noble gas radon constitute the most important natural source of ionizing radiation. Another important isotope is ^{40}K, which is incorporated along with stable K isotopes into the human body. Due to regional variations of the geological composition of the earth's crust, as well as to an increase of the cosmic radiation with height above sea level

Table 5.9: Average effective dose values for common radiological examinations [49].

radiological examination	effective dose E [mSv]
dental radiograph	< 0.01
chest radiograph	0.01
mammogram	0.4
head CT	2
chest CT	7
abdomen CT	8

and towards the earth's poles, the natural radiation background can vary considerably [57].

By far the largest fraction of the man made radiation exposure of the average individual, at least in industrialized countries, is due to the medical use of x-rays. In Germany, it makes up the vast part of the effective dose from man made sources and amounts to about the same effective dose as the average natural radiation exposure. Typical effective doses of different radiological procedures are given in Tab. 5.9. Note that a large part of the population is not exposed to any medical radiation sources at all, while some individuals receive much larger doses than the average value of 1.9 mSv. In particular, a single CT scan of the thorax or abdomen easily exceeds the natural annual effective dose by a factor of up to 5.

References

[1] S. P. Ahlen. Theoretical and experimental aspects of the energy loss of relativistic heavily ionizing particles. *Rev. Mod. Phys.*, 52(1):121–173, 1980.

[2] A. Ahnesjö, P. Andreo, and A. Brahme. Calculation and Application of Point Spread Functions for Treatment Planning with High Energy Photon Beams. *Acta Oncologica*, 26(1):49–56, 1987.

[3] A. Ahnesjö and M. M. Aspradakis. Dose calculations for external photon beams in radiotherapy. *Phys. Med. Biol.*, 44(11):R99, 1999.

[4] B. Alberts, A. Johnson, J. Lewis, M. Raff, K. Roberts, and P. Walter. *Molecular Biology of the Cell.* Garland Science, 5th edition, 2008.

[5] P. Andreo. Monte Carlo techniques in medical radiation physics. *Phys. Med. Biol.*, 36(7):861, 1991.

[6] F. H. Attix. *Introduction to radiological physics and radiation dosimetry.* Wiley VCH, 1986.

[7] S. M. Bentzen, L. S. Constine, J. O. Deasy, A. Eisbruch, A. Jackson, L. B. Marks, R. K. Ten Haken, and E. D. Yorke. Quantitative Analyses of Normal Tissue Effects in the Clinic (QUANTEC): An Introduction to the Scientific Issues. *International Journal of Radiation Oncology Biology Physics*, 76(3):S3–S9, 2010.

[8] H. Bethe. Zur Theorie des Durchgangs schneller Korpuskularstrahlen durch Materie. *Ann. Physik*, 5:325–400, 1930.

[9] H. Bethe. Bremsformel für Elektronen relativistischer Geschwindigkeit. *Z. Physik*, 76:293–299, 1932.

[10] W. Böcker, H. Denk, and P. U. Heitz. *Pathologie.* Urban und Fischer Verlag/Elsevier GmbH, 5th edition, 2012.

[11] N. Bohr. On the theory of the decrease of velocity of moving electrified particles on passing through matter. *Phil. Mag.*, 25(6):10–31, 1913.

[12] T. Bortfeld. An analytical approximation of the Bragg curve for therapeutic proton beams. *Med. Phys.*, 24:2024–2033, 1997.

[13] T. Bortfeld. IMRT: A review and preview. *Physics in Medicine and Biology*, 51(13):R363–R379, 2006.

[14] T. Bortfeld and W. Schlegel. An analytical approximation of depth-dose distributions for therapeutic proton beams. *Phys. Med. Biol.*, 41:1331–1339, 1996.

[15] D. J. Brenner, R. Doll, D. T. Goodhead, E. J. Hall, C. E. Land, J. B. Little, J. H. Lubin, D. L. Preston, R. J. Preston, J. S. Puskin, E. Rone, R. K. Sachs, J. M. Samet, R. B. Setlow, and M. Zaider. Cancer

risks attributable to low doses of ionizing radiation: Assessing what we really know. *Proc. Nat. Acad. Sci. USA*, 100:13761–13766, 2003.

[16] K. H. Chadwick and H. P. Leenhouts. A molecular theory of cell survival. *Phys. Med. Biol.*, 18:78–87, 1973.

[17] International Commission on Radiation Units and Measurements. ICRU Report 16: Linear Energy Transfer, 1970.

[18] International Commission on Radiation Units and Measurements. ICRU Report 36: Microdosimetry, 1983.

[19] International Commission on Radiation Units and Measurements. ICRU Report 37: Stopping Powers for Electrons and Positrons, 1984.

[20] International Commission on Radiation Units and Measurements. ICRU Report 49: Stopping Powers and Ranges for Protons and Alpha Particles, 1993.

[21] International Commission on Radiation Units and Measurements. ICRU Report 85: Fundamental Quantities and Units for Ionizing Radiation (revised). *Journal of the ICRU*, 11(1), 2011.

[22] International Commission on Radiological Protection. ICRP Publication 103: The 2007 recommendations of the international commission on radiological protection. *Ann. ICRP*, 37, 2007.

[23] U. Fano. Penetration of Protons, Alpha Particles, and Mesons. *Ann. Rev. Nucl. Sci.*, 13:1–66, 1963.

[24] Bundesamt für Strahlenschutz (BfS) und Bundesministerium für Umwelt und Naturschutz und Reaktorsicherheit (BMU). Umweltradioaktivität und Strahlenbelastung im Jahr 2013: Unterrichtung durch die Bundesregierung (Parlamentsbericht 2013). 2015. Available online at https://doris.bfs.de/jspui/handle/urn:nbn:de:0221-2015072412951.

[25] E. H. Hall and A. J. Giaccia. *Radiobiology for the Radiologist*. Lippincott Williams and Wilkins, 7th edition, 2012.

[26] D. Hanahan and R. A. Weinberg. Hallmarks of cancer: The next generation. *Cell*, 144:646–674, 2011.

[27] R. K. Hobbie and B. J. Roth. *Intermediate Physics for Medicine and Biology*. Springer, 4th edition, 2007.

[28] J. D. Jackson. *Klassische Elektrodynamik*. De Gruyter, 3rd edition, 2002.

[29] R. Jeraj and P. Keall. The effect of statistical uncertainty on inverse treatment planning based on Monte Carlo dose calculation. *Physics in Medicine and Biology*, 45(12):3601–3613, 2000.

[30] D. Jette and W. Chen. Creating a spread-out Bragg peak in proton beams. *Phys. Med. Biol.*, 56:N131–N138, 2011.

[31] M. W. K. Kan, P. K. N. Yu, and L. H. T. Leung. A Review on the Use of Grid-Based Boltzmann Equation Solvers for Dose Calculation in External Photon Beam Treatment Planning. *BioMed Research International*, 2013.

[32] C. A. Kelsey, P. H. Heintz, D. J. Sandoval, G. D. Chambers, N. L. Adolphi, and K. S. Paffett. *Radiobiology of Medical Imaging*. John Wiley and Sons, 2014.

[33] F. M. Khan and J. P. Gibbons. *Khan's The Physics of Radiation Therapy*. Lippincott Williams and Wilkins, 5th edition, 2014.

[34] G. F. Knoll. *Radiation Detection and Measurement*. John Wiley & Sons, 1986.

[35] F. Kohlrausch. *Praktische Physik*. B. G. Teubner, Stuttgart, 24th edition, 1996.

[36] H. Krieger. *Grundlagen der Strahlungsphysik und des Strahlenschutzes*. Springer Spektrum, 4th edition, 2012.

[37] H. Krieger. *Strahlungsmessung und Dosimetrie*. Vieweg+Teubner Verlag, 2nd edition, 2013.

[38] H. Krieger. *Strahlungsquellen für Technik und Medizin*. Springer Spektrum, 2nd edition, 2013.

[39] T. Krieger and O. A. Sauer. Monte Carlo- versus pencil-beam-/collapsed-cone-dose calculation in a heterogeneous multi-layer phantom. *Physics in Medicine and Biology*, 50(5):859–868, 2005.

[40] C. Leroy and P. G. Rancoita. *Priciples of Radiation Interaction in Matter and Detection*. World Scientific, 3rd edition, 2012.
[41] M.P. Little, R. Wakeford, E.J. Tawn, S.D. Bouffler, and A. Berrington de Gonzalez. Risks Associated with Low Doses and Low Dose Rates of Ionizing Radiation: Why Linearity May Be (Almost) the Best We Can Do. *Radiology*, 251(1):6–12, 2009.
[42] H. Lodish, A. Berk, C. A. Kaiser, M. Krieger, M. P. Scott, and A. Bretscher. *Molecular Cell Biology*. DeGruyter, 6th edition, 2007.
[43] M. E. Lomax, L. K. Folkes, and P. O'Neill. Biological Consequences of Radiation-induced DNA Damage: Relevance to Radiotherapy. *Clinical Oncology*, 25(10):578–585, 2013. Advances in Clinical Radiobiology.
[44] L. Lu. Dose calculation algorithms in external beam photon radiation therapy. *International Journal of Cancer Therapy and Oncology*, 1(2), 2013.
[45] L. Marcu, E. Bezak, and B. Allen. *Biomedical Physics in Radiotherapy for Cancer*. Springer, 1st edition, 2012.
[46] L. A. Mathews, S. M. Cabarcas, and E. M. Hurt. *DNA Repair of Cancer Stem Cells*. Springer, 2013.
[47] P. N. McDermott. *Tutorials in Radiotherapy Physics – Advanced Topics with Problems and Solutions*. CRC Press, 2016.
[48] P. Metcalfe, T. Kron, and P. Hoban. *The Physics of Radiotherapy X-Rays And Electrons*. Medical Physics Publishing, 2007.
[49] F.A. Mettler, W. Huda, T.T. Yoshizumi, and M. Mahesh. Effective Doses in Radiology and Diagnostic Nuclear Medicine: A Catalog. *Radiology*, 248(1):254–263, 2008.
[50] R. Milo and R. Phillips. *Cell Biology by the numbers*. Garland Science, 2016.
[51] W. D. Newhauser and R. Zhang. The physics of proton therapy. *Phys. Med. Biol.*, 60:R155–R209, 2015.
[52] United Nations Scientific Committee on the Effects of Atomic Radiation. UNSCEAR 2012 Report: Sources, Effects and Risks of Ionizing Radiation. 2015. Available online at http://www.unscear.org/unscear/en/publications.html.
[53] S. F. C. O'Rourke, H. McAneney, and T. Hillen. Linear quadratic and tumour control probability modelling in external beam radiotherapy. *Journal of Mathematical Biology*, 58(4):799–817, 2008.
[54] H. Owen, D. Holder, J. Alonso, and R. Mackay. Technologies for delivery of proton and ion beams for radiotherapy. *International Journal of Modern Physics A*, 29(14):1441002, 2014.
[55] N. Papanikolaou and S. Stathakis. Dose-calculation algorithms in the context of inhomogeneity corrections for high energy photon beams. *Medical Physics*, 36(10):4765–4775, 2009.
[56] E. B. Podgorsak. *Radiation Oncology Physics: A Handbook for Teachers and Students*. International Atomic Energy Agency (IAEA), Vienna, 2005.
[57] M. Reiser, F.-P. Kuhn, and J. Debus. *Duale Reihe Radiologie*. Thieme, 3rd edition, 2011.
[58] N. Reynaert, S. C. van der Marck, D. R. Schaart, W. Van der Zee, C. Van Vliet-Vroegindeweij, M. Tomsej, J. Jansen, B. Heijmen, M. Coghe, and C. De Wagter. Monte Carlo treatment planning for photon and electron beams. *Radiation Physics and Chemistry*, 76(4):643–686, 2007.
[59] D. W. O. Rogers. Fifty years of Monte Carlo simulations for medical physics. *Phys. Med. Biol.*, 51(13):R287, 2006.
[60] H. H. Rossi and M. Zaider. *Microdosimetry and its Applications*. Springer, 1996.
[61] M. Spurio S. Braibant, G. Giacomelli. *Particles and Fundamental Interactions – An Introduction to Particle Physics*. Springer, 2009.
[62] A. Sancar, L. A. Lindsey-Boltz, K. Ünsal-Kacmaz, and S. Linn. Molecular Mechanisms of Mammalian DNA Repair and the DNA Damage Checkpoints. *Annual Review of Biochemistry*, 73(1):39–85, 2004. PMID: 15189136.

[63] S. M. Seltzer. Calculation of photon mass energy-transfer and mass energy-absorption coefficients. *Radiation Research*, 136(2):147–170, 1993.
[64] C. Tomasetti and B. Vogelstein. Variation in cancer risk among tissues can be explained by the number of stem cell divisions. *Science*, 347:78–81, 2015.
[65] M. Tubiana, L.E. Feinendegen, C. Yang, and J.M. Kaminski. The Linear No-Threshold Relationship Is Inconsistent with Radiation Biologic and Experimental Data. *Radiology*, 251(1):13–22, 2009.
[66] E. A. Uehling. Penetration of Heavy Charged Particles in Matter. *Ann. Rev. Nucl. Sci.*, 4:315–350, 1954.
[67] W. Ulmer. Theoretical aspects of energy-range relations, stopping power and energy straggling of protons. *Radiation Physics and Chemistry*, 76:1089–1107, 2007.
[68] W. Ulmer and E. Matsinos. Theoretical methods for the calculation of Bragg curves and 3D distributions of proton beams. *Eur. Phys. J. Special Topics*, 190:1–81, 2010.
[69] O. N. Vassiliev, T. A. Wareing, J. McGhee, G. Failla, M. R. Salehpour, and F. Mourtada. Validation of a new grid-based Boltzmann equation solver for dose calculation in radiotherapy with photon beams. *Physics in Medicine and Biology*, 55(3):581, 2010.
[70] F. Verhaegen and J. Seuntjens. Monte Carlo modelling of external radiotherapy photon beams. *Phys. Med. Biol.*, 48(21):R107, 2003.
[71] J. Z. Wang, Z. Huang, S. S. Lo, W. T. C. Yuh, and N. A. Mayr. A generalized linear-quadratic model for radiosurgery, stereotactic body radiation therapy, and high dose rate brachytherapy. *Science Translational Medicine*, 2:39–48, 2010.
[72] S. Webb. The physical basis of IMRT and inverse planning. *The British Journal of Radiology*, 76(910):678–689, 2003.
[73] S. Webb and A. E. Nahum. A model for calculating tumour control probability in radiotherapy including the effects of inhomogeneous distributions of dose and clonogenic cell density. *Physics in Medicine and Biology*, 38(6):653–666, 1993.
[74] H. Willers, J. Dahm-Daphi, and S. N. Powell. Repair of radiation damage to DNA. *British Journal of Cancer*, 90:1297–1301, 2004.
[75] K. Zink. Einführung in die Strahlentherapie und Therapie mit offenen Nukliden (Vorlesungsskript Fachhochschule Gießen-Friedberg), 2004.

Symbols and abbreviations used in Chapter 5

A	atomic mass number (nucleon number)
b	impact parameter
BGC	Bragg–Gray conditions
$\beta = \frac{v}{c}$	ratio of particle velocity and speed of light
c	speed of light in vacuum
CPE	charged particle equilibrium
CSDA	continuous slowing down approximation
$D(\underline{r})$	absorbed dose at $\underline{r}$
DNA	deoxyribonucleic acid
DSB	DNA double strand break
$\bar{D}_T$	mean absorbed dose in organ/tissue T
eV	electron volt

ϵ_0	vacuum permittivity
F_C	Coulomb force
FSU	functional subunit
Φ	fluence
γ	Lorentz factor
$H_T = w_R \bar{D}_T$	equivalent dose in organ/tissue T
$h_{point}(\underline{r})$	point energy deposition kernel
$h_{pencil}(\underline{r})$	pencil energy deposition kernel
$\bar{I}$	mean excitation energy
ICRU	International Commission on Radiation Units and Measurement
ICRP	International Commission on Radiological Protection
K	kerma (kinetic energy released per unit mass)
LET	linear energy transfer
LNT	linear nonthreshold
LQ	linear quadratic
LBTE	linear Boltzmann transport equation
MLC	multi-leaf collimator
μ	linear attenuation coefficient
μ_{en}	energy absorption coefficient
μ_{tr}	energy transfer coefficient
N_A	Avogadro's number
NTCP	normal tissue complication probability
m_0	rest mass of a particle
m_e	electron rest mass
m_p	proton rest mass
OAR	organ at risk
$\underline{p}$	momentum
PDD(z)	percentage depth dose curve
PSF	point-spread function
PTV	planning target volume
$\underline{r}$	position vector
R_{CSDA}	CSDA range
R_p	projected range
RBE	relative biological effectiveness
ROS	reactive oxygen species
ρ	mass density
ρ_e	electron density
S_{el}	electronic stopping power
S_{nuc}	nuclear stopping power
S_{rad}	radiative stopping power
SSB	DNA single strand break

SSD	source-to-surface distance
T	terma (total energy released per unit mass)
TCP	tumor control probability
$\bar{W}_{\text{air}}$	mean ionization energy of air
w_R	radiation weighting factor
w_T	tissue weighting factor
Z	atomic charge number (proton number)

6 Phase contrast radiography

Despite the enormous success and widespread use of radiography for over 100 years, projection imaging based on differential absorption of x-rays in matter is subject to a number of severe limitations and shortcomings:

- The contrast of soft and weakly absorbing tissue is too low. As discussed in Chapter 3, the photoelectric effect is the dominant interaction for light elements in the diagnostic energy range up to $\sim$ 100 keV, resulting in a particular scaling of the linear absorption coefficient μ with elemental composition (atomic number Z), density ρ and photon energy E, approximately as

  $$\mu \propto \rho \frac{Z^4}{E^3} \,. \tag{6.1}$$

  Hence, absorption of x-rays provides excellent contrast between bone (high content of Ca and P with relatively high Z), and soft (unmineralized) tissue. The absorption coefficient of soft tissues is in the range of water or lower (for example, in adipose tissue or the lung). Imaging with absorption contrast is, therefore, severely limited in view of important applications, such as detection and delineation of metastases.
- Absorption contrast vanishes at small length scale, typically in the micron range. If the typical extent d of an object to be imaged is *small* compared to the inverse of μ, we have

  $$I(d) = I_0 \exp(-\mu d) \approx I_0 \underbrace{(1 - \mu d)}_{\approx 1} \lesssim I_0 \,, \qquad d \ll \mu^{-1} \,, \tag{6.2}$$

  and, therefore, insufficient absorption contrast. This contrast related resolution limit is even more pronounced for internal tissue structures, where the difference in absorption coefficient enters $\mu \to \Delta\mu$. Nanoscale resolution, therefore, requires a different contrast mechanism.
- The geometrical optical model breaks down at high resolution. As we have seen, tomography is based on *projections* along a given direction $\underline{n}_\theta$, e.g., in the formulation of the Radon transform

  $$\mathcal{R}f(\underline{n}_\theta, s) = \int \mathrm{d}^2 x\, f(\underline{x})\, \delta(\underline{x} \cdot \underline{n}_\theta - s) \,. \tag{6.3}$$

  When the resolution approaches length scales where diffraction effects become apparent, a tomography concept based solely on geometrical optics must break down. The validity of the geometrical optical concept should be investigated based on the more suitable wave optical description.

In this chapter, we discuss radiography with contrast mechanisms beyond absorption, mostly known as phase contrast x-ray imaging, sensitive to absorption **and**

DOI 10.1515/9783110426694-006

phase contrast. We also discuss a further contrast mechanism known as scattering contrast. Phase contrast radiography can be implemented based on very different imaging modalities and setups, but all require the wave optical concepts presented in the first sections of the chapter, before the different experimental implementations are addressed. Phase contrast has not yet reached clinical practice, but is currently under active development and at the stage of preclinical and clinical testing. First applications can be expected for mammography. A crucial step was the translation of the method from synchrotron beams to laboratory radiation sources, with the associated significant decrease in partial coherence. In this chapter, however, we do not focus on medical diagnostic imaging. Phase contrast imaging is discussed more broadly, also in view of high resolution imaging of tissues for biomedical research or diagnostic purposes based on biopsies. Here, the prospect of phase contrast imaging in combination with tomographic reconstruction is the development of a virtual 3d histology beyond the conventional techniques based on thin slices, which often make it difficult to establish the true 3d connectivity of tissue.

6.1 Radiography beyond geometrical optics

6.1.1 Limits of geometrical optics

Let us consider the limits of the geometric optical description of x-ray imaging, by addressing the conditions for observing diffraction and refraction effects of x-rays. Diffraction manifests itself after propagation of the wavefield over some distance, the propagation distance, see Figure 6.1. If not mentioned otherwise, we assume coherent propagation in this chapter. Many of the conclusions, however, carry over to the more realistic case of partial coherence, provided the degree of coherence is sufficiently high.

- **Diffraction:** Figure 6.2 shows the near field diffraction of a sharp edge as a paradigmatic example of diffraction effects. Following [54], we consider the path length difference between rays $\overline{AC}$ and $\overline{BC}$

$$\overline{AC} = z\,, \qquad \overline{BC} = \sqrt{z^2 + a^2} \tag{6.4}$$

$$\Rightarrow \Delta s = |\overline{AC} - \overline{BC}| = \sqrt{z^2 + a^2} - z \overset{a \ll z}{\approx} z\left(1 + \frac{a^2}{2z^2}\right) - z = \frac{a^2}{2z}\,. \tag{6.5}$$

We have destructive interference if $\Delta s = \frac{\lambda}{2}$ or, equivalently, $\frac{a^2}{z} = \lambda$. This illustrates how the appearance of diffraction effects is controlled by the Fresnel number

$$F = \frac{a^2}{\lambda z}\,, \tag{6.6}$$

as a unitless combination of structure size a, propagation distance z and wavelength λ. For $F \ll 1$ (the contact regime) we encounter absorption contrast only.

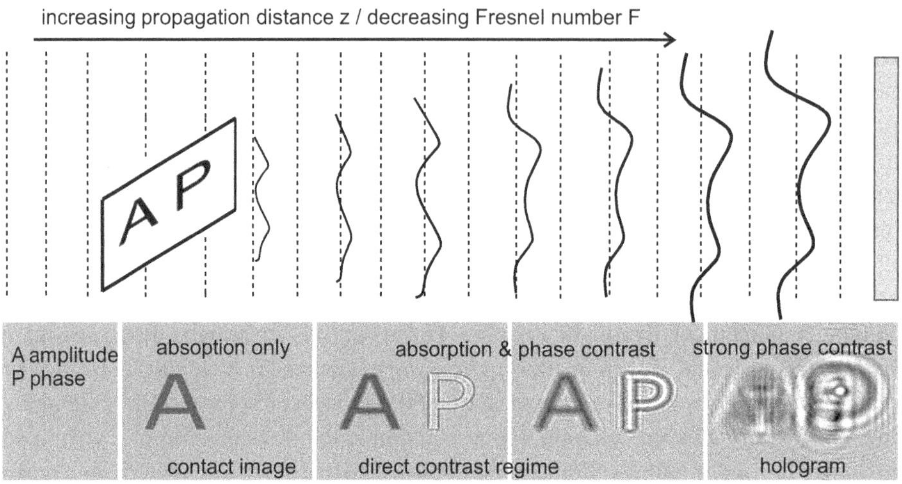

Fig. 6.1: Principle of phase contrast arising from free propagation. An object consisting of structural elements with (A) amplitude and (P) phase contrast is illuminated by a plane wave. Phase contrast arises by self interference of the diffracted waves with the primary beam behind the object plane, effectively converting phase shifts into measurable intensity fringes. The different imaging regimes are characterized by the respective Fresnel numbers F, decreasing with the propagation distance (also called the defocus distance). The contact regime behind the object (pure absorption contrast) is followed by the direct contrast regime, and finally by the fully holographic regime.

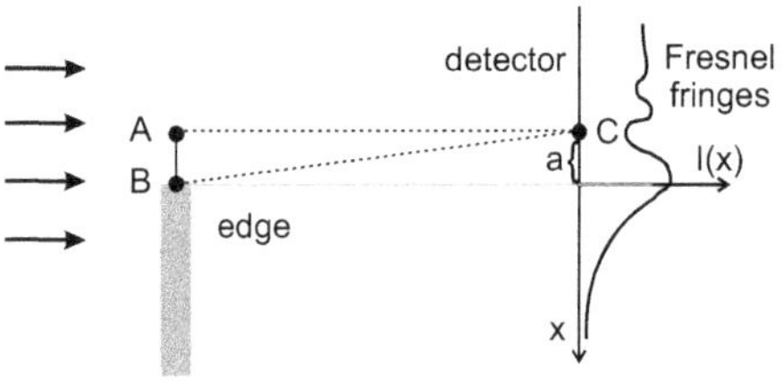

Fig. 6.2: Near field diffraction at an edge, leading to characteristic fringes behind the object. This well known wave optical effect creates visibility for tissue interfaces by edge enhancement. At the same time the fringes lead to a blurring in the projection image. Diffraction is hence a blessing (visibility, contrast) and a curse (blurring) for imaging. With suitable phase retrieval algorithms, the blurring can be inverted to yield a sharp reconstruction, even of a completely nonabsorbing object (pure phase contrast).

For increasing the propagation distance z approaching $F \approx 1$ (Fresnel diffraction regime), diffraction effects in the form of intensity oscillations or fringes become more and more visible. Let us take $\lambda \approx 10^{-11}$ m and $z \approx 10^{-1}$ m for a typical radiographic recording. Diffraction effects will manifest themselves only for small structure sizes a in the micron range and below.

$$F \approx 1 \qquad \Leftrightarrow \qquad a \leq \sqrt{10^{-11}\,\mathrm{m} \cdot 10^{-1}\,\mathrm{m}} = 10^{-6}\,\mathrm{m}\,. \tag{6.7}$$

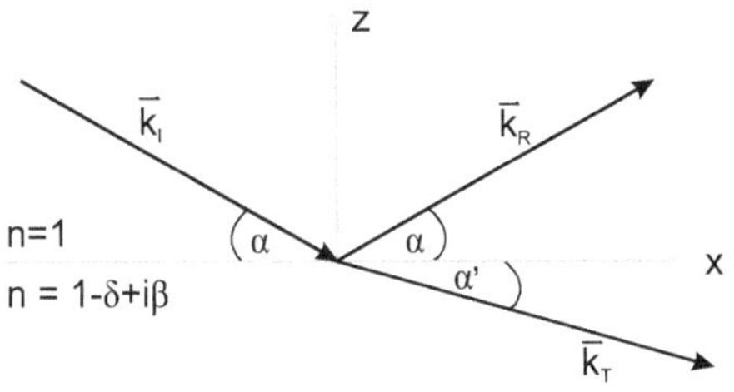

Fig. 6.3: Refraction and reflection of x-rays at an interface between air and condensed matter. Total external reflection is observed for rays reaching the interface from the lower density medium under grazing angles.

- **Reflection:** As we know from Chapter 3, the real part of the x-ray index of refraction $n = 1 - \delta$ is asymptotically close to unity for high photon energies, say for the matter of concreteness that we have a tissue with $\delta \approx \mathcal{O}(10^{-6})$ (in Section 6.1.2 we will discuss this in more detail). For perpendicular incidence of the x-ray beam onto an air/tissue interface, the reflection coefficient R (for intensities) can be calculated from the Fresnel equations

$$R = \left(\frac{n_1 - n_2}{n_1 + n_2}\right)^2 = \left(\frac{\delta}{2}\right)^2 \approx \mathcal{O}(10^{-12}) , \tag{6.8}$$

showing that reflection is completely negligible. Snell's law with notation as in Figure 6.3

$$n_1 \cos \alpha_1 = n_2 \cos \alpha_2 \tag{6.9}$$

gives a critical angle of total external reflection α_c of (using the Taylor approximation $\cos x \approx 1 - \frac{x^2}{2}$, $x \ll 1$)

$$\cos \alpha_c = 1 - \delta \tag{6.10}$$

$$1 - \frac{\alpha_c^2}{2} = 1 - \delta \tag{6.11}$$

$$\Rightarrow \alpha_c = \sqrt{2\delta} = \mathcal{O}(10^{-3}\,\text{rad}) . \tag{6.12}$$

With a maximum angular offset α_c of an x-ray by the presence of the interface, the smallest structure to be resolved in projection radiography is limited by $a = \alpha_c z$, which for a minimum propagation distance between body and detection plane of $y \simeq 0.1$ m gives a resolution limit of only 100 µm. Luckily, in practice, this estimate would be overly restrictive, since typical interfaces are neither flat nor sharp.

6.1.2 The x-ray index of refraction and phase shift in matter

As discussed in Section 3.2.4, the x-ray index of refraction is typically written as

$$n = 1 - \delta + i\beta , \tag{6.13}$$

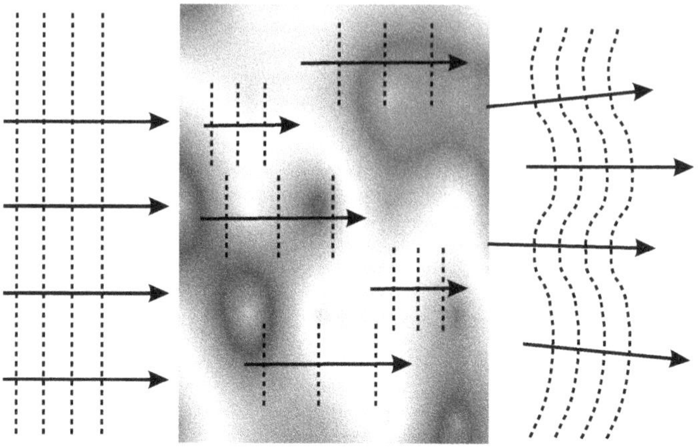

Fig. 6.4: Propagation of x-ray waves through a medium with varying index of refraction.

and determines the refraction of x-rays at interfaces, as well as absorption and phase shift when the wave penetrates a slab of matter, i.e., a medium with index n, see Figure 6.4. If a plane wave e^{ikz} with wavenumber $k = \frac{2\pi}{\lambda}$ transits from vacuum ($n = 1$) to a medium with $n = 1 - \delta + i\beta$, it changes its wavenumber

$$e^{ikz} \xrightarrow{k \to k'} e^{iknz} , \tag{6.14}$$

according to the respective difference in the phase velocity

$$k' = \frac{2\pi}{\lambda'} = \frac{2\pi}{\lambda} n . \tag{6.15}$$

The x-ray index of refraction is a continuum property of matter, but can be traced back quantitatively to atomic properties, notably elastic scattering of bound electrons present in the medium at a certain density, see Section 3.2.4. For large diffraction angles (wide angle diffraction and Bragg peaks) the continuum description of the x-ray index of refraction breaks down, but for forward directed beam propagation and diffraction at small angles, the optics are insensitive to the interatomic distances, and matter is probed only in the form of its local averaged response of elastic (Thomson) scattering. The optical response of a single atom is described by the atomic form factor as a function of the scattering vector

$$f(\underline{q}, \hbar\omega) = f_0(\underline{q}) + f'(\hbar\omega) + if''(\hbar\omega) . \tag{6.16}$$

The first term

$$f_0(\underline{q}) = \int \rho_e(\underline{r})\, e^{i\underline{q}\cdot\underline{r}} d^3r = \begin{cases} Z, & |\underline{q}| \to 0 , \\ 0, & |\underline{q}| \to \infty \end{cases} \tag{6.17}$$

corresponds to the Fourier transform of the electron density distribution ρ_e of the atom, as given by the orbital wave functions. For $\underline{q} \to 0$ (forward scattering), we have $f_0(\underline{q}) \to Z$ for nonionized atoms or $f_0(\underline{q}) \to (Z - I)$ for ions of ionization state I. The real and imaginary corrections f' and if'' not only depend on $\underline{q}$ but also on $\hbar\omega$, reflecting the resonance properties of atomic orbitals, peaking in magnitude at the absorption edges. Form factors for each element are tabulated in the International Tables for Crystallography and are also available online.[1] Often, data have been acquired by precision measurements of the absorption cross section to determine f'', and then using the Kramers–Kronig relation to determine f' from f'' [1]. With this notation of the atomic form factor, the dispersion term δ of the x-ray refractive index becomes

$$\delta = \frac{r_0 \lambda^2}{2\pi} \rho_a(\underline{r})[Z + f'(\omega)] , \tag{6.18}$$

where ρ_a is the number density of atoms, $r_0 = 2.82 \cdot 10^{-15}$ m the Thomson scattering length and $f'(\omega)$ the real part of the correction. For the imaginary part β of the index of refraction we have

$$\beta = \frac{r_0 \lambda^2}{2\pi} \rho_a(\underline{r}) f''(\omega) . \tag{6.19}$$

After a propagation distance Δ through a homogeneous medium with index n, the transmitted plane wave

$$e^{ik\Delta(1-\delta+i\beta)} = e^{ik\Delta} e^{-ik\delta\Delta} e^{-k\beta\Delta} \tag{6.20}$$

has a reduced amplitude and a retarded phase given with respect to the (reference) vacuum wave $e^{ik\Delta}$. The phase retardation $\Delta\varphi$, is proportional to the traversed length of material and to δ

$$e^{-ik\delta\Delta} \equiv e^{-i\Delta\varphi} . \tag{6.21}$$

The decrease in amplitude is given by the factor

$$e^{-k\beta\Delta} , \tag{6.22}$$

resulting in a decrease of intensity by

$$\frac{I}{I_0} = \left| e^{-k\beta\Delta} \right|^2 = e^{-2k\beta\Delta} = e^{-\mu\Delta} , \tag{6.23}$$

which shows that the linear attenuation coefficient μ and β are related by

$$\beta = \frac{\mu}{2k} = \frac{\mu\lambda}{4\pi} . \tag{6.24}$$

1 http://it.iucr.org/Cb/ch4o2v0001/sec4o2o6/ or http://henke.lbl.gov/optical_constants/

For a material composed of several elements, we have

$$\delta = \frac{r_0 \lambda^2 N_A}{2\pi} \rho \sum_j \frac{[Z_j + f_j']}{A_j} w_j \tag{6.25}$$

$$\beta = \frac{r_0 \lambda^2 N_A}{2\pi} \rho \sum_j \frac{f_j''}{A_j} w_j \,, \tag{6.26}$$

with Avogadro's constant N_A, mass density ρ ($\rho_a = \rho \frac{N_A}{A}$), atomic mass A_j and mass fraction w_j of element j. For biological matter (dominated by light elements C, H, O, N, S, P) we have in good approximation

$$\frac{Z_j}{A_j} \approx \frac{1}{2} \quad \Rightarrow \quad \delta = \frac{r_0 \lambda^2 N_A}{4\pi} \rho \tag{6.27}$$

(except for H with $Z = A = 1$). Inserting representative values $\lambda = 1\,\text{Å}$ and $\rho = 1\,\text{g/cm}^{-3}$ (density of water) we obtain $\delta \approx 1.3 \cdot 10^{-6}$. For a biological cell we can expect $\delta = 1.7 \cdot 10^{-6}$ and $\beta = 2.5 \cdot 10^{-9}$ at $\lambda = 1\,\text{Å}$ [27]. Importantly, the optical constants underlying the phase shift dominate over the imaginary part (absorption) by an overwhelming factor, here

$$\frac{\delta}{\beta} = \mathcal{O}(10^3) \,. \tag{6.28}$$

Phase contrast methods should, therefore, enable a fundamentally improved resolution and contrast for a given dose, or at given contrast and resolution a significantly reduced dose. The only challenge is how to measure the phase retardation in a suitable manner. In order to present the currently most promising approaches, we first need to consider some fundamentals of (x-ray) wave propagation in matter.

6.2 Wave equations and coherence

Before addressing phase contrast imaging, some optical foundations are required, which we present in this section in a condensed manner.

6.2.1 The stationary wave equation and diffraction integrals

For clarity of notation, let us first rewrite the four Maxwell equations, in the absence of free charges and currents as

$$\nabla \times \underline{E} = -\partial_t \underline{B} \tag{6.29}$$

$$\nabla \cdot \underline{D} = 0 \tag{6.30}$$

$$\nabla \times \underline{H} = \partial_t \underline{D} \tag{6.31}$$

$$\nabla \cdot \underline{B} = 0 \tag{6.32}$$

in terms of the electric field $\underline{E}$, the magnetic (magnetizing) field $\underline{H}$, the electric displacement $\underline{D} = \epsilon_0\epsilon_r\underline{E}$ and the magnetic induction $\underline{B} = \mu_0\mu_r\underline{H}$. The electrical permittivity and the magnetic permeability are denoted by $\epsilon = \epsilon_0\epsilon_r$ and $\mu = \mu_0\mu_r$, respectively. To derive the wave equation, we take the curl on both sides of equation (6.29), which gives

$$\nabla \times (\nabla \times \underline{E}) = \nabla(\nabla \cdot \underline{E}) - \nabla^2\underline{E} = -\nabla^2\underline{E} = -\partial_t(\nabla \times \underline{B}) \tag{6.33}$$

where we use $\nabla \cdot \underline{E} = \nabla \cdot \underline{D}/\epsilon = 0$ for homogeneous media.[2] Inserting equation (6.31) into equation (6.33) yields the wave equation for the electric field $\underline{E}$

$$-\nabla^2\underline{E} = -\mu\partial_t(\nabla \times \underline{H}) = -\mu\partial_t^2\underline{D} = -\epsilon\mu\partial_t^2\underline{E} \quad \Rightarrow \quad \nabla^2\underline{E} - \frac{n^2}{c^2}\partial_t^2\underline{E} = 0\,, \tag{6.34}$$

and similarly for the magnetic field. In summary, using the vacuum speed of light $c = \frac{1}{\sqrt{\epsilon_0\mu_0}}$ and the index of refraction $n = \sqrt{\epsilon_r\mu_r}$, we can write for any component of the fields

$$\nabla^2\Psi - \frac{n^2}{c^2}\partial_t^2\Psi = 0 \qquad \forall\, \Psi \in \{E_x, E_y, E_z, H_x, H_y, H_z\}\,. \tag{6.35}$$

Note that the components are, in general, not decoupled, since the Maxwell equations not used in the derivation also constrain the fields. Only if polarization effects are negligible, as for the forward directed diffraction of x-rays, can equation (6.35) be applied quasi independently for each polarization direction orthogonal to the optical axis, i.e., the equations reduce to a scalar wave equation. Further, the scalar field Ψ can be spectrally decomposed,

$$\Psi = \Psi(\underline{r}, t) = \frac{1}{\sqrt{2\pi}}\int_0^\infty \Psi_\omega(\underline{r})\mathrm{e}^{-i\omega t}\,d\omega\,, \tag{6.36}$$

i.e., it can be written as a superposition of monochromatic fields Ψ_ω. Simple stationary solutions of the wave equation can be constructed from such time harmonic waves

$$\Psi(\underline{r}, t) = \Psi_\omega(\underline{r})\mathrm{e}^{-i\omega t}, \quad \Psi_\omega(\underline{r}) \in \mathbb{C}\,. \tag{6.37}$$

Inserting equation (6.37) into the wave equation yields the stationary wave equation or Helmholtz equation, which is a differential equation for the spatially varying complex amplitude $\Psi_\omega(\underline{r})$

$$(\nabla^2\Psi_\omega(\underline{r}))\mathrm{e}^{-i\omega t} + \frac{n^2}{c^2}\omega^2\Psi_\omega(\underline{r})\mathrm{e}^{-i\omega t} = 0 \tag{6.38}$$

$$\Rightarrow \quad \nabla^2\Psi_\omega(\underline{r}) + n^2(\underline{r})k_0^2\,\Psi_\omega(\underline{r}) = 0 \qquad \text{(Helmholtz Equation)}\,, \tag{6.39}$$

with the vacuum wave number $k_0 = \frac{\omega}{c} = \frac{2\pi}{\lambda}$. Alternatively, one could write $k(\underline{r}) = n(\underline{r})k_0$ with $k(\underline{r})$ the wavenumber in the propagation medium of the wave. It is immediately evident that the plane wave $\Psi_\omega(\underline{r}) = \mathrm{e}^{i\underline{k}\cdot\underline{r}}$ satisfies the Helmholtz equation. In

2 It is sufficient to have a piecewise homogeneous medium, with appropriate boundary conditions.

the following, we will write k instead of k_0 and refer to k as the vacuum wave number or the modulus of the wave vector in vacuum. Further, we will only consider quasi monochromatic waves, keeping in mind that any realistic (broad bandpass) spectrum can always be written as a suitable superposition of corresponding solutions. For notational simplicity, we will also use $\Psi \rightarrow \Psi_\omega$ and $n \rightarrow n_\omega$, remembering that the index will differ for the respective spectral components. Before turning to solutions of the Helmholtz equation in terms of diffraction integrals, we want to point out the formal identity of the wave equation and the time independent Schrödinger equation for a particle without spin

$$\frac{\hbar^2}{2m}\nabla^2\Psi + V\Psi = 0\,, \tag{6.40}$$

obtained for the variable replacement $n^2k^2 \rightarrow 2m/\hbar^2 V$, with the conventional symbols for the potential V, mass m and (reduced) Planck's constant $\hbar$.

Diffraction integrals

For use in biomedical imaging, we need formal solutions of the stationary wave equation, for example, explicit formulas to calculate how a field propagates through an object or in free space. This seems like a formidable task, and no generality can be expected. The solution depends on the precise 3d arrangement of the index of refraction $n(\underline{r})$ and the boundary conditions specifying the radiation going into and out of – say – a bound region containing the object. In fact, explicit solutions without approximations can be expected only for very generic objects and geometries, such as, for example, reflecting surfaces (Fresnel equations), spherical objects (Mie scattering) [56] or stratified media supporting bound modes and guided wave solutions [52]. As a partial differential equation, the Helmholtz equation is elliptic in nature [75], requiring knowledge of the field and/or its derivative all over a closed surface around the object, comparable to the Poisson equation. There are, in principle, two general strategies to deal with this challenge, which are depicted in Figure 6.5 that illustrates the paradigms of the far field diffraction and optical propagation, respectively. Far field diffraction is treated based on the Born series, and in practice is almost always in the form of the first Born approximation, to compute the amplitude and intensity diffracted from an object. The scattering process is treated as weak, neglecting multiple scattering. The diffracted field is simply computed by summing over an assembly of scattering centers (atoms, molecules, or more generally scattering constituents) placed at position vectors $\underline{r}_n$ or a corresponding continuous density of scattering centers ρ_f

$$A_s \propto \sum_{n=0}^{N-1} f_i\, e^{i\underline{q}\cdot\underline{r}_n}\,, \qquad A_s \propto \int d\underline{r}\,\rho_f\, e^{i\underline{q}\cdot\underline{r}_n}\,, \tag{6.41}$$

with the measurable diffraction intensity computed by $I \propto |A_s|^2$ and the scattering vector $\underline{q} := \underline{k}_f - \underline{k}_i$ defined as the wave vector difference of the outgoing and incoming

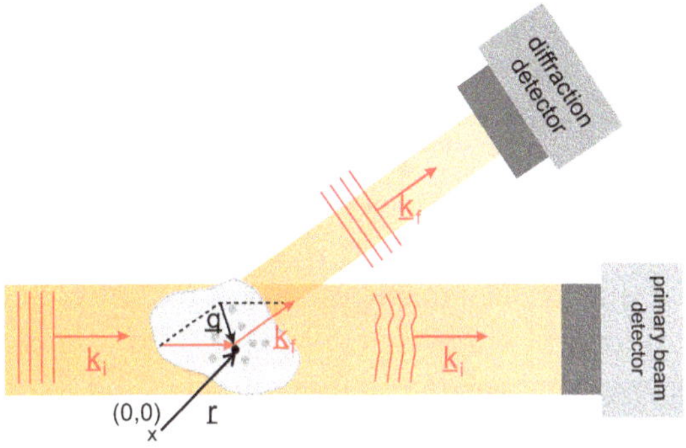

Fig. 6.5: Far field diffraction versus full field imaging in the near field. Far field diffraction: A plane wave is scattered from an assembly of scatterers, and the resulting interference is detected in the (Fraunhofer) far field. At detection distances that are large compared to all length scales of the object and the beam cross section, only scattered intensity is recorded without contributions from the primary beam. Even when the detector is placed on the optical axis, the primary beam typically covers only a few central detector pixels and is blocked by a beam stop. Contrarily, (Fresnel) near field diffraction measures the interference of scattered and primary beam, with holographic encoding of the phase. Typically, a detector with small pixel sizes is required to detect the holographic fringes. In practice, the different regimes are realized by changing either propagation distance, beam divergence or detector type. For structure analysis by x-ray diffraction, the paradigmatic far field scheme prevails with the scattering vector defined as $\underline{q} := \underline{k}_f - \underline{k}_i$, where $\underline{k}_i$ is the wavevector of the incoming and $\underline{k}_f$ of the outgoing wave with $|\underline{k}_i| = |\underline{k}_f| = 2\pi/\lambda$. This is a powerful concept and correctly takes into account the dimensionality (3d) of the object. However, is not well suited to treat x-ray imaging, which requires the Fresnel (near field) wave optical description. Unfortunately, the near field case has to be formulated in terms of diffraction planes, which are often less suitable for the 3d nature of objects. Therefore, additional assumptions such as thin object, projection approximation, etc., are required.

plane wave. The density ρ_f is proportional to the decrement δ of the index of refraction $Re(n) = 1 - \delta$, see the discussion of the x-ray index of refraction in Section 6.1.2.

An important restriction is the following: Not only must the scattering be weak, but the incoming and outgoing (diffracted) waves must be plane waves in order to keep the expression as simple and tractable as in equation (6.41). This necessitates that the detection device be placed in the optical far field (Fraunhofer diffraction). Only more recently has the restriction to plane wave illumination been lifted by ptychographic wavefield reconstruction [86]. On the other hand, the merit of this classical approach of x-ray crystallography or x-ray diffraction is that it links the diffraction pattern directly to the 3d assembly of scatterers. The diffraction intensity is calculated in 3d reciprocal space by ways of a simple 3d Fourier transform, followed by taking the modulus squared. Contrarily, in near field diffraction, the detection plane is typically not a subspace of a 3d reciprocal space, and recordings for different object orientations can only under certain assumptions be assembled in a 3d measurement space [79]. Far field x-ray diffraction treated in the first Born approximation has been an extremely successful concept, which is surprising even for an approach that violates such basic laws as energy conversion. A particular advantage is that there is no necessity to limit the structure to a thin diffraction mask or to define the wavefronts over planes perpendicular to the optical axis. Further, no paraxial approximation is required; only the detection distance has to be much larger than the size of the object. This is clearly not the typical situation that we encounter in biomedical imaging, where the objects are extended and interact more strongly. Moreover, in the diffraction approach, the diffracted radiation is treated without interference with the primary beam, since diffracted waves and primary beam separate in the optical far field.

For full field imaging we, therefore, need a setting with an extended beam, where the downstream field (e.g., in the detection plane) is to be computed from an upstream field (e.g., in the source plane) with interference of scattered and transmitted waves, and with the concept of an optical axis, by which upstream and downstream can be defined. The Fresnel–Kirchhoff diffraction integrals provide such a framework.[3] As a mathematical formulation of the Huygens principle, the integrals can be derived from the Helmholtz equation using Green's formalism. Forward propagation is calculated by a sum over source points; back propagation or reflection is suppressed by suitable boundary conditions. Hence, the stationary wavefield in the detection plane $u(x_2, y_2, z)$ at a distance z downstream from the source plane is calculated from the Fresnel–Kirchhoff diffraction integral

$$u(x_2, y_2, z) = \frac{1}{i\lambda} \iint dx_1 dy_1 \, u(x_1, y_1, z = 0) \, \frac{e^{ikr_{1,2}}}{r_{1,2}} \cos(\theta) \,, \tag{6.42}$$

where $\cos\theta = \hat{r}_{1,2} \cdot \underline{e}_z$ is the Stokes inclination factor (obliquity factor) and $r_{1,2}$ is the distance between a point in the source plane and the point in the detection plane, at which the field is to be calculated. The distance $r_{1,2}$ can be written as

$$r_{1,2} = \sqrt{(x_2 - x_1)^2 + (y_2 - y_1)^2 + z^2} \approx z + \frac{(x_2 - x_1)^2}{2z} + \frac{(y_2 - y_1)^2}{2z} \qquad (x, y \ll z) \,, \tag{6.43}$$

where the paraxial approximation holds if all contributing rays between the two planes propagate at small angles with respect to the optical axis. While this second order expansion for the distance $r_{1,2}$ is inserted in the argument of the exponential in equation 6.42, the prefactor describing the amplitude decrease is approximated more coarsely by $r_{1,2}^{-1} \simeq z^{-1}$. With $\cos(\theta) \simeq 1$ we finally obtain for the paraxial field propagation

$$u(x_2, y_2, z) = \underbrace{\frac{e^{ikz}}{i\lambda z} \, e^{i\frac{\pi}{\lambda z}(x_2^2 + y_2^2)}}_{\text{compl. prefactor}} \underbrace{\iint dx_1 dy_1 \, u(x_1, y_1) \, e^{i\frac{k(x_1^2 + y_1^2)}{2z}} \, e^{\frac{-ik}{z}(x_1 x_2 + y_1 y_2)}}_{\mathcal{F}[u(x,y) e^{\frac{ik}{2z}(x_1^2 + y_1^2)}]} \,. \tag{6.44}$$

3 Since the integration kernels can differ depending on the choice of boundary conditions and approximations, it is more appropriate to speak of diffraction integrals in plural.

Hence, the downstream field is given by the Fourier transform of the "decorated" upstream field. The input of the Fourier transformation is not the bare input field, but the field multiplied by a complex exponential function, often denoted as a chirp function. It is only for very large distances z that the phase variation of the chirp function becomes negligible, and the field in the detection plane is proportional to the Fourier transform of the "naked" input field. If one is only interested in the measurable intensity distribution (diffraction pattern), the complex prefactor in front of the Fourier integral vanishes; this is sometimes also denoted as the detector chirp factor. In fact, the Fourier operator maps the input to an output field, as function of the spectral coordinates $(kx_2/z, ky_2/z)$. If the downstream field is plotted in these variables of spatial frequencies rather than in positions in the detector plane, the detector chirp factor reduces to 1. The discussion of the chirp factor in the integral leads to the definition of the Fraunhofer and the Fresnel regimes

$$\text{Fraunhofer (far field) regime:} \quad z \gg \frac{x^2 + y^2}{\lambda} \tag{6.45}$$

$$\text{Fresnel (near field) regime:} \quad z \lesssim \frac{x^2 + y^2}{\lambda} \tag{6.46}$$

where the respective regimes and the entire propagation can be described by the unitless Fresnel number $F = \frac{a^2}{\lambda z}$ given in equation (6.6). For discrete (numerical) problems we will identify a with the pixel size, for practical purposes, keeping in mind that large structures will then still exhibit a near field appearance even at $F \ll 1$.

Angular spectrum approach

We have seen above how propagation of a wavefield in free space can be calculated by the Fresnel–Kirchhoff diffraction integral, for arbitrary propagation distance or, equivalently, for arbitrary Fresnel number F. In this paragraph, we restrict ourselves to the regime where the diffracted beams interfere with the primary beam, i.e., the optical near field defined with respect to beam size b and Fresnel numbers $F = b^2/(\lambda z) \gg 1$. In this regime, an extremely elegant description of propagation is possible in reciprocal space, which is known as the angular spectrum decomposition. This is extremely useful for numerical implementation of the propagation problem. The angular spectrum approach is discussed in detail in the textbook by Paganin [69] and in [45], which are the main references for the condensed presentation here. The idea of the angular spectrum approach starts with the observation that plane waves $u^{(PW)}(x, y, z) = \mathrm{e}^{i(k_x x + k_y y + k_z z)}$ with the components of the wavevector $\underline{k}$ fulfilling $k_x^2 + k_y^2 + k_z^2 = k^2$, are elementary solutions of the Helmholtz equation (6.39). In fact, they form a suitable basis for the entire space of solutions, since the propagation of any plane wave $u^{(PW)}(x, y, z)$ that propagates at angles $\arctan(k_x/k)$ and $\arctan(k_y/k)$ in the xz and yz planes is simply given by

$$u^{(PW)}(x, y, z) = \mathrm{e}^{i(k_x x + k_y y)} \mathrm{e}^{iz\sqrt{k^2 - k_x^2 - k_y^2}}, \tag{6.47}$$

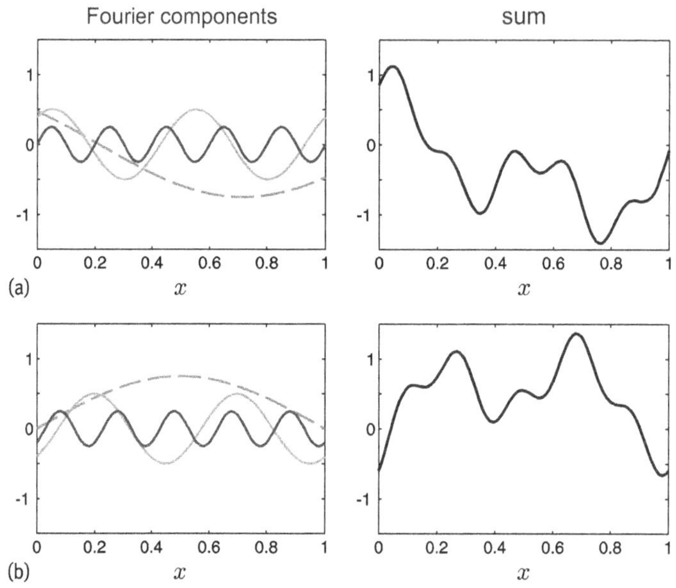

Fig. 6.6: Illustration of Fourier synthesis in 1d, showing Fourier components (*left*) for the same amplitude but different phases in (a) and (b), respectively. The phase corresponds to shifts in the sine functions. Fourier synthesis, i.e., summing up the components to a total signal (*right*), yields very different results. This illustrates the challange of structure determination without phase information. The problem can only be tackled by using prior information (constraints).

where we have expressed the solution in terms of the lateral components k_x and k_y. This means that a field that varies in the source plane as $u^{(PW)}(x, y, z = 0) = e^{i(k_x x + k_y y)}$ can be propagated analytically along the optical axis z by multiplication with the free space propagator $e^{iz\sqrt{k^2-k_x^2-k_y^2}}$. Hence, we can use this method to expand arbitrary wavefields in the source plane $u(x, y, z = 0)$ in terms of $u^{(PW)}$, which amounts to taking the 2d Fourier transform of the unpropagated wavefield

$$u(x, y, z = 0) = \frac{1}{2\pi} \iint \tilde{u}(k_x, k_y, z = 0)\, e^{i(k_x x + k_y y)}\, dk_x\, dk_y\,, \tag{6.48}$$

where $\tilde{u}(k_x, k_y, z = 0)$ is the Fourier transform of $u(x, y, z = 0)$. We can then simply compute the propagated wavefield by multiplication of the plane wave components by the free space propagator

$$u(x, y, z = \Delta) = \frac{1}{2\pi} \iint \tilde{u}(k_x, k_y, z = 0)\, e^{i\Delta\sqrt{k^2-k_x^2-k_y^2}}\, e^{i(k_x x + k_y y)}\, \mathrm{d}k_x \mathrm{d}k_y\,. \tag{6.49}$$

In compact form we can write this as a free space diffraction operator

$$\mathcal{D}_\Delta = \mathcal{F}^{-1} \exp[i\Delta\sqrt{k^2 - k_x^2 - k_y^2}]\, \mathcal{F}\,, \tag{6.50}$$

and if the paraxial approximation is warranted

$$\mathcal{D}_\Delta \simeq \mathcal{D}_\Delta^{\mathrm{F}} = \exp[ikz]\, \mathcal{F}^{-1} \exp[-i\Delta(k_x^2 + k_y^2)/(2k)]\, \mathcal{F}\,, \tag{6.51}$$

where $\mathcal{D}_\Delta^F$ is known as the Fresnel diffraction operator. Note that backpropagation of the field is simply carried out by changing the sign of propagation $\Delta \to -\Delta$. As we will see later, this is an important step of many phase retrieval algorithms. According to equation (6.51), propagation amounts to a multiplication of the wavefield in Fourier space by the Fresnel propagator; hence, it can also be formulated as a convolution in real space

$$u(x,y,z=\Delta) = u(x,y,z=0) * h(x,y,\Delta)\,. \tag{6.52}$$

The convolution kernel $h(x,y,\Delta)$ is the real space representation of the Fresnel propagator given by

$$h(x,y,\Delta) \equiv \frac{1}{2\pi} e^{ik\Delta} \mathcal{F}^{-1} e^{\frac{-i\Delta(k_x^2+k_y^2)}{2k}} = -\frac{ike^{ik\Delta}}{2\pi\Delta} e^{\frac{ik(x^2+y^2)}{2\Delta}}\,. \tag{6.53}$$

Numerical implementation of propagation

Using the angular spectrum approach, numerical propagation is carried out by three numerically efficient steps: (1) Fourier transformation of the unpropagated wavefield, yielding $\tilde{u}(k_x, k_y, z = 0)$, (2) multiplication by $e^{i\Delta\sqrt{k^2-k_x^2-k_y^2}}$ and (3) inverse Fourier transformation, yielding the propagated field $u(x_2, y_2, z)$, again in real space. For data on an equidistant grid, this reduces to [2]

$$\begin{aligned} &u(n_x\Delta x, n_y\Delta y, \Delta) \\ &= \exp[ik\Delta] \cdot \mathrm{iDFT}\left[\exp\left(\frac{-i\Delta(n_x^2\Delta k_x^2 + n_y^2\Delta k_y^2)}{2k}\right)\mathrm{DFT}[u(n_x\Delta x, n_y\Delta y, 0)]\right]\,, \end{aligned} \tag{6.54}$$

where Δx, Δy, Δk_x and Δk_y are the equidistant sampling intervals in real and in reciprocal space, and DFT and iDFT the discrete forward and inverse Fourier transformation. The DFT/iDFT are mostly carried out by the fast Fourier transform (FFT) algorithm, provided by many platforms and packages for data analysis. In the DFT/iDFT, the sampling intervals mutually fulfill the Shannon sampling criteria with

$$\Delta k_x \Delta x = 2\pi/N_x\,, \tag{6.55}$$

and the equivalent expression for y. Interestingly, the paraxial approximation of the discrete Fresnel propagator can be expressed by only a single unitless parameter, the Fresnel number $F = \frac{(\Delta x)^2}{\lambda z}$, eventually defined separately as F_x and F_y, in the case that $\Delta_x \neq \Delta_y$. Propagation can, hence, be computed in pixel units with F as the only control parameter, if the paraxial form of the Fresnel propagator is used. This means that the propagated field is only determined by the known combination (from the Fresnel number) of sampling interval, propagation distance and wavelength.[4] Contrarily, if the

4 Note that the plane wave prefactor requires the product of wavenumber and propagation distance separately, but this global phase is not relevant for the field distribution. Furthermore, it cancels out in cyclic phase retrieval algorithms.

square root in the exponential was kept, the parameter wavelength and propagation distance enter separately. Numerically, the complex exponential in equation (6.54) is, hence, simply evaluated as [2]

$$\exp\left(\frac{-iz(n_x^2\Delta k_x^2 + n_y^2\Delta k_y^2)}{2k}\right) = \exp\left(-\frac{i\pi n_x^2}{N_x^2 F_x}\right)\exp\left(-\frac{i\pi n_y^2}{N_y^2 F_y}\right) . \tag{6.56}$$

Numerical implementation of wave propagation is treated in several helpful textbooks on the DFT [9, 12]. In particular, for the implementation of free space propagation using Matlab [37], we refer to [25, 38, 91]. Issues of numerical sampling of the chirp functions can be found in [91, 92].

6.2.2 The optical far field and the phase problem

In classical diffraction studies used for structure analysis, the intensity of a diffracted wave is recorded in the optical far field or Fraunhofer regime. Hence, the amplitude of the diffraction pattern is available, but the phase is not. The corresponding information deficit is illustrated in Figure 6.7, where far field phase and amplitude of two test objects (portraits of Röntgen and Fresnel) are exchanged, before inverse Fourier transformation. For the 'reconstructed' objects the phase (which is typically lost) appears as the dominating information.

Autocorrelation, oversampling and support

We may ask: How much information and what kind of information on the object wavefield ψ[5] is contained in the far field intensity distribution $I(\underline{q})$? According to the autocorrelation theorem the inverse Fourier transform of the measured intensities is equivalent to the autocorrelation of the field in the object plane

$$\mathcal{F}^{-1}[2\pi I(q)] = \int \psi(x)\,\psi^*(x'+x)\,dx' = \psi(x)\otimes\psi(x) =: \Gamma(r)\,, \tag{6.57}$$

where * denotes the complex conjugate, $\otimes$ the autocorrelation operation, and Γ the autocorrelation function in the object plane. As before, the diffraction pattern intensity is $I = \tilde{\psi}(q)\tilde{\psi}^*(q)$, and the Fourier transform is denoted by $\mathcal{F}[\psi](q) = \tilde{\psi}(q)$. Therefore, we will address under which conditions and constraints it is possible to determine a function already from its autocorrelation function. In this case a diffraction pattern can be inverted, i.e., the undiffracted (unpropagated) wave or correspondingly the object

5 For a diffraction aperture, we define the exit wavefield as the field directly behind the aperture, which, hence, includes all the information that the aperture has imparted on the wave. For extended objects, the exit field is calculated by integration over the object thickness and refers to the downstream end of the object.

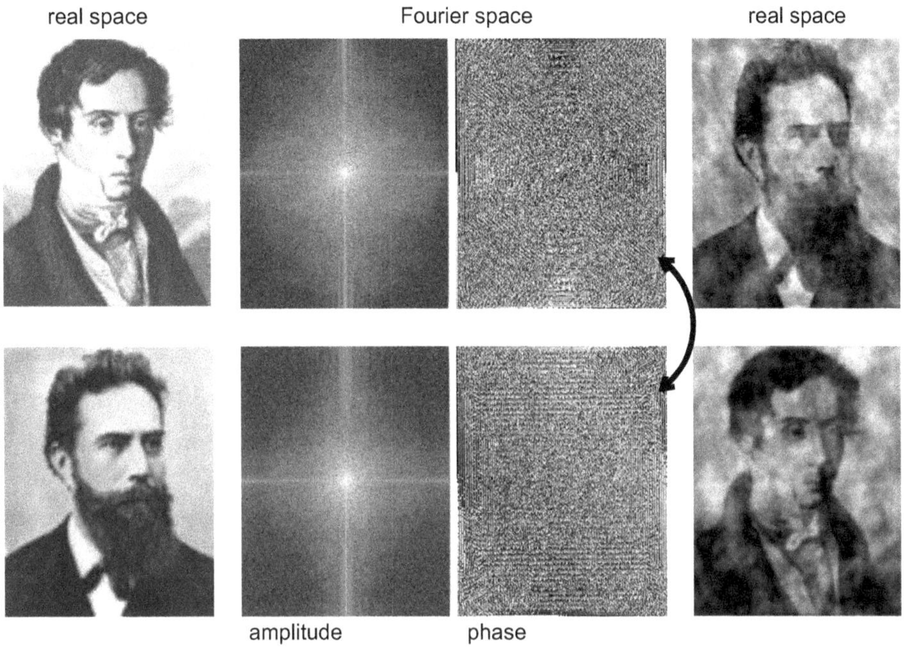

Fig. 6.7: Portraits of A. J. Fresnel and W. C. Röntgen, pioneers of Fourier optics and x-ray imaging, respectively, are shown along with grayscale representations of the corresponding amplitudes and phases, calculated from a DFT. To illustrate the relevance of the information contained in amplitudes (measurable) and phase (nonmeasurable), the two phases are exchanged, followed by inverse Fourier transformation. Interestingly, the 'phases of Fresnel' combined with the 'amplitudes of Röntgen' bear more similarity to Fresnel's than to Röntgen's portrait, highlighting the importance of the phase (and hence of the phase problem) for perception of images or identification of structures.

function can be reconstructed. However, first we must know how finely the diffraction pattern has to be sampled in order to fully capture the autocorrelation function. Assume that the object's transmission function $o(x)$ has a finite support $L = N\Delta x$. Then, the support of the autocorrelation of $o(x)$ is twice as large as the support of $o(x)$ itself. According to the sampling theorem (Section 2.8), discrete sampling of a function is equivalent to periodizing its Fourier transform. Sampling the diffraction pattern (e.g., by the finite pixel size of a detector) results in periodizing the object's autocorrelation function in real space. Hence, here we must sample the intensities finely enough so that the periodized autocorrelations do not overlap, as illustrated in Figure 6.8.

Contrarily, if we were able to measure the complex valued diffracted wave, the inverse Fourier transformation would yield the object function itself, and owing to the smaller support of the object compared to the support of the autocorrelated object, sampling at $\Delta k_{\text{Bragg}} = \frac{2\pi}{L}$ would be sufficient. Sampling at $\Delta k_{\text{Bragg}} = \frac{2\pi}{L}$ is also called

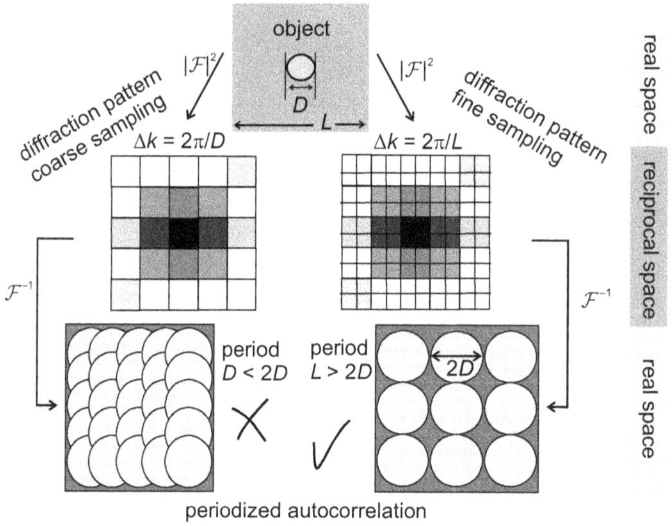

Fig. 6.8: Inverse Fourier transformation of the discretely sampled diffraction pattern (measured intensities) results in a periodized version of the object autocorrelation function in real space. Since the support of the autocorrelation is twice the support of the object, the pixel size in the detection plane has to be chosen small enough so that the periodic copies of the autocorrelation function in real space do not overlap.

Bragg sampling. For a given experimental sampling rate Δk_{exp}, the oversampling rate σ is defined as

$$\sigma = \frac{\Delta k_{\text{Bragg}}}{\Delta k_{\text{exp.}}} \,. \tag{6.58}$$

The sampling condition for complete assessment of the autocorrelation, hence, is

$$\sigma \geq 2 \,.$$

Further increase of σ does not provide more information on the autocorrelation. Note that if the support of o is not finite, no sampling – however fine the grid – is sufficient. The sampling condition can also be expressed in terms of the FOV in real space (i.e., the FOV of the computational reconstruction), which is

$$D_{\text{FOV}} = N\Delta x = 2L \,,$$

with $\Delta k_{\text{exp}} \Delta x = \frac{2\pi}{N}$ and $\Delta k_{\text{exp}} = \frac{\pi}{L}$. The FOV, therefore, has to be larger than the object by at least a factor of 2 (along both directions) in order to correctly sample the autocorrelation function. Hence, the diffraction data allows us to reconstruct the autocorrelation $\Gamma(\underline{r}_\perp)$ of the object, given sufficient sampling. However, can we obtain enough independent data points to retrieve $o(\underline{r}_\perp)$ itself? In general, we can not. Consider a detector with $N_x \cdot N_y = N$ pixels, yielding N intensity values. If $o(\underline{r}_\perp)$ is complex, i.e., an object with phase and absorption contrast, we have $2N$ unknowns in real space,

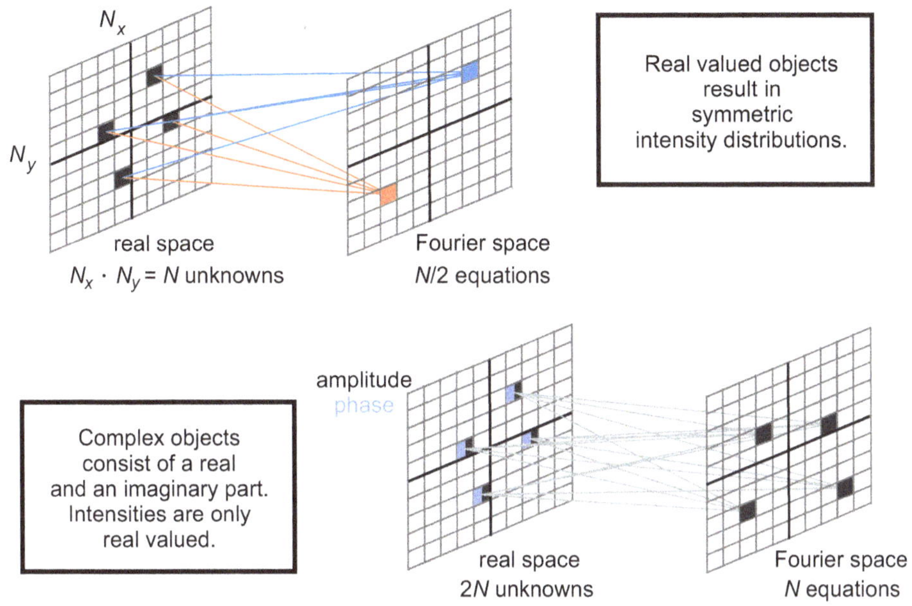

Fig. 6.9: If we want to reconstruct the wavefield in the object plane, we face the following dilemma: In either case (real or complex object) the number of known information in reciprocal space exceeds the number of unknowns by a factor of 2. For object reconstruction by inverse Fourier transform, we ought to be measuring complex diffraction amplitudes, while we can measure only real valued intensities. Making the object real valued does not solve the information deficit, since now the diffraction data becomes centrosymmetric.

as we have to determine phase and amplitude for each point in the object plane $\underline{r}_\perp$. If we were to reconstruct a real valued object, we would not be better off. As the Fourier transform of real valued objects and, hence, also the diffracted intensities are centrosymmetric, we have only $N/2$ equations at hand from the data compared to the N unknowns for the object. Hence in either case, the number of unknowns exceeds the number of independent data points by a factor of 2. This is illustrated in Fig. (6.9). To compensate for the lack of data, however, we could decrease the number of unknowns of the object; for example, if we had knowledge about a certain number of object pixels. In the simplest case, we can use the information that the object is nonzero only within a known (compact) support, as illustrated in Fig. (6.10). If the support is sufficiently small, we have a unique or even over determined set of equations. Solving the corresponding equations still poses a significant challenge, since the square operation in the intensities results in a nonlinear problem. For far field diffraction, the field of coherent diffractive imaging (CDI) has been opened by the algebraic approach of budgeting the information with respect to the available data, which was the key to solving the phase problem. To this end, a generalized oversampling ratio (also valid

Fig. 6.10: The object is constrained to a finite support (*green pixels*). Outside the support region, the object is zero valued (*gray pixels*). The support includes N_s pixels, whereas the total amount of pixels is N.

for more than two dimensions) was introduced by J. Miao [55]

$$\sigma' = \frac{\text{total number of pixels}}{\text{unknown valued pixel number}} = \frac{A_{\text{FOV}}}{A_{\text{support}}} . \tag{6.59}$$

Obtaining only half of the required information from the diffraction pattern, the other half has to be provided by knowing the object's support. Hence, we require

$$\sigma' \geq 2 . \tag{6.60}$$

A further increase in the number of known pixels (a decrease of the support) would not necessarily deliver more information, but can still be a good choice experimentally, e.g., in the presence of additional uncertainties or experimental errors. Of course, a compact and known support poses an experimental restriction. We have to constrain the unknown object to a small region in the object plane. For example, consider a detector with pixel size Δx_d. With equation (6.55), the pixel size in reciprocal space is $\Delta k = \Delta x_2 \frac{2\pi}{\lambda z}$, with distance z between detector and sample. Bragg sampling with $\Delta k_{\text{Bragg}} = \frac{2\pi}{L}$, and requiring an oversampling condition $\sigma \geq 2$ (equation (6.58)) yields a maximum object cross section of

$$L \leq \frac{\lambda z}{2\Delta x_d} . \tag{6.61}$$

For a detector pixel size of $\Delta x_d = 172\,\mu\text{m}$ (PILATUS, Dectris), wavelength $\lambda = 10^{-10}$ m and propagation distance $z = 5$ m, we see that the object diameter has to be smaller than 1.5 µm. It would be impossible to reconstruct any extended object, which rules out this type of coherent imaging for most applications of biomedical imaging. Some of these stringent constraints have been lifted by ptychographic CDI, after the realization that compactness of the object can be exchanged against compactness of the illumination function [86]. Hence, larger objects can be imaged by scanning of a (compactly supported) beam. However, in phase contrast radiography and tomography for biomedical imaging, we are primarily interested in full field techniques, facing extended objects with extended beams, such as in radiography. The sampling criteria and support constraints, first introduced in far field CDI, also bear some relevance for imaging in the near field. Moreover, the iterative algorithms discussed below were first used in far field CDI and later generalized to near field imaging, where they have proven extremely useful.

6.2.3 Coherence

Phase contrast radiography requires a sufficiently high partial coherence of the probing x-rays. The temporal and spatial variation of the phases in the illuminating wavefront must be small enough not to impede the interpretation of the phase shift imparted by the object. As in many other areas of physics, coherence describes the ability of a wave to show interference effects. Formation of interference patterns is typically subject to two conditions. Firstly, the radiation has to be of sufficiently small bandwidth, since the interference pattern would wash out if the radiation becomes too 'white'. Information initially contained in the pattern is lost, when minima and maxima begin to cancel. Secondly, the phase variations of two neighboring points in space must be correlated, in order to enable coherent superposition of radiation, which 'has taken different paths'. The length scale over which the field at different points on the wavefront are correlated is denoted as the **lateral coherence length** $\xi_\perp$, the most important parameter for describing the **spatial coherence** properties. Contrarily, the relative wavelength spread $\Delta\lambda/\lambda$ determines the time scale τ over which two wave trains can be kept registered in phase, yielding the **longitudinal coherence length** $\xi_\parallel = c\tau$, the coherence length scale parallel to the propagation direction (optical axis). This parameter relates to the **temporal coherence** properties. In particular it sets the maximum tolerable path length difference $\Delta s \leq \xi_\parallel$ between two radiation paths in order to observe interference. Here, interference always means superposition of field amplitudes, instead of superposition of intensities, which is what remains if coherence is lost. For x-ray radiation, coherent optics and coherence properties are treated in [69] and are also reviewed in [63] in view of coherent imaging techniques.

Mutual coherence function

In order to quantify the coherence properties of a wavefield, we consider the **mutual coherence function** $\Gamma_{12} = \langle u_1(t_1)u_2^*(t_2)\rangle$ [96], which describes the correlation between two wavefields u probed at two points (1) and (2), and at times t_1 and t_2, respectively. Γ is also denoted as a **complex degree of coherence**. Normalized to the intensities, the mutual coherence function is written as

$$\gamma_{1,2}(t_1, t_2) = \frac{\langle u_1(t_1)u_2^*(t_2)\rangle}{\sqrt{\langle I_1\rangle\langle I_2\rangle}} \,, \tag{6.62}$$

where $I_i = \langle u_i u_i^*\rangle$ is the intensity at the two points and $\langle\ldots\rangle$ the statistical average. Assuming ergodicity, the temporal and ensemble averages are considered equal. For fields with stationary coherence properties, the correlations depend only on the *relative* time difference $\tau = t_2 - t_1$, and we have

$$\gamma_{1,2}(\tau) = \frac{\langle u_1(t)u_2^*(t+\tau)\rangle}{\sqrt{\langle I_1\rangle\langle I_2\rangle}} \,. \tag{6.63}$$

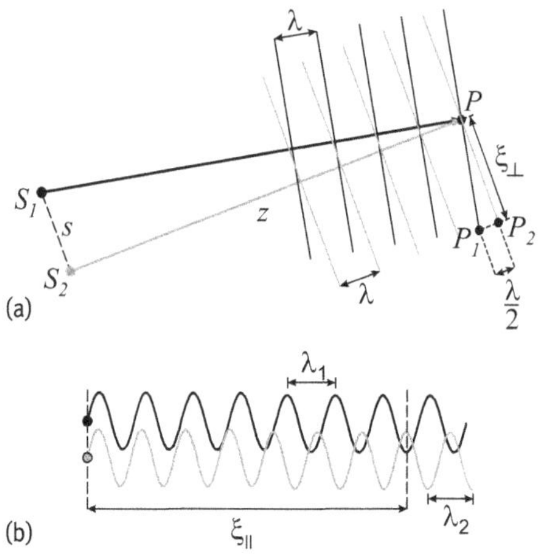

Fig. 6.11: Coherence properties of a beam can be described based on two length scales. (a) The lateral length over which a wavefront is coherent is called lateral coherence length $\xi_\perp$. Even for an incoherently emitting source, a finite $\xi_\perp$ arises solely by propagation. To illustrate this, consider two point sources S_1 and S_2 at a distance s, each emitting a monochromatic wave with wavelength λ, but with random phase shift between the two. A point P in an observation plane perpendicular to the propagation direction will receive signal from both sources. When moving across the wavefront, this superposition will remain stationary until the tilts of the wavefront manifest themselves, i.e., when the deviation between the *black* and the *gray* planes reach $\lambda/2$. Since the triangles $\overline{S_1S_2P}$ and $\overline{P_1P_2P}$ are similar, we can infer $\xi_\perp = \lambda z/(2s)$. (b) The longitudinal coherence length $\xi_\parallel = \lambda^2/(2\Delta\lambda)$ describes the distance parallel to the propagation direction, over which two wave trains with a wavelength difference $\lambda_1 - \lambda_2 = \Delta\lambda$ 'detune' in phase.

For this case, we can also define the **cross spectral density** as the Fourier transform of $\Gamma_{1,2}$ with respect to τ

$$W_{1,2}(\omega) = \int d\tau\, \Gamma_{1,2}(\tau)\, \exp(-i\omega\tau)\,. \tag{6.64}$$

If, in addition to being stationary, the field is also monochromatic, Γ is constant in time (up to a trivial harmonic term). This is directly evident for the case that the two probing points coincide spatially, which gives $\gamma(\tau) = \exp(-i\omega\tau)$, a function of constant magnitude in time.

Wiener–Khinchin theorem and longitudinal coherence

Next, we consider the more realistic case of quasi monochromatic fields, i.e., radiation with a small bandwidth $\Delta\omega$ centered around a mean angular frequency $\omega = 2\pi\nu$, or correspondingly, a wavelength band $\Delta\lambda$ around an average λ. For the mutual intensity function, taken at identical points but at a time difference τ, we equally have unit magnitude

$$\gamma(\tau) = \frac{\langle u(t)u^*(t+\tau)\rangle}{\sqrt{\langle I^2\rangle}} \overset{\tau\Delta\nu \ll \lambda/(2\Delta\lambda)}{\simeq} e^{i\omega\tau}\,, \tag{6.65}$$

for short time scales $\tau \ll \xi_{\|}/c$ with longitudinal coherence length defined by $\xi_{\|} = \lambda^2/(2\Delta\lambda)$. Contrarily, for larger time difference $\tau \gg \xi_{\|}/c$, γ decreases to zero by negative interference of the different frequencies, which run out of phase. For an x-ray beam of central frequency 10^{18} Hz and a relative bandwidth of 10^{-4}, we have a correlation time of $\tau \sim 10^{-14}$ s, which is well below the integration time of any detector. This shows that any realistic measurement (apart from single shot free electron laser (FEL) pulses [6]) will to an excellent degree correspond to a temporal ensemble average with respect to the source (not necessarily to the object of course). The precise functional form of $\gamma(\tau)$ is related to the PSD of the field. After all, $\gamma(\tau)$ is a second order correlation function of the field in time, and hence must be related to the PSD of the signal, see equation (2.124). In optics, the fact that the temporal coherence function is related to the spectrum of radiation frequencies, is known as the **Wiener–Khinchin theorem**

$$\gamma(\tau) = \int d\omega \, |\tilde{u}(\omega)|^2 \, e^{i\omega\tau} \,. \tag{6.66}$$

In practice, the finite longitudinal or temporal coherence length $\xi_{\|}$ simply limits the maximum path length difference. Consider interference between any two points in the object separated by $\Delta \underline{r}$. The path length is then controlled by the maximum momentum transfer $\underline{q}$. For all interfering points we must have

$$\underline{q} \cdot \underline{r} = \frac{4\pi}{\lambda} \sin\theta \, r \leq \frac{\xi_{\|}}{\lambda} \,, \tag{6.67}$$

with θ the maximum scattering angle or numerical aperture. In terms of the maximum (half period) resolution $\Delta = \pi/q$, we obtain the intriguing relation that the relative bandwidth has to be smaller than the ratio between resolution and object size L, if all points in the object $r \leq L$ are to interfere, with

$$2\pi \, \frac{\Delta\lambda}{\lambda} \leq \frac{\Delta}{L} \,. \tag{6.68}$$

Note that the prefactor 2π depends on the precise definition of $\xi_{\perp}$, which requires a criterion for interference visibility (such as a certain percentage of the Michelson contrast), as well as some definition work concerning $\Delta\lambda$ with respect to the functional form of the spectrum. Let us consider a micro-CT source operating with Mo-K_α radiation, with corresponding photon energy $E \simeq 17.48$ keV and 105 eV the difference between the $K_{\alpha,1}$ and $K_{\alpha,2}$ lines, which cannot be separated by filters. The corresponding inverse relative bandwidth is $E/\Delta E \simeq 166$, which has to be compared to the number of resolution elements in the FOV. Note that the requirement that rays through all points

6 FEL single shot experiments are nearly fully coherent, i.e., the pulse is shorter than the coherence time $\xi_{\|}/c$, and the beam cross section is smaller than the lateral coherence length $\xi_{\perp}$. By using appropriate 'seeding', the coherence extends even over a train of pulses, i.e., the phases of individual pulses are no longer uncorrelated [23].

in the reconstruction volume must support interference with each other is overly restrictive. This would only be the case in the fully holographic regime. Furthermore, for $\lambda = 0.0708\,\text{nm}$, we have $\xi_\parallel \simeq 5.9\,\text{nm}$, which sounds very small, but it would already allow us to observe up to 116 interference orders for a grating!

Spatial coherence and radiation modes

Importantly, if these criteria are granted, it is sufficient to consider only the mutual coherence function at equal time, the so called mutual intensity function, describing only the spatial dependence of the coherence function. We denote the mutual intensity function by $j = \gamma_{1,2}(0)$, and by $J = \Gamma_{1,2}(0)$, for the corresponding quantities before normalization. These functions are also sometimes denoted as equal time complex degree of coherence. Hence, in terms of the field u we now have to consider

$$j(x_1, x_2) = \frac{\langle u(x_1)u^*(x_2)\rangle}{\sqrt{\langle I(x_1)\rangle\langle I(x_2)\rangle}} . \tag{6.69}$$

The limiting cases are a fully coherent ($|j| = 1$) and fully incoherent ($j = 0$) field, which are both idealizations. Real fields always exhibit **partial coherence** $0 < |j| < 1$. Often it is helpful to decompose the field into radiation modes

$$u(x) = \sum_n w_n\, u_n(x) , \tag{6.70}$$

where w_n is a real valued weight function of the mode n. The resulting intensities are $I = |u|^2 = |\sum_n w_n u_n(x)|^2$ and $I = \sum_n |w_n u_n(x)|^2$, for coherent and incoherent superposition of modes, respectively. In most cases, the concept of radiation modes is used in such a way that the different modes are assumed to be completely incoherent with respect to each other, while a mode is by definition a coherent wavefield. Partial coherence is modeled by considering an ensemble of stochastic field realizations [66]

$$u(x) = \sum_n w_n c_n^{\text{rand}} u_n(x) , \quad c_n^{\text{rand}} = a_n e^{i\varphi_n} , \tag{6.71}$$

with $a_n = \text{rand}(0, 1)$ and $\varphi_n = \text{rand}(0, 2\pi)$, and average field $\langle u\rangle = \langle \sum_n w_n c_n^{\text{rand}} u_n(x)\rangle$. For example, $u_n(x)$ can be used to describe the fields emitted by a collection of point sources. For each realization, these fields have a random, but fixed phase relation, given by random coefficients c_n. The intensities or mutual intensities can then be computed from many such realizations, by simple averaging or correlation, respectively. This approach has been used to model an incoherent undulator source by a set of distributed point sources, each emitting a mode u_n that can be propagated coherently. The partial coherence in any plane is then simply calculated by the incoherent superposition of these modes (basis functions), see Figure 6.12 [66]. The mutual intensity function then can be written as

$$j(x_1, x_2) = \frac{\sum_n w_n^2\, u_n(x_1)u_n^*(x_2)}{\sqrt{\sum_n w_n^2\, |u_n(x_1)|^2 |u_n(x_2)|^2}} . \tag{6.72}$$

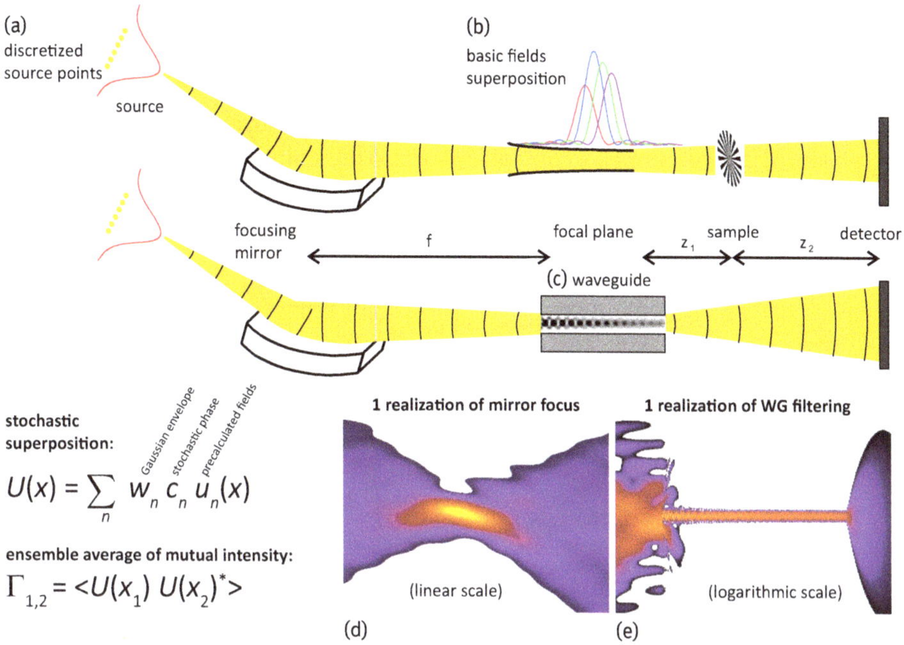

Fig. 6.12: Numerical simulation of partial coherence for x-ray propagation imaging. An extended (incoherent) source is described by independent point sources (a), each emitting a mode which can be propagated coherently through the optical system. In this example, focused synchrotron radiation is simulated for (b) a focusing mirror and (c) a mirror/waveguide compound optical system. The sample is placed in the defocus z_1 to achieve high magnification. Importantly, the random superposition of modes can be carried out in any plane of observation. The average intensity which would be recorded by a detector corresponds to the ensemble average over many statistical realizations. For illustration purposes, single field realizations of (d) the mirror focus and (e) a field propagation through the waveguide are shown. In this manner, and by running the simulation over many realizations, the mutual intensity function can be determined. From [66].

Van Cittert–Zernike theorem and lateral coherence length

If x_1 and x_2 denote two points on an observation screen, i.e., on a surface perpendicular to the line of observation, and $\underline{r}_{1,2}$ the corresponding difference vectors to a point within the source, the mutual intensity function by free propagation in the observation plane and at distance z from the source is given by an integral over all points of the source [96]

$$J(x_1, x_2) = \frac{1}{\lambda^2} \int_S ds\, I(s) \frac{\exp(ik(r_1 - r_2))}{r_1 r_2} \cos\theta_1 \cos\theta_2 \,, \tag{6.73}$$

where S denotes the source area, and $\cos\theta_{1,2} = \hat{r}_{1,2} \cdot \underline{e}_z$ are inclination factors as in equation (6.42). This formulation of the **van Cittert–Zernike theorem** is valid for the case of an incoherent source. Importantly, a finite (spatial) coherence length $\xi_\perp$ in the observation plane emerges solely by wave propagation. For a given intensity distribution $I(s)$ of the source[7], equation (6.73) gives a function J, which decreases as a function of separation $\Delta x = |x_2 - x_1|$ between the source points. The coherence length $\xi_\perp$ then denotes a characteristic decay length, taken in the observation plane perpendicular to the propagation direction, i.e., it describes over which lateral length the phases of the wavefront are correlated; see also the simple geometric construction in Figure 6.11. Assuming that the observation plane is in the far field of the source, the van Cittert–Zernike theorem gives

$$j(\Delta x) = \frac{\exp[i\psi] \int ds\, I(s)\, \exp[i\frac{2\pi}{\lambda z}\,\Delta xs]}{\int ds\, I(s)}\,, \tag{6.74}$$

for the normalized mutual intensity function. If we are just interested in the absolute value, i.e., in the degree of coherence $|j(\Delta x)|$, the phase factor $\psi = \pi/(\lambda z)(x_2^2 - x_1^2)$ becomes irrelevant, and we have

$$|j(\Delta x)| = \frac{1}{I_0} \int ds\, I(x)\, \exp\left[i\frac{2\pi}{\lambda z}\Delta xs\right]\,, \tag{6.75}$$

where I_0 denotes the integrated intensity of the source. For the example of a Gaussian intensity profile $I(s) = \exp(-x^2/2\sigma^2)$, we hence obtain

$$|j(\Delta x)| = \exp\left[-(\Delta x)^2/2\xi_\perp^2\right], \quad \xi_\perp = \frac{\lambda z}{2\pi\sigma}\,. \tag{6.76}$$

Note that in the correlation length $\xi_\perp$ and the width of the source were defined as rms values of the respective functions. For the full width at half maximum (FWHM) convention, the relation would be $\xi_\perp = 0.883\ \lambda z/S$, with S and $\xi_\perp$ denoting the FWHM of the source profile and the function $|j|$, respectively. This rationalizes the conventional use of

$$\xi_\perp \approx \frac{\lambda z}{S}\,. \tag{6.77}$$

Hence, the lateral coherence length is proportional to the propagation distance and inversely proportional to the source size. Consider a micro-CT source operating with Mo-K_α radiation with $\lambda = 0.0708$ nm and a spot size of $S = 1\ \mu$m. At a distance of $z = 0.1$ m from the source, we would have $\xi_\perp \simeq 6\ \mu$m, which is large enough to cover a few neighboring pixels in the object.

Coherence characterization

As a certain degree of spatial coherence is essential for phase contrast propagation, coherence diagnostics is, to some extent, indispensable. In lucky cases, the mere observation of some phase contrast may be regarded as sufficient. In more sophisticated

7 For simplicity, we now only consider a 1d source; the generalization to 2d is straightforward.

approaches, in particular when the coherence function should be included in the reconstruction, more quantitative characterization tools are required. For the longitudinal (temporal) properties, it is directly clear that the energy spectrum needs to be recorded, for example, by a suitable (single crystal) spectrometer. For the lateral (spatial) coherence properties, we can use different methods, most notably grating interferometry as discussed for imaging in Section 6.4.2. For example, in [80] grating interferometry was used to determine the mutual intensity function in the imaging planes of a synchrotron nano focus setup. To this end, gratings of 500 nm, 200 nm and 50 nm lines and spaces were scanned longitudinally in a range of defocus distances between 1 mm to 15 mm behind the focal plane of the elliptic mirror system. The increase of the coherence length $\xi_\perp$ with distance from the source was found to be in perfect agreement with propagation theory, see Figure 6.13, and numerical coherence simulations assuming an incoherent undulator source. Further it was shown that x-ray waveguides

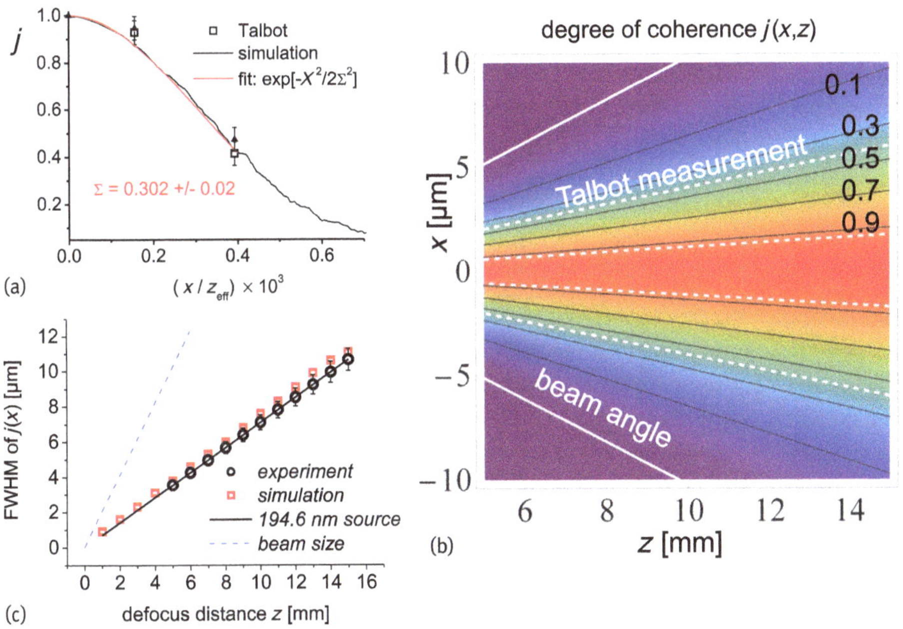

Fig. 6.13: Characterization of spatial coherence properties by Talbot (grating) interferometry. (a) Experimentally determined degree of coherence $|j|$ in the imaging plane, compared to simulations. (b) The mutual coherence function $|j|$ evaluated for each point separated from the optical axis by a distance x (vertical axis), with respect to the central point on the optical axis and as a function of distance from the source z (horizontal axis). Roughly, the beam cross section was about three times larger than the lateral coherence length $\xi_\perp$ (c) The increase of the lateral coherence length $\xi_\perp$ as a function of defocus distance z, comparing experimental values (*open circles*), numerical simulations (*solid red squares*), as well as the coherence length corresponding to a fully incoherent source of size s, observed at distance z_1. From [80].

inserted in the focal plane can be used as additional coherence filters and to generate spatially fully coherent illumination wavefronts.

6.2.4 Paraxial wave equation (parabolic wave equation)

We have seen that the Fresnel–Kirchhoff diffraction integrals provide convenient formulations of the Huygens principle and allow us to treat wave propagation in free space, along an optical axis. The downstream (output) field is only determined by the upstream (input) field. Strictly speaking, the Helmholtz equation does not support this kind of *causality*; it was imposed onto its solutions by the concept of diffraction masks forcing the field to vanish outside a bounded region within the source plane. It is easy to see that the concept of a source plane breaks down for the case where a beam is back reflected. Even if only forward propagation is admitted, there is a significant limitation in the concept of propagating a field between 2d planes in ignoring the 3d nature of an object, i.e., when the object is not well approximated by a 'diffraction mask'. Finally, for far field diffraction it is more suitable to ask for the field as a function of angular coordinates instead of real space detector coordinates. In view of all these limitations, it can be more appropriate to replace the Helmholtz equation by an equation that has a 'causal' optical axis 'built in' and which is often more amenable to 3d distributions of $n(\underline{r})$. The paraxial wave equation (PWE) provides such a framework. It is also known as the parabolic wave equation, reflecting its parabolic nature as a partial differential equation. The PWE is treated in many advanced optics textbooks, e.g., in [28, 82]. First devised to treat propagation of radar and radio waves by the Russian scientists LEONTOVICH and FOCK in the 1940s [47], it was introduced to x-ray optics by the Moscow x-ray optics group around VINOGRADOV [40]. In fact, the PWE turned out to be extremely useful for numerical x-ray optics, allowing efficient forward stepping by finite difference (FD) schemes [21, 75].

The PWE is obtained from the Helmholtz equation by separating the propagating wave in terms of a rapidly oscillating plane wave term along the optical axis z and a smoothly varying envelope function $A(\underline{r})$

$$u(\underline{r}) = A(\underline{r})e^{ikz} . \tag{6.78}$$

This is particularly useful when we can demand that $A(\underline{r})$ is a slowly varying function with respect to z, i.e., $\frac{\partial A}{\partial z} \ll kA$ and $\frac{\partial^2 A}{\partial^2 z} \ll k^2 A$; see Figure 6.14). This turns out to be a perfect approximation for x-rays propagating in media that change on scales much larger than the x-ray wavelength. For numerical work, it is obvious that in view of node points, it is much more efficient to solve a differential equation for A than for u. Inserting equation (6.78) into the Helmholtz equation (6.39) yields

$$\nabla^2 \left(A(\underline{r})e^{ikz}\right) + n^2(\underline{r})k^2 A(\underline{r})e^{ikz} = 0 \tag{6.79}$$

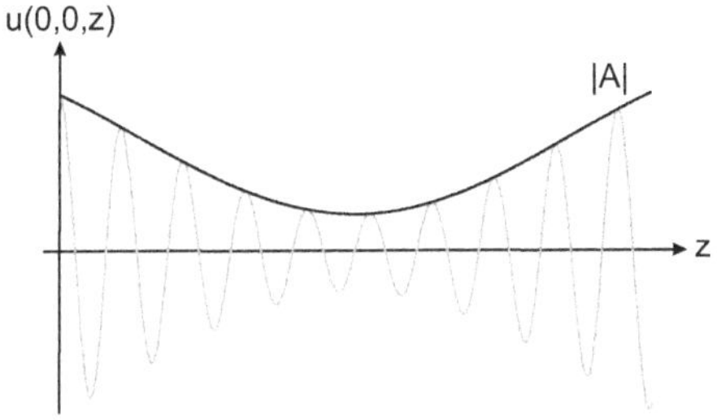

Fig. 6.14: Numerical simulations of a propagating x-ray wave with its rapidly varying field on sub-nm scales would require a tremendous number of node points. If, however, the harmonic term e^{ikz} describing the oscillations along the optical axis z is factored out, the envelope $A(\underline{r})$ (*black*) of the wave (*gray*), which varies much more slowly, can be efficiently treated. The parabolic wave equation (PWE) is a partial differential equation for the slowly varying envelope function of a beam directed along the optical axis z. Note that the variation of A along the optical axis is small on the scale of the 'carrier wave'. The PWE is well suited to treat x-ray propagation in objects and in free space by efficient numerical forward stepping.

which by developing the Laplacian in lateral and longitudinal derivatives gives

$$\begin{aligned}\nabla_\perp^2 A(\underline{r})e^{ikz} + \partial_z\left(\partial_z A(\underline{r})e^{ikz} + ike^{ikz}A(\underline{r})\right) = \\ \nabla_\perp^2 A(\underline{r})e^{ikz} + (\partial_z^2 A) + 2(\partial_z A(\underline{r}))ike^{ikz} - k^2 e^{ikz}A(\underline{r})\,,\end{aligned} \tag{6.80}$$

and by eliminating e^{ikz}, gives the following differential equation for $A(\underline{r})$

$$\left[\nabla_\perp^2 + \partial_z^2 + 2ik\partial_z + k^2(n^2(\underline{r}) - 1)\right]A(\underline{r}) = 0\,. \tag{6.81}$$

This equation is now subject to further approximation. The second order derivative in z can be neglected for paraxial beams, as $\partial_z^2 A \ll k^2 A$, leading to the paraxial (or parabolic) wave equation

$$\left[\nabla_\perp^2 + 2ik\partial_z + k^2(n^2(\underline{r}) - 1)\right]A(\underline{r}) = 0\,. \tag{6.82}$$

The paraxial approximation holds, as long as the relative change of the envelope along z is small on the scale of the x-ray wavelength. A further discussion of the validity of this approximation can be found in [21]. In free space, the PWE reduces to

$$\left[\nabla_\perp^2 + 2ik\partial_z\right]A(\underline{r}) = 0\,. \tag{6.83}$$

As simple as it looks, the last equation has important solutions, for example, the parabolic beam, centered at $z = 0$,

$$A(\underline{r}) = \frac{A_1}{z}e^{-ik\frac{\rho^2}{2z}} \qquad \text{with} \qquad \rho^2 = x^2 + y^2\,, \tag{6.84}$$

and constant A_1. A more general solution is

$$A(\underline{r}) = \frac{A_1}{q(z)}e^{-ik\frac{\rho^2}{2q(z)}}\,, \tag{6.85}$$

which for $q(z) = z + iz_0$ is reduced to the Gaussian beam, which is well known to model optical beams [82]. The radius of curvature $R(z)$ of the wave fronts and the beam waist $W(z)$ are related to q by [82]

$$\frac{1}{q(z)} = \frac{1}{R(z)} - i\frac{\lambda}{\pi W(z)} . \tag{6.86}$$

For $q(z) = z - \xi$, equation (6.85) again describes a parabolic beam, centered at ξ.

The intensity of a Gaussian beam

The optical intensity $I(\underline{r}) = |A(\underline{r})|^2$ is a function of the axial position z and the radial position $\rho = \sqrt{x^2 + y^2}$.

$$I(\rho, z) = I_0 \left[\frac{W_0}{W(z)}\right]^2 \exp\left[-\frac{2\rho^2}{W^2(z)}\right] , \tag{6.87}$$

where $I_0 = |A|^2$ and W_0 parameterizes the waist of the beam in the focal plane. For each value of z, the intensity is a Gaussian function of the radial distance ρ, with its maximum centered on the z axis at $\rho = 0$. The beam width $W(z)$ increases with increasing distance to the focus. Along the axis of propagation at $\rho = 0$ the intensity varies as (Figure 6.15)

$$I(0, z) = I_0 \left[\frac{W_0}{W(z)}\right]^2 = \frac{I_0}{1 + (z/z_0)^2} , \tag{6.88}$$

with the maximum I_0 at $z = 0$ and a power law decay with $|z|$. The intensity is reduced to half its maximum at $z = \pm z_0$.

The power of a Gaussian beam

The total optical power in an arbitrary plane perpendicular to the direction of propagation z is given by the integral of the optical intensity in this plane

$$P = \int_0^\infty I(\rho, z)\, 2\pi\rho\, d\rho = \frac{1}{2} I_0 (\pi W_0^2) . \tag{6.89}$$

The power of a Gaussian beam, thus, is half of its intensity multiplied by the size of the central illuminated area. As required by flux conservation, the power does not depend on z.

The cross section of a Gaussian beam

The dependence of the beam width W on distance z is

$$W(z) = W_0 \sqrt{1 + \left(\frac{z}{z_0}\right)^2} , \tag{6.90}$$

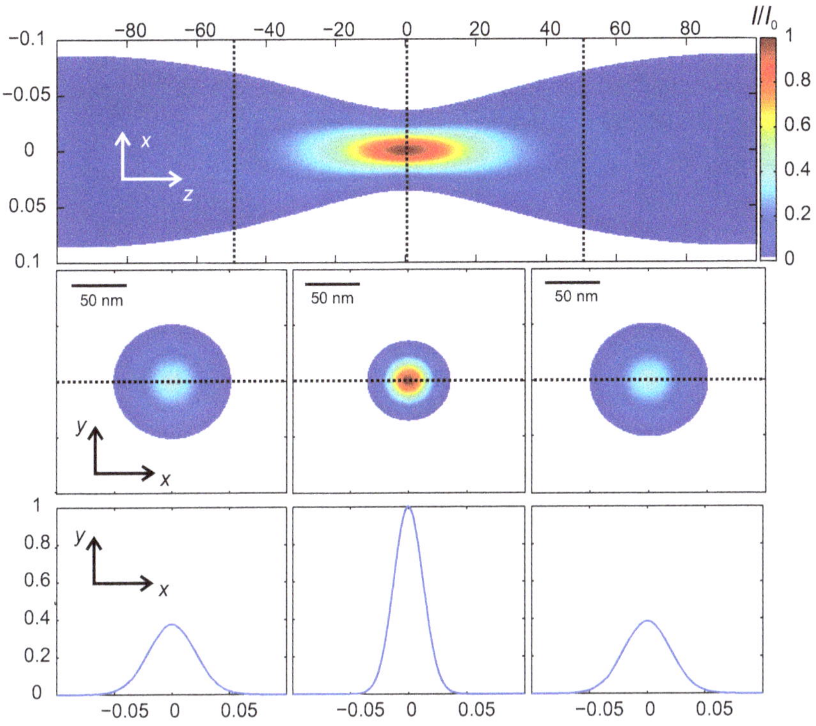

Fig. 6.15: A Gaussian beam as an important solution of the parabolic wave equation. The intensity in the lateral planes is always given by a Gaussian distribution. Approaching the focal plane $z < 0$, the intensity increases (converging) and decreases (diverging) for $z > 0$, accompanied by a change in the beam width. This simple analytical solution can also serve to test numerical propagation methods.

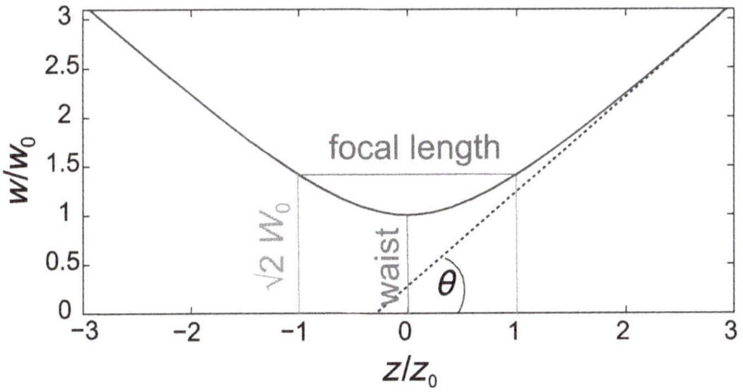

Fig. 6.16: The diameter of the beam reaches its minimal value at the beam waist. At the distance of the Rayleigh length, the diameter of the beam is increased by a factor of $\sqrt{2}$ with respect to the size of the beam waist.

which is illustrated in Figure 6.16. The diameter reaches a minimal value W_0 in the focal plane at $z = 0$. The beam waist is also called the spot size. The diameter of the beam increases monotonically for increasing $|z|$ and reaches a size of $\sqrt{2}W_0$ at $|z| = z_0$. The distance z_0 is called the Rayleigh length.

6.2.5 Projection approximation of the object transmission function

The next step is to treat the interaction of the wave with an object. The simplest, and at the same time most versatile, concept to do this is to consider the effect of a sufficiently thin object on the wave, since any extended object may be constructed from (virtual) thin slices. The propagation in arbitrary objects can then be treated by incrementally repeated steps of thin object transmission and propagation. To derive the complex valued object transmission function, we start with the PWE in matter. Following [25], we will neglect the second derivative in x, y in addition to the second derivative in z (paraxial approximation). This essentially 'turns off' any diffraction and is a rather crude approximation for thin objects, which are thus reduced to quasi two-dimensional diffraction apertures. Approximating $n^2 \approx 1 - 2\delta + 2i\beta$, the parabolic wave equation (6.82) becomes

$$\left[\frac{1}{k^2}\nabla_\perp^2 + \frac{1}{k^2}\partial_z^2 + \frac{2i}{k}\partial_z\right] A(\underline{r}) = \left[2\delta(\underline{r}) - 2i\beta(\underline{r})\right] A(\underline{r}) \,. \tag{6.91}$$

For $\frac{\lambda}{2\pi}\partial_z^2 A \ll \partial_z A$ and $\frac{\lambda}{2\pi}\partial_{x,y}^2 A \ll \partial_z A$ we further find

$$\partial_z A(\underline{r}) = -ik(\delta(\underline{r}) - i\beta(\underline{r}))A(\underline{r}) \,. \tag{6.92}$$

After a propagation distance Δ along the optical axis z, we obtain

$$A(\underline{r}, z + \Delta) = A(\underline{r}, z) \exp\left[-ik\int_0^\Delta \left(\delta(\underline{r}) - i\beta(\underline{r})\right) \mathrm{d}z\right] , \tag{6.93}$$

for the envelope of the wavefield. Accordingly, the exit wavefield is given by

$$u(\underline{r}, z + \Delta) = \mathrm{e}^{ik\Delta}\, u(\underline{r}, z) \underbrace{\exp\left[-ik\int_0^\Delta \left(\delta(\underline{r}) - i\beta(\underline{r})\right) \mathrm{d}z\right]}_{\exp[-ik\Delta\langle\delta\rangle_z]\,\exp[-k\Delta\langle\beta\rangle_z]} \,. \tag{6.94}$$

This is also known as the projection integral in geometrical optics, as the mean refractive index along z is taken into account. The object transmission function describing the interaction with matter is then defined as

$$\tau(\underline{r}_\perp) = \exp\left[-ikz\langle\delta\rangle_z\right] \exp\left[-kz\langle\beta\rangle_z\right] , \tag{6.95}$$

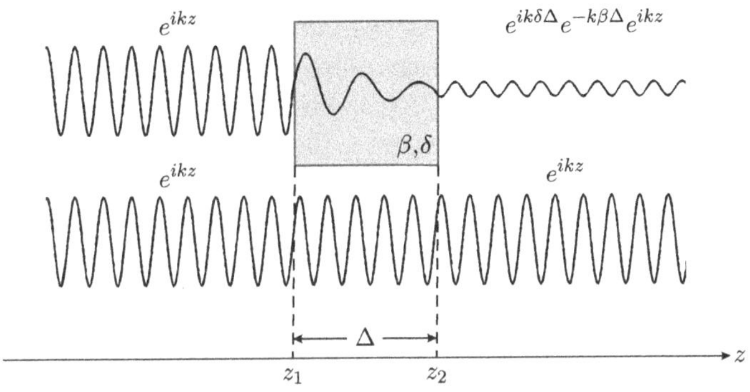

Fig. 6.17: A homogeneous slab with index of refraction $n = 1 - \delta + i\beta$ and thickness Δ illuminated by a plane wave. The wave fronts of the incident wave are modified with respect to those propagating in free space in two ways: (1) damping of the amplitude and (2) increase of the wavelength resulting in a decrease of the phase in the exit plane. While (1) depends on β and is the contrast mechanism of conventional radiography, (2) depends on δ and is exploited by phase contrast radiography.

which relates the beam u_E that exits the object (exit wave) to the incident beam (illumination function) u_I by

$$u_E(\underline{r}_\perp, z + \Delta) = u_I(\underline{r}_\perp, z + \Delta)\, \tau(\underline{r}_\perp)\,. \tag{6.96}$$

For a homogeneous slab of thickness Δ and plane wave illumination e^{ikz} incident at $z = z_1$ (Figure 6.17), we have again $e^{ikn\Delta}e^{ikz_1}$ in the exit plane $z_2 = z_1 + \Delta$, resulting in

$$\tau = \mathrm{e}^{ik\Delta(-\delta+i\beta)} = \mathrm{e}^{-ik\delta\Delta}\,\mathrm{e}^{-k\beta\Delta}\,, \tag{6.97}$$

where $\mathrm{e}^{-ik\delta\Delta}$ describes the phase shift and $\mathrm{e}^{-k\beta\Delta}$ the absorption of the wave. For an inhomogeneous medium and if the object is thin enough so that diffraction in the specimen can be neglected, the exit wavefield is

$$u_E = \mathrm{e}^{ikz}\,\mathrm{e}^{ik\langle\delta\rangle_z\Delta}\,\mathrm{e}^{-k\langle\beta\rangle_z\Delta}\,. \tag{6.98}$$

This approximation is called **projection approximation,** for which one assumes that the value of the wavefield is entirely determined by the phase and amplitude shifts, which are accumulated along streamlines of the unscattered beam. The validity of approximating the object to be "sufficiently thin" depends on the resolution. For a minimum feature size a to be detected, the first order diffraction imparts at an angle $\theta \simeq \lambda/a$. This diffraction signal will have reached the next resolution element (pixel) for a critical thickness $\Delta = a^2/\lambda$, resulting in the condition

$$\Delta\lambda/a^2 \leq 1\,. \tag{6.99}$$

In other words, the Fresnel number (defined for a single pixel and the object thickness) must be larger than 1 for the projection approximation to hold. If $\Delta \geq a^2/\lambda$, one can resort to multislice approaches, with iterative steps of projection and propagation advancing through the object. In these iterations propagation is treated as in free space. The above validity criterion for the projection approximation does not explicitly depend on the refractive index. This is in contrast to the expectation that weakly scattering objects should be more benevolent with respect to neglecting diffraction in the object.

Born and Rytov approximation

This raises the interesting question, by which approach beyond the projection approximation, wave propagation in the object can be suitably described. The parabolic wave propagation can be calculated efficiently by finite differences [22]. Hence, numeric computation is always an option if the assumptions for paraxial propagation are met. For the more general case of the Helmholtz equation $\nabla^2 u(\underline{r}) + k^2\, u(\underline{r}) = 0$ (equation 6.39), integral equations for the field can be obtained from the method of Green's functions. These analytical solutions provide deeper insight into the relationship between object and scattered near field wave. However, in order to turn the implicit integral equations into a closed form, further approximations are required, such as the Born or Rytov approximations, see, for example, [83] for a general introduction, [32, 84] for a discussion with respect to near field phase retrieval or [15] for earlier work considering far field diffraction. While the well known Born ansatz is based on a small additive correction of the scattering contribution to the primary wave

$$u(\underline{r}) = u_0(\underline{r}) + u_s(\underline{r}) \,, \tag{6.100}$$

the Rytov ansatz is to write the solution in a multiplicative form as

$$u(\underline{r}) = u_0(\underline{r}) \, \exp(\Psi_s(\underline{r})) \,, \tag{6.101}$$

where $\Psi_s(\underline{r})$ is the complex valued phase of the scattered field and $u_0(\underline{r})$ the primary wave amplitude $u_0(\underline{r})$, without perturbation by the object. The object is described by its index or scattering potential with $O(\underline{r}) = k_0^2(1 - n(\underline{r}))$. With the Green's function $G(\underline{r}) = \exp(ikr)/(4\pi r)$, the solution to the Helmholtz equation is obtained as [84]

$$u_0 \Psi_s(\underline{r}) = \int d^3r' \; G(\underline{r} - \underline{r}')\, u_0 \left((\nabla \Psi_s(\underline{r}'))^2 - O(\underline{r}') \right) . \tag{6.102}$$

Rytov's approximation requires $|(\nabla\Psi_s)^2| \ll |O|$, turning this implicit integral solution to an explicit closed form [84]

$$\Psi_s(\underline{r}) = -\frac{1}{u_0} \int d^3r' \; G(\underline{r} - \underline{r}')\, u_0 \, O(\underline{r}') \,, \tag{6.103}$$

which in frequency space gives [84]

$$\tilde{\Psi}_s(k_x, k_y, z) = \frac{\exp\left[iz\left(\sqrt{k_0^2 - k_x^2 - k_y^2} - k_0 \right) \right]}{2i\sqrt{k_0^2 - k_x^2 - k_y^2}} \, \tilde{O}\left(k_x, k_y, \sqrt{k_0^2 - k_x^2 - k_y^2} - k_0 \right) . \tag{6.104}$$

Furthermore, if the paraxial approximation

$$\sqrt{k_0^2 - k_x^2 - k_y^2} \simeq k_0 \left(1 - \frac{k_x^2}{2k_0^2} - \frac{k_y^2}{2k_0^2}\right) \tag{6.105}$$

and the projection approximation[8]

$$\sqrt{k_0^2 - k_x^2 - k_y^2} - k_0 \ll \frac{\pi}{\Delta} \tag{6.106}$$

hold, we obtain

$$\tilde{\Psi}_s(k_x, k_y, z) = \frac{1}{2ik_0} \exp\left[-\frac{i\lambda z}{2}(k_x^2 + k_y^2)\right] \tilde{O}(k_x, k_y, 0) \,. \tag{6.107}$$

By inserting this result in equation (6.101), and expanding the exponential, the first Born approximation is obtained. The advantage of the Rytov approximation is that it requires only the phase shifts over one wavelength to be small and it is, hence, also suitable for describing refraction effects in larger (but weakly varying) objects, while the Born approximation requires the total accumulated phase to be small [15]. The transition to geometrical optics and the eikonal equation is also better accomplished by Rytov formalism.

6.3 Propagation imaging

Following the basic considerations on the wave equation, coherence and the object transmission function, we now first turn to image formation by free propagation and then to phase retrieval and image reconstruction. However, before we do so, we still need some prerequisites concerning the illuminating wavefunction.

6.3.1 The illumination wavefront: cone beam and plane wave

So far we have tacitly assumed plane wave illumination of the object. We now address the question of how to calculate the propagated field behind a sample that is illuminated by a divergent beam, in particular a spherical wave emitted by a hypothetical point source.[9] The spherical beam geometry is important because it enables geometric magnification and hence extends phase contrast imaging to phase contrast microscopy. The geometry is also denoted as cone beam in the literature. While the terms

8 This expression can be regarded as a definition of the projection approximation, and when inserting the maximum lateral momentum transfer $k_{x,y} = 2\pi/a$ it directly gives equation (6.99).

9 Strictly speaking, point beam emitters do not exist; we should, therefore, speak of quasi point beam sources. X-ray waveguides [5] or high resolution Fresnel zone plate lenses [17, 18] provide reasonable realizations of such nanoscale quasi point sources, but in contrast to optical fields, x-ray sources will always be much larger than λ and, hence, emit only with correspondingly limited numerical aperture.

spherical wave and point source imply coherence, cone beam also includes beams of low coherence.

Importantly, there is a close analogy between parallel beam (plane wave) imaging and point source (spherical beam) imaging. According to the **Fresnel scaling theorem** [69], a Fresnel diffraction pattern generated by point source illumination, is equivalent to plane wave illumination, up to a transform of variables, involving the (de)magnification M of detector pixels, and a transform of the propagation distance $z \rightarrow z_{\text{eff}}$ to an effective value. To this end, we reconsider the cone beam geometry already encountered in Chapter 4, with the geometric magnification determined by the source object distance z_1 and the object detector distance z_2

$$M = \frac{z_1 + z_2}{z_1} , \tag{6.108}$$

as sketched in Figure 6.18, resulting in the effective propagation distance

$$z_{\text{eff}} = \frac{z_1 \, z_2}{z_1 + z_2} . \tag{6.109}$$

Note that contrary to Chapter 4, we here use 'lighter' labeling $z_{01} \rightarrow z_1$ and $z_{12} \rightarrow z_2$, in order to simplify the equations. For cone beams, z_1 is also called the "defocus distance", which also explains the terminology "effective defocus distance" for z_{eff}. From the above, it follows that

$$\frac{1}{z_1} + \frac{1}{z_2} = \frac{1}{z_{\text{eff}}} = \frac{M}{z_2} . \tag{6.110}$$

We now use these relations to rephrase the Fresnel–Kirchhoff integral, for the field u_2 in the detection plane, as a function of the field in the object exit plane u_1. We assume the projection approximation to hold $u_1 = \tau u_I$, with the illumination field u_I, and the complex valued object transmission function τ. In paraxial approximation (Eq. (6.44)), we have

$$u_2(x_2, y_2) = A \iint dx_1 dy_1 \; u_I(x_1, y_1) \, \tau(x_1, y_1) \, e^{i\frac{k(x_1^2+y_1^2)}{2z_2}} \, e^{\frac{-ik}{z_2}(x_1x_2+y_1y_2)} , \tag{6.111}$$

where A is the complex prefactor defined in equation (6.44). Two special cases of the incident illumination u_I can be distinguished:

(i) For a plane wave illumination (Figure 6.18(b)), u_I is given by e^{ikz_1}.

(ii) For a point source illumination (Figure 6.18(a)), u_I can be expressed by $e^{ik\sqrt{x_1^2+y_1^2+z_1^2}} \approx e^{ikz_1} e^{\frac{ik}{2z_1}(x_1^2+y_1^2)}$. Note that the amplitude decrease $\frac{1}{r} \simeq \frac{1}{z}$ is already absorbed in the prefactor A.

Inserting (ii) for u_I in the Fresnel–Kirchhoff integral yields

$$u_2(x_2, y_2) = A \iint dx_1 dy_1 \; \tau(x_1, y_1) \, e^{ikz_1} \underbrace{e^{\frac{ik}{2z_1}(x_1^2+y_1^2)} e^{i\frac{k(x_1^2+y_1^2)}{2z_2}}}_{e^{\frac{ik}{2}(\frac{1}{z_2}+\frac{1}{z_1})(x_1^2+y_1^2)}} \, e^{\frac{-ik}{z_2}(x_1x_2+y_1y_2)} , \tag{6.112}$$

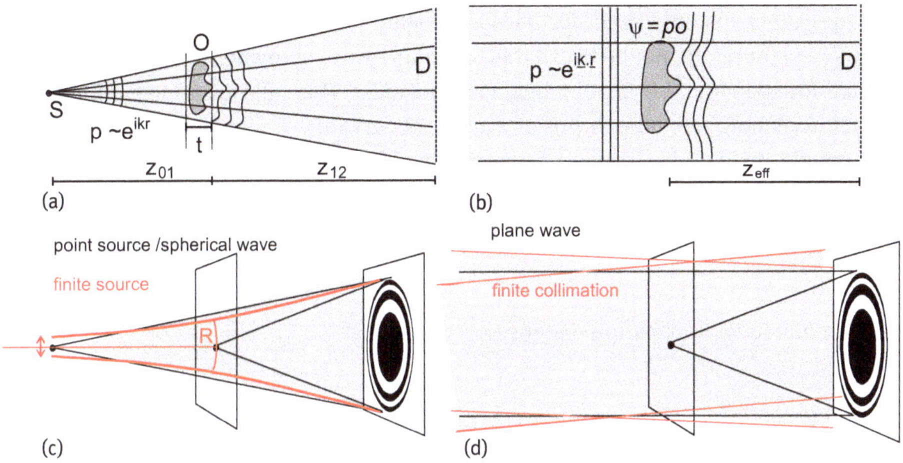

Fig. 6.18: According to the Fresnel scaling theorem, the two geometries of (a) a point source and (b) the plane wave illumination of an object are equivalent. Within the paraxial and the projection approximation, the diffraction patterns are identical up to a variable transformation involving the source object distance z_1 and object detector distance z_2. In essence, the magnification $M = (z_1 + z_2)/z_1$ of geometric optics is preserved in the wave optical treatment. Fresnel diffraction patterns recorded in divergent geometry can be understood as recordings of a miniaturized detector (easily achieving nanoscale effective pixel sizes) in a parallel beam (plane wave) setup with the effective propagation distance $z_{\text{eff}} = z_2/M$. (c,d) One can show that the hologram of a point in the object is equivalent in both geometries, and by linearity it follows that this is true for the holograms of any object, which can be constructed as a superposition of point sources. For finite source size, the scheme has to be slightly modified (indicated in *red*) and z_1 has to be replaced by the radius of curvature R.

where A accounts for the amplitude. Using equation (6.108) and equation (6.110), it follows that

$$u_2(x_2, y_2) = A \iint dx_1 dy_1 \, \tau(x_1, y_1) \, e^{ikz_1} \, e^{\frac{ik}{2z_{\text{eff}}}(x_1^2+y_1^2)} e^{-2\pi i\left[\left(\frac{x_2}{M}\right)\left(\frac{x_1}{\lambda z_{\text{eff}}}\right)+\left(\frac{y_2}{M}\right)\left(\frac{y_1}{\lambda z_{\text{eff}}}\right)\right]} . \quad (6.113)$$

Hence, up to the coordinate transformation, the results of the Fresnel–Kirchhoff integrals are equal for the cases of (i) and (ii). This is an important result, since it allows us to calculate image formation in a divergent beam in terms of an equivalent parallel beam setup, which is much easier for numerical implementation. The same, similar or related derivations are found in many references; we recommend [69] as the primary textbook reference, as well as a number of recent monographs on x-ray imaging, presenting and using the Fresnel scaling theorem. In particular, the prefactors to scale the field amplitudes correctly with respect to the pixel sizes are given in [25], an explicit derivation for a Gaussian beam illumination is presented in [21] and implementations for cone beam recordings at multiple distances are given in [41, 76]. Two words of caution: First, it should not escape our attention that – while speaking of spherical beams – we have used the paraxial approximation in the derivation, hence the equivalence is only exact for parabolic beams. High numerical aperture experiments can, therefore, not be translated to a plane wave equivalent geometry. Second, as mentioned in footnote 9, point beams do not exist. This calls for some modification. In fact, replacing z_1 by the radius of wavefront curvature R gives a set of transformation equations that also work for finite source size [94]:

$$M = 1 + \frac{z_2}{R} , \tag{6.114}$$

and the effective propagation distance

$$z_{\text{eff}} = \frac{z_1 z_2}{R + z_2} . \tag{6.115}$$

For point sources, these more general equations obviously again give equation (6.108) and equation (6.110). We can see that the maximum magnification is limited by the inverse Rayleigh length of a focused beam. The case of finite source size is sketched in red in Figure 6.18. Let us close with a remark on other illumination functions, which are neither described by plane waves nor by spherical waves. Of course, the Fresnel scaling theorem can still be used to factor out an average spherical or more precisely parabolic curvature term of u_I. The respective wave front deviations (aberrations) are simply shifted from a spherical to a parallel beam geometry. Importantly, one way or another, the exact wavefield u_I has to be known to predict image formation or to reconstruct the object (the inverse image formation). This poses an experimental challenge. In particular highly coherent and focused beams are typically full of different 'imprints' of optical elements used to focus and shape the beam somewhere upstream from the sample. It seems like an impossible mission to factor out all these features in order to separate the pure object function. Yet, this can be achieved if images are recorded at multiple lateral and longitudinal distances by advanced ptychographic algorithms, as demonstrated in [77]. If extended data can be recorded, clean illumination functions are, hence, no prerequisite. However, in view of dose and speed, single exposures are preferred, and in practice a much more simple raw data correction scheme is used: The measured defocus (near field) intensity is divided by the empty beam intensity. This ignores the propagation of the illumination u_I itself [21, 25]. The corresponding approximations and errors were investigated by analytical means in [36], and numerically in [34], showing that in phase retrieval, a flawed empty beam correction can induce an effective resolution loss. Therefore, clean wavefronts are clearly of advantage for phase contrast x-ray imaging. This is particularly important for nanoscale resolution. In contrast to high resolution synchrotron work, for micronscale recordings at laboratory sources, the empty beam approximation is often well justified, given the smoothness of typical partially coherent micro-CT wavefronts.

6.3.2 Contrast transfer by free space propagation

In this section, we address an analytic form for the contrast transfer function (CTF) of free space propagation. This formalism is extremely useful not only for an understanding of image formation, but also as a starting point of object reconstruction (CTF based reconstruction). The concept has been developed in electron microscopy but is now well established for x-ray imaging. Here, we follow the presentations of [21, 64, 69] and [81], where the validity of the formalism was verified experimentally. We start with considering the propagated field behind a weak phase object. According to Equation (6.52), in Fourier space, the propagated field is given by

$$\tilde{u}(\underline{k}_\perp, z) = \tilde{u}_E(\underline{k}_\perp)\, \tilde{h}(\underline{k}_\perp, z)\,, \tag{6.116}$$

where $\tilde{h}$ is the Fourier transform of h given in equation (6.53). For brevity, we use the notation of $(x, y) = \underline{r}_\perp$ and $(k_x, k_y) = \underline{k}_\perp$. In operator notation (Equation (6.50) and (6.51)), and for an object illuminated by a plane wave, we have

$$\begin{aligned} u(\underline{r}_\perp, z) &= \mathcal{D}_z\left[u_E(\underline{r}_\perp, z)\right] \\ &= \mathcal{D}_z\Big[\tau(\underline{r}_\perp)\ \underbrace{u_I(\underline{r}_\perp)}_{\text{plane wave}}\ \Big]\,. \end{aligned} \tag{6.117}$$

The object transmission function τ in real space and Fourier space is given by

$$\tau(\underline{r}_\perp) = \mathrm{e}^{-i\overbrace{k\Delta z\bar{\delta}(\underline{r}_\perp)}^{\varphi(\underline{r}_\perp)}}\,\mathrm{e}^{-\overbrace{\Delta z\bar{\mu}(\underline{r}_\perp)}^{\mu(\underline{r}_\perp)}/2} \simeq 1 - i\varphi(\underline{r}_\perp) - \mu(\underline{r}_\perp)/2\,, \tag{6.118}$$

$$\tilde{\tau}(\underline{k}_\perp) = \delta_D(\underline{k}_\perp) + i\tilde{\varphi}(\underline{k}_\perp) - \tilde{\mu}(\underline{k}_\perp)/2\,, \tag{6.119}$$

where the weak object approximation has been used; $\bar{\delta}$ and $\bar{\mu}$ denote the mean phase and amplitude shifts, imparted on the wavefield along the sample, $\mu = 2k\beta$ the linear attenuation coefficient and δ_D is the Dirac delta function. As usual, the tilde denotes Fourier transformation. The convolution kernel can be rewritten as

$$\begin{aligned} \tilde{h}(\underline{k}_\perp) &= \mathrm{e}^{ikz}\ \exp\left[-iz(k_x^2 + k_y^2)/2k\right] \\ &= \mathrm{e}^{ikz}\Big[\cos\underbrace{\frac{z}{2k}\left(k_x^2 + k_y^2\right)}_{=:\chi} - i\,\sin\frac{z}{2k}\left(k_x^2 + k_y^2\right)\Big]\,, \end{aligned} \tag{6.120}$$

where $\sqrt{\chi}$ is a generalized (unitless) spatial frequency. The intensity is proportional to the square of the absolute value of the amplitude, and in Fourier representation we obtain

$$I(\underline{r}_\perp) = |u|^2 \quad \Leftrightarrow \quad \tilde{I}(\underline{k}_\perp) = \tilde{u} * \tilde{u}\,. \tag{6.121}$$

Together with the transmission function, we obtain

$$\begin{aligned} \tilde{u} &= (\delta_D + i\tilde{\varphi} - \tilde{\mu}/2)\,(\cos\chi - i\,\sin\chi) \\ &= \delta_D + \tilde{\varphi}\,\sin\chi - \tilde{\mu}/2\,\cos\chi + i(\tilde{\varphi}\,\cos\chi + \tilde{\mu}/2\,\sin\chi)\,. \end{aligned} \tag{6.122}$$

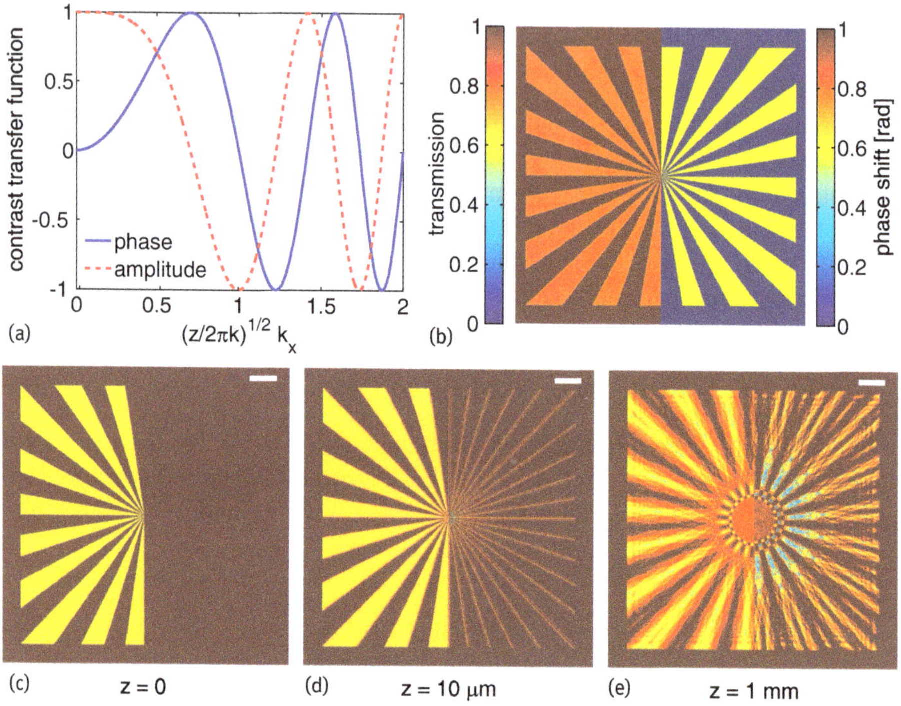

Fig. 6.19: (a) Amplitude and phase CTFs $\cos\chi$ and $\sin\chi$. (b) Simulation of a Siemens star test object [45]. The Siemens star is composed of an amplitude object with 80% transmission (*left*) and 0.6 rad phase shift (*right*). (c–e) Fresnel diffraction patterns of the Siemens star simulated for 10 keV; scale bar 1 μm. (c) In the contact plane only the amplitude object is visible. (d) With increasing distance, the outline of the phase object becomes visible and Fresnel fringes appear at the edges. (e) Upon further increase of z, the interference fringes become more pronounced, starting with the high spatial frequencies. Direct resemblance to the object is lost, as more and more feature sizes reach the Fresnel regime, followed by the regime of CTF oscillation. Experimentally, the validity of the CTF formalism was first verified by nanoscale the Siemens star test pattern in [81] and with cleaner illumination in [42]. Figure from [45].

Hence, the intensity in Fourier space $\tilde{u} * \tilde{u}$ developed in lowest order in $\tilde{\varphi}$ and $\tilde{\mu}$ is given by

$$\tilde{I}(\underline{k}_\perp) \approx \delta_D(\underline{k}_\perp) + 2\tilde{\varphi}(\underline{k}_\perp)\,\sin\chi - \tilde{\mu}(\underline{k}_\perp)\,\cos\chi\,. \tag{6.123}$$

Thus, along the propagation axis we obtain an oscillatory contrast in phase and amplitude, and the intensity distribution varies, depending on the different transmitted spatial frequencies. In particular, not all of the spatial frequencies are transmitted, due to the zeros in the respective CTFs $\sin\chi$ and $\cos\chi$ (Figure 6.19). This characteristic feature of propagation imaging is discussed in detail in [25] in view of single distance phase retrieval, and more recently by [2, 41] in view of multidistance phase retrieval. A very elegant extension of CTF formalism to 3d tomographic datasets is presented in [78]. For small propagation distances $\chi \ll \pi/2$ and by setting absorption to zero $\mu = 0$, we have for a pure phase object

$$\tilde{I}(\underline{k}_\perp) \approx \delta_D(\underline{k}_\perp) + 2\tilde{\varphi}(\underline{k}_\perp)\sin\chi \approx \delta_D(\underline{k}_\perp) + \frac{z}{k}(k_x^2 + k_y^2)\tilde{\varphi}(\underline{k}_\perp), \tag{6.124}$$

where in the last step we used the small angle approximation for χ. In real space, we equivalently obtain

$$I(\underline{r}_\perp) = 1 - \frac{z\lambda}{2\pi}\nabla_\perp^2\varphi(\underline{r}_\perp). \tag{6.125}$$

This last expression for pure phase contrast holds beyond the weak phase approximation. In fact, it can be directly derived from the Transport of Intensity Equation (TIE, Section 6.3.3) without the approximation of small phase shifts or weakly varying phase. The only requirement is that the propagation distance be sufficiently small (or equivalently F sufficiently large), as is shown further below. This indicates the importance of linearization, either of the propagation distance or the object's phase shift. By using linearization in one or the other, the nonlinear near field phase retrieval problem becomes a linear problem, with correspondingly much larger leverage for solvers. Three different ingredients can make the problem very challenging: strong objects, small F, or a mixed object composed of arbitrary absorbing and phase shifting components. For the simple case of pure phase objects, equation (6.125) and equation (6.123) are excellent starting points to discuss the different regimes of propagation imaging, see also [21]:

- **Contact regime:** Directly behind the object at $\chi \simeq 0$, diffraction is not yet visible, and the wavefield shows a sharp imprint of the object's absorption profile. Phase effects have not yet been transformed into measurable intensities. In practice, the range of the contact regime depends strongly on the feature size. For macroscopic radiographic imaging, the detection may be spaced many centimeters apart from the body, and the image formation may still be characterized as the contact regime. In this respect, not only feature size but also vanishing coherence suppress the visibility of diffraction effects.
- **Direct contrast regime:** For increasing normalized spatial frequencies $\sin\chi \simeq \chi$ (or equivalently decreasing Fresnel number) we turn to equation (6.125). The contrast is governed by the Laplacian of the phase shift or, equivalently, the projected object density. Edges are strongly enhanced in the form of a double fringe. Hence, the visibility is linked to the second order of spatial derivative. Furthermore, as the wave front curvature $\nabla_\perp^2\varphi$ determines the intensity variations, we see that a positive wave front curvature leads to a decrease in the propagated intensity, while a negative wave front curvature implies an increase in the propagated intensity. Equation (6.125) also shows that the signal is contained in the deviation of intensity from unity, after dividing by the empty beam (flat field or empty beam correction). The amplitude of the signal is proportional to λ and z, and hence it can be easily 'tuned' experimentally. The simple proportionality to λ indicates that this direct contrast regime is compatible with a broader bandpass, in contrast to the

holographic regime. The regime is, hence, very suitable for laboratory sources, including low coherence conditions.

- **Holographic regime:** As the oscillations of the CTF reach $\chi \geq \pi$, the Fresnel diffraction pattern loses resemblance with the object. Many oscillations interfere, leading to a fully developed inline holographic image, in the sense of Gabor's original holography concept. Phase retrieval algorithms beyond linearization in the defocus distance are required. For weak objects, the analytic knowledge of the CTF provides such an ansatz, see Section 6.3.5. For more general cases, one has to resort to iterative algorithms. For macroscopic objects, the holographic regime can easily require several meters of propagation distance, depending on feature size and photon energy. Contrarily, for nanoscale objects (and medium photon energies) distances of a few mm may already be sufficient with the holographic regime. In fact, in cone beam geometry with $z_{\text{eff}} \simeq z_1$, decreasing the defocus distance also decreases the Fresnel number, since the effective pixel size enters F quadratically. Images thus become increasingly holographic when z_1 is reduced.

6.3.3 Transport of intensity equation (TIE)

Above, we saw that the approximation of a weak object already leads to a very tractable expression of the contrast transfer, providing an ideal starting point for reconstruction, which will be further pursued in Section 6.3.5. Instead of linearizing the object's optical properties, we now turn to a formalism in which linearization with respect to the propagation distance is used, first to understand image formation, and later again as a starting point for reconstruction. The strategy is to rewrite the PWE in terms of measurable quantities, the intensity and its first derivative along z. Subsequent approximation of the differential quotient by the difference quotient, then paves the way to reconstruction. We start by reconsidering the PWE (6.83)

$$(2ik\partial_z + \nabla_\perp^2)u(\underline{r}_\perp, z) = 0\,. \tag{6.126}$$

By writing the wave function in terms of its intensity, we obtain

$$u(\underline{r}_\perp, z) = \sqrt{I(\underline{r}_\perp, z)}\, \mathrm{e}^{i\varphi(\underline{r}_\perp, z)}\,. \tag{6.127}$$

Inserting the right hand side of equation (6.127) into the PWE yields

$$ik\frac{1}{\sqrt{I}}\,\mathrm{e}^{i\varphi}\partial_z I - 2k\sqrt{I}\,\mathrm{e}^{i\varphi}\partial_z\varphi + \nabla_\perp\cdot\left(\frac{1}{2\sqrt{I}}\,\mathrm{e}^{i\varphi}\,\nabla_\perp I + i\sqrt{I}\,\mathrm{e}^{i\varphi}\,\nabla_\perp\,\varphi\right) = 0\,. \tag{6.128}$$

We can further simplify the last equation and use a separation in terms of real and imaginary parts, which leads (for the imaginary part) to

$$ik\partial_z I + \frac{i}{2}\nabla_\perp I\cdot\nabla_\perp\varphi + \frac{i}{2}\nabla_\perp I\cdot\nabla_\perp\varphi + iI\nabla_\perp^2\varphi = 0\,. \tag{6.129}$$

Using the properties of the ∇ operator and rearranging the terms yields

$$\nabla_\perp \cdot [I(\underline{r}_\perp, z)\, \nabla_\perp \varphi(\underline{r}_\perp, z)] = -k\, \frac{\partial I(\underline{r}_\perp, z)}{\partial z} \,. \tag{6.130}$$

This continuity equation is known as the Transport of Intensity Equation (TIE) [33, 74, 85]. Hence, intensity variations along the propagation direction are determined by the curvature of the wave fronts (divergent beam) and the intensity gradient in the plane of the propagated wavefield. In view of the desired use for phase retrieval, we have to approximate $\partial_z I$ by a differential quotient, which can be computed from the measurements

$$\partial_z I \simeq \frac{I(\underline{r}_\perp, z + \Delta) - I(\underline{r}_\perp, z)}{\Delta} \,. \tag{6.131}$$

Inserting this into the TIE equation yields

$$I(\underline{r}_\perp, z + \Delta) \simeq I(\underline{r}_\perp, z) - \frac{\Delta}{k}\, \nabla_\perp \cdot \left(I(\underline{r}_\perp, z)\, \nabla_\perp \varphi(\underline{r}_\perp, z) \right) . \tag{6.132}$$

Rearranging this equation and applying the approximation

$$\nabla_\perp \cdot [I\, \nabla_\perp \varphi] = I\, \nabla_\perp^2 \varphi + \nabla_\perp I \cdot \nabla_\perp \varphi \;\simeq\; I\, \nabla_\perp^2 \varphi \tag{6.133}$$

(which holds if intensity variations are small compared to phase variations) leads to

$$I(\underline{r}_\perp, z) \simeq I(\underline{r}_\perp, 0) \left(1 - \frac{z}{k}\, \nabla_\perp^2 \varphi(\underline{r}_\perp, 0) \right) , \tag{6.134}$$

holding for small propagation distances $z = \Delta$ behind a plane taken to be at the origin $z = 0$. TIE based phase retrieval – in its simplest form – is the inversion of this equation, and hence computing φ from $I(\underline{r}_\perp, z)$. To this end, the numerical inversion of the Laplacian is required, which can be achieved based on the Fourier representation of the $\nabla_\perp$ operator

$$\nabla_\perp^{-2} = -\mathcal{F}^{-1}\, \frac{1}{k_x^2 + k_y^2}\, \mathcal{F} \,, \tag{6.135}$$

$$\nabla_\perp = -\mathcal{F}^{-1}\, \underline{k}_\perp\, \mathcal{F} \,. \tag{6.136}$$

When boundary conditions are not important, the phase φ can be solved by implementation of FFT filtering. Alternatively, for special requirements regarding boundary conditions, multigrid algorithms (finite element calculation on increasingly finer grids) can be used to solve such elliptical partial differential equations.

6.3.4 Propagation imaging in the TIE regime

Propagation based phase contrast can be implemented in any spectral range of electromagnetic waves or also particle waves (electrons, neutrons). For multi-keV x-rays it has become particularly attractive over the last two decades, since suitable lenses for full field imaging are lacking.[10] In view of this deficit in optical components of suitable specifications, a simple lensless radiography setup with an extended x-ray beam and no additional optical elements, apart from source, object (sample) and detector is appealing, also in view of practical implementation and compatibility with conventional radiography. Furthermore, without any optical element between object and detector, propagation imaging is comparatively dose effective.

In order to exploit the phase contrast, the simplistic geometric optical approximation used in the hospital has to be extended to a wave optical description. After significant research efforts with synchrotron radiation sources over the last decades, x-ray propagation imaging has now matured [13, 63, 69]. As we saw above, contrast for both absorbing and phase shifting objects is generated by the inherent near field diffraction of the wave transmitted through the object, while propagating in free space to the detector. In the following two sections, we now turn from image formation to object reconstruction, i.e., to the inverse problem of propagation imaging. We concentrate on the direct contrast regime at small defocus distances, where the TIE formalism provides a fast and robust phase retrieval approach, and then turn to the holographic regime. In this summary of phase retrieval based on the TIE, we follow the presentation of [2].

The goal of object reconstruction is to retrieve the complex valued object transmission function $\tau(\underline{r}_\perp)$ from the near field diffraction pattern $I(\underline{r}_\perp, z \geq 0)$ recorded downstream from the object. As in the previous section, we use $\underline{r}_\perp = (x, y)$ for lateral coordinates and z for the optical axis. The transmission function τ defined by the projection integrals over the object's optical indices (see equation (6.95)) can be written in terms of amplitude and phase $\tau(\underline{r}_\perp) = a(\underline{r}_\perp)e^{i\varphi(\underline{r}_\perp)}$. For weak objects, we can further linearize $\tau(\underline{r}_\perp) = 1 + o(\underline{r}_\perp)$ with $o(\underline{r}_\perp) = -k\bar{\beta}(\underline{r}_\perp)t - ik\bar{\delta}(\underline{r}_\perp)t$, where t is the object thickness and $\bar{\delta}$ and $\bar{\beta}$ are the respective averages of the optical constants over the line integral through t for each lateral position $\underline{r}_\perp$. Then $\varphi := -k\bar{\delta}(\underline{r}_\perp)t$ is the phase in the object exit plane in radian. The goal of phase contrast radiography is to retrieve φ from intensity measurements.[11]

The Bronnikov formula

The required data is provided by the downstream Fresnel diffraction pattern $I(\underline{r}_\perp, z)$ and a second measurement recorded directly behind the object $I(\underline{r}_\perp, 0)$ (which could

---

10 This is not quite the case for the soft and medium x-ray spectral range, where Fresnel zone plates and compound refractive lenses are more routinely available for x-ray microscopy. Contrarily, lenses for x-ray photon energies used in biomedical imaging on the tissue, organ and body scale are lacking.

11 The use of the term phase retrieval is sometimes ambiguous. It can designate the phase of the object exit plane or the phase of the wavefield in the detection plane. Of course, when the latter has been retrieved, the first can be obtained by a straightforward Fresnel backpropagation.

be alternatively provided as prior information). Under the condition that the absorption image does not "diffract", i.e., for sufficiently slowly varying absorption, phase is calculated based on the TIE formalism, yielding [2, 10]

$$\varphi(\underline{r}_\perp) \simeq -2\pi F\, \mathcal{F}\left[\frac{\mathcal{F}[I(\underline{r}_\perp, z)/I(\underline{r}_\perp, 0) - 1]}{\underline{q}_\perp^2}\right], \tag{6.137}$$

written here in form of the unitless (single pixel) Fresnel number, i.e., in pixel coordinates; $\underline{q}_\perp$ denotes the pixel indices of the DFT grid.

Paganin–Nugent formula

In Bronnikov's formula the difference between $I(\underline{r}_\perp, 0)$ and $I(\underline{r}_\perp, z)$ is only attributed to φ, since the absorption profile is not subject to diffraction itself. This can be justified based on the initial regime of the CTF, but requires the assumption of the object's intensity profile to be slowly varying. This is no longer necessary in the extension of Paganin and Nugent, who derived the reconstruction formula [68, 69]

$$\varphi(\underline{r}_\perp) \simeq -k\nabla_\perp^{-2}\left[\nabla_\perp\cdot\left(\frac{1}{I(\underline{r}_\perp, 0)}\nabla_\perp\left[\nabla_\perp^{-2}\left(\frac{I(\underline{r}_\perp, z) - I(\underline{r}_\perp, 0)}{z}\right)\right]\right)\right]. \tag{6.138}$$

In the derivation, a scalar potential is used, which exists in the absence of phase vortices [68]. The inverse Laplacians can be implemented numerically based on Fourier operators (or multigrid approaches) as detailed before. This equation is also the starting point for the so called Holo-TIE reconstruction scheme, see Section 6.3.5.

Modified Bronnikov algorithm (MBA)

Both of the above formula require knowledge about the object's absorption $I(\underline{r}_\perp, 0)$. If this is not available, and at the same time the absorption contrast is assumed negligible, one may be tempted to set $I(\underline{r}_\perp, 0) = 1$. However, this pure phase limit of the TIE formalism is numerically unstable, due to the singularity at $\underline{q} \to 0$ (i.e., the $1/q^2$ correction in equation (6.137)). This calls for a regularization with an adjustable unitless regularization parameter α, provided by the MBA reconstruction formula [30, 31]

$$\varphi(\underline{r}_\perp) \simeq -2\pi F\, \mathcal{F}^{-1}\left[\frac{\mathcal{F}[I(\underline{r}_\perp, z) - 1]}{\underline{q}_\perp^2 + \alpha}\right], \tag{6.139}$$

again written in pixel coordinates as a function of F as in [2]. In practice, α is found empirically. The parameter is increased until edge enhancement is compensated, but kept below the values at which the reconstructed image becomes blurred. Since $\underline{q}_\perp$ is given in inverse pixel coordinates, there is a simple interpretation for $\sqrt{\alpha}$ as a critical spatial frequency, at which the power law filter crosses over to a plateau (regularization). In practice, there may often be residual absorption in the image, which provides some physical basis for regularization. In other words, when a contact image $I(\underline{r}_\perp, 0)$ is not available, the strategy assumption of a pure phase object (even if, strictly speaking, not justified) can be permissible if a regularized parameter α is used. In many cases, absorption provides even a physical justification for the parameter, and as starting value for empirical optimization one can take $\alpha = 4\pi F\delta/\beta$ [2], where the ratio $\kappa = \delta/\beta$ is initialized according to the expected average of optical constants.

Paganin's single material formula

The last approach already points to the relevance of the ratio $\kappa = \delta/\beta$. However, above it was used in a rather coarse grained way for an empirical regularization parameter. This brings up the issue of a more rigorous derivation, for the case of a fixed coupling of optical constants. If a material has fixed stoichiometry, and only the overall density varies (but at fixed composition), this extra knowledge can be employed in phase retrieval as an important prior information. Solving the TIE for this case, Paganin obtained an expression, which by now has found widespread application [67]

$$\varphi(\underline{r}_\perp) = \frac{\kappa}{2} \cdot \ln\left(\frac{4\pi F\beta}{\delta} \mathcal{F}^{-1} \left[\frac{\mathcal{F}[I(\underline{r}_\perp, z)]}{\underline{q}_\perp^2 + 4\pi F/\kappa} \right] \right), \tag{6.140}$$

again written in convenient pixel coordinates [2]. In practice, the approach yields results that are not too different from MBA [8, 11]. Both approaches fail, if the variation in $\kappa = \delta/\beta$ becomes too significant, yielding blurry images. If the variation in κ is sufficiently small, i.e., when the object is nearly single material, the following strategy may be useful. For ideal imaging conditions, for example, using monochromatic and highly coherent synchrotron radiation, Paganin's single material approach should be used with prescribed (fixed) κ. Contrarily, with laboratory sources, where even for a single material κ enters as a bandpass weighted parameter, it is more suitable to choose κ or, equivalently, α in the MBA by variation and visual inspection.

Bronnikov aided correction (BAC)

In practice, when reconstructing data of low coherence obtained at laboratory sources, one can easily fall into either of two traps: reconstruction with small regularization parameter leads to low frequency artifacts, reconstruction with high regularization parameter leads to blurring. For nonideal data, there does not seem to be an ideal value, and it is often assumed that phase reconstruction under these conditions does not work. However, as shown in [2, 3] and [41] the additional strategy known as **Bronnikov aided correction** (BAC) introduced by [95] works surprisingly well for laboratory sources. It is based on two steps. First, an approximative phase distribution $\phi^{\text{app}}(\underline{r}_\perp)$ is reconstructed by MBA, almost as if it were a pure phase object. In fact, this step should just compensate for the edge enhancement. From this phase, the intensity pattern in the exit plane is computed in a second step

$$I(\underline{r}_\perp, 0) = \frac{I(\underline{r}_\perp, z)}{1 - \gamma \nabla_\perp^2 \phi^{\text{app}}(\underline{r}_\perp)}, \tag{6.141}$$

where γ is a further variable parameter, which depends on the regularization parameter α (used to find ϕ^{app} in the first step), and which has to be determined by visual inspection. The result is an effective absorption image with contributions of both absorption and phase contrast. The BAC approach achieves an effective inversion of diffraction induced blurring. It is built on the assumption that the phase image diffracts, while the absorption image is stationary with z, which is approximately true for small χ in the CTF. By selecting γ and α, the measured signal $I(\underline{r}_\perp, z)$ is attributed partly to phase (high frequencies), partly to absorption (low frequency), in a way that preserves sharpness, while compensating the edge effect. This seems an inferior approach, lacking rigorous justification, and one would certainly not resort to this approach for ideal data, such as data recorded with synchrotron radiation. For laboratory data, however, BAC is an enabling algorithm that provides stunning results [87, 88]. More effort should be put into better understanding why this is the case.

Summary

The TIE based approach to phase retrieval is based on a near field diffraction pattern recorded in the direct contrast regime along with an estimate of the derivatives of the field with respect to z. The obvious way to estimate the derivative is to use a contact image at z_1 (which is only sensitive to absorption) and to approximate the respective differential quotient by the difference quotient, computed from the two images. This can be regarded as a linearization of the field evolution over the propagation distance $\Delta z = z_2 - z_1$. An important aspect for practical relevance is the fact that the contact image can be made obsolete by prior information, so that TIE phase retrieval is compatible with single distance recordings. The strategy is to use additional knowledge of the object's optical properties, for example, the fact that the object has vanishing absorption (pure phase object), coupled phase and absorption contrast (single material object) or – in a more approximate fashion – small absorption, which affects only the small spatial frequencies. TIE based phase retrieval is compatible with broad bandpass, i.e., with sources exhibiting a broad spectrum of x-rays. An insightful comparison of MBA, BAC, and Paganin's approach for laboratory based phase retrieval methods is given in [89].

6.3.5 Propagation imaging in the holographic regime

The TIE based approach presented in the preceding section is limited, since it is restricted to small propagation distances, where the phase signal (i.e., contrast in the Fresnel diffraction pattern) is not yet fully developed. In practice, the downstream image has to be recorded at z_2 high enough to achieve reasonable contrast, but low enough to keep the error due to linearization small.

For purposes of resolution and contrast, it is of advantage to increase the propagation distance further than the TIE range. In terms of the CTF (equation (6.123) and Figure 6.19), the TIE regime corresponds to the initial parabolic increase of the phase CTF or the plateau of absorption CTF, respectively. Contrarily, the holographic regime corresponds to the rapidly oscillating regime of the CTF. Here, the challenge is not the image formation itself, but the decoding of the near field pattern in this regime, i.e., the inverse problem. One strategy is to replace the linearization with respect to the propagation distance, by a linearization regarding the optical properties. To this end, we can use the knowledge of the nature of the CTF for phase retrieval. Alternatively, one can leave behind linearizations altogether, and use iterative projection algorithms that are suitable for solving the nonlinear problem based on the data *and* prior knowledge. Furthermore, iteratively regularized Gauss–Newton (IRGN) methods have recently been devised for holographic phase retrieval. In the following, we present these approaches, one by one, starting with a method that extends the TIE concept to the full range of the CTF.

Holo-TIE

The idea is simple. The TIE concept is extended to larger propagation distances by taking the difference quotient for the diffraction pattern recorded at z_1 and $z_2 = z_1 + \Delta$. We no longer compute the phase in the object's exit plane, but in the detection plane, by a 'differential measurement', recording two closely spaced Fresnel diffraction images. From the mean intensity and the TIE reconstructed phase, we obtain the field in the center of the diffraction plane (midplane of the two recordings), followed by backpropagation of the field to the object's exit plane [42]. This simple concept yields very convincing reconstructions and does not require any further assumptions on the object, i.e., it is also suitable for mixed phase and amplitude objects. Residual low frequency artifacts arising from noise in computing the difference quotient are easily removed by additional application of iterative algorithms, or by variation of Δ [42, 44].

CTF based reconstruction

For a weak object, where the exponential factors of amplitude and phase can be linearized, we have seen that the CTF is described in reciprocal space (2d Fourier transform of near field images) by a simple oscillatory function, the CTF of equation (6.123). The significance of this simple CTF is the following: For an object of either pure phase or pure absorption contrast, the Fourier transforms of the optical indices $\tilde{\varphi}(\underline{k}_\perp)$ or $\tilde{\mu}(\underline{k}_\perp)$ are directly imprinted in the Fourier transform of the near field image, multiplied with an additional radially symmetric function of $\sin\chi$ or $\cos\chi$. Phase reconstruction, therefore, amounts to simply dividing by this function with the Fresnel number F as the only parameter, followed by inverse Fourier transform. In practice, however, things are not that easy, even for pure phase or amplitude contrast objects. Division by zero and the suppression of certain spatial frequencies by the zero crossings of the CTF lead to artifacts. The remedy to this problem is to combine images recorded at several Fresnel numbers in order to achieve redundancy in the spatial frequencies which are transferred and to use regularization techniques in order to prevent division by zero. CTF based reconstruction was pioneered by P. CLOETENS and coworkers [13], who additionally proposed a simple scheme to combine the information from a defocus series with m images (in most applications $m = 4$). The idea is to minimize the deviation of the data from the expected CTF behavior, which for a pure phase object amounts to minimizing

$$E = \frac{1}{N} \sum_{m=1}^{N} \int \mathrm{d}\underline{k}_\perp \left| \tilde{I}^{(\mathrm{exp})}(\underline{k}_\perp, z_m) - 2\tilde{\varphi}(\pi z k^2_{\perp/k}) \sin\left[\pi z k_\perp^2/k\right] - \delta_\mathrm{D}(\underline{k}_\perp) \right|^2 . \quad (6.142)$$

Taking the derivative $\partial E/\partial\tilde{\varphi} = 0$ yields the phase distribution φ, which minimizes the error E

$$\tilde{\varphi}(\underline{k}_\perp) = \frac{\sum_{m=1}^{N} \tilde{I}^{(\mathrm{exp})}(\underline{k}_\perp, z_m) \sin\left[\pi z k_\perp^2/k\right]}{\sum_{m=1}^{N} 2\sin^2\left[\pi z k_\perp^2/k\right]} . \quad (6.143)$$

To prevent division by zero, the expression has to be regularized by a frequency dependent regularization parameter, as introduced in [97] and reviewed in [2]: $\alpha(\underline{q}_\perp) = \alpha_1 \cdot f(\underline{q}_\perp) + \alpha_2 \cdot (1 - f(\underline{q}_\perp))$ with a suitable switch function f.[12] The idea is to have different regularization parameters for the TIE and the holographic regime of spatial frequencies, accounting for the different nature of zero crossings. It is commonly assumed that this form of the CTF calculated for monochromatic radiation is incompatible with large bandpass. However, a controlled bandpass may also help to effectively regularize the zero crossings. In future, it may be of advantage to extend the scheme to finite bandpass. An alternative way to deal with the zero crossings essentially by masking the data of the corresponding rings in Fourier space is presented in [59]. The CTF approach, at least in the present form, generally fails for mixed absorption and phase objects, except for the fixed material case with parameter $\kappa = \delta/\beta$ [90]

$$\varphi(\underline{r}_\perp) = \mathcal{F}^{-1}\left[\frac{\sum_{m=1}^{N} \mathcal{F}[I^{(\mathrm{exp})}(\underline{r}_\perp, z_m) - 1] \cdot \left(\sin\left[\underline{q}_\perp^2/4\pi F\right] + \frac{1}{\kappa}\cos\left[\underline{q}_\perp^2/4\pi F\right]\right)}{\sum_{m=1}^{N} 2\left(\sin\left[\underline{q}_\perp^2/4\pi F\right] + \frac{1}{\kappa}\cos\left[\underline{q}_\perp^2/4\pi F\right]\right)^2 + \alpha(\underline{q}_\perp)}\right] , \quad (6.144)$$

again written as in [2]. The case of pure phase contrast is readily obtained from the above formula in the limit $\kappa \to \infty$, by which the cos terms drop out. The CTF approach has been successfully applied to a number of high resolution tomography studies on different tissues, ranging from the nervous system [4] to the lungs [43], see also Figure 6.20.

12 Again, $q_\perp$ is used to denote the spatial frequency in units of inverse image pixel.

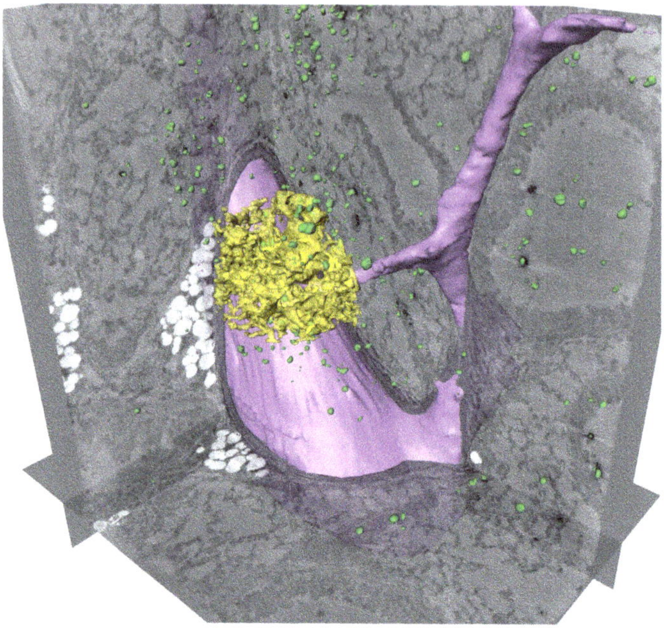

Fig. 6.20: CTF based reconstruction of lung tissue from a mouse from cone beam tomography scans recorded at the European Synchrotron Radiation Facility [43]. The 3d rendering shows three orthoslices together with barium labeled macrophage cells (*green*), and alveolar walls rendered in a small ROI (*yellow*), and a blood vessel marked semiautomatically (*purple*). Voxel size $p = 430$ nm. From [43].

6.3.6 Iterative algorithms for reconstruction

As we discussed above in Section 6.2.2, the phase problem can be solved by reducing the number of unknowns in the object plane, most easily by requiring the object to have a sufficiently small and compact support. Many other constraints can be formulated to enable solutions to the inverse problem of diffraction. Here, the object is reconstructed from diffraction data **and** the prior information provided by the constraints. The need for further constraints expresses the fact that measured diffraction patterns record only intensities that are real valued. In contrast, the original wavefield consists of amplitudes and phases, i.e., of complex values. We now face the problem that a multitude of complex valued functions could have produced the measured data. In order to find a complex valued wavefield that most likely produced the recorded intensities we can use the measured data, additional information (constraints) and iterative projection algorithms.

Projection algorithms

When the problem of near field imaging can be linearized, either by considering weakly interacting samples, or by linearization with respect to the propagation distance, solutions based on the TIE or the CTF formalism are most appropriate and yield deterministic one step solutions. For the deeply holographic regime and for stronger samples, however, linearizations are not appropriate, and the nonlinear aspect of the problem becomes more apparent. Projection algorithms are then the method of choice, cycling iteratively between the detection plane (to account for the data) and the object plane (to formulate additional constraints). Issues to be addressed concern the initialization (e.g., by a random initial guess), convergence, numerical efficiency, and eventually stagnation.

The concept of projection algorithms is illustrated in Figure 6.21. Let X denote a vector space containing all possible solutions ψ of the wave equation. A subset of these solutions ψ_i fulfills certain constraints in real space – the a priori information. For example, ψ_i could correspond to all solutions with a finite predefined support in the object plane. A second subset of solutions ψ_j corresponds to those wavefields in the object plane that are consistent with the measured intensity data in the detection plane. For example, in a far field setting, we could demand $|\mathcal{F}[\psi_j]|^2 = I$, or – by exchange of the propagator – $|\mathcal{D}_z[\psi_j]|^2 = I$ for data recorded in the optical near field (cf. Section 6.2.1). The intersection of both sets contains all ψ_s that are possible solutions of the phase problem. Based on experimental design, the sets are chosen such

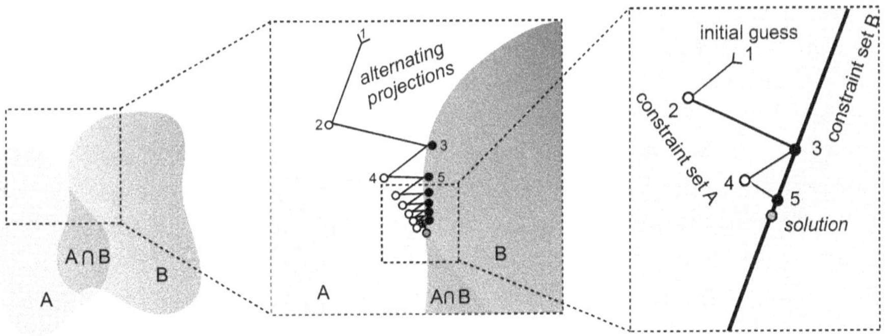

Fig. 6.21: Illustration of alternating projections to find the solution of the phase problem. Let X denote a vector space of functions ψ that fulfill the wave equation. A subset $A \subset X$ of these solutions is consistent with the measured data. A second subset $B \subset X$ denotes all elements of X that fulfill certain constraints. These constraints are considered as prior information when solving the inverse problem of diffraction. The intersection of both sets contains the solution. Alternating projections are applied to iteratively find the intersection starting from an initial guess. For convex constraint sets they will finally converge to the solution (illustrated here as two intersecting lines). In reality, not all constraint sets are convex, but eventually one can find an initial guess that is sufficiently close to the locally convex topology of the sets. If the problem is too challenging, one should reformulate the constraints or change the experimental design.

that this intersection ideally has a unique solution (up to perhaps symmetries, considered irrelevant for object reconstruction). If it turns out that an intersection of data and constraints comprises (too) many different solutions, the constraints have to be 'sharpened' or data recording has to be extended, for example, by longer acquisitions (less noise) or finer sampling, by adding more detection planes, or more generally by recording data with additional diversity (variation of experimental parameters such as wavelength, defocus, translations, etc.).

As illustrated in Figure 6.21, the nonlinear problem of finding the intersection of the two sets can be solved iteratively by alternating projections. First, an initial guess ψ is defined. The shortest distance (i.e., the projection) of ψ onto the elements in the first set is calculated, resulting in an element ψ_i inside the first set and closest to the initial guess. This element ψ_i is projected onto the second set resulting in ψ_j. By iterations of these alternating projections, the intersection – and hence the solution – may be found, at least in fortunate cases when the constraint sets are convex (i.e., every point on a straight line between two elements of a set is also element of the set). However, not all of the relevant constraint sets are convex. Most importantly, the magnitude constraint set is not convex [49]. From a mathematical point of view, existence uniqueness and convergence are central issues when studying alternating projections. Experimentally, the main approach is to study these issues empirically by numerical simulations of the reconstruction from artificial data and then to decide whether the experimental design has to be changed. Three important aspects have to be considered: the prior information at hand, the data at hand and, finally, the algorithmic engine to find a solution. Often, much emphasis is put in finding a suitable type of algorithm. For a good reason: Small changes in the algorithms can lead to very different dynamic behavior (convergence versus stagnation, local versus nonlocal search, numerical complexity). What is often neglected, however, is the experimental design. The experimentalist has many parameters at hand to design the type of problem: near field versus far field, single plane versus multiplane, consideration of the object with respect to the optical indices, variation of photon energy, considerations of sparsity, etc. In the following, however, we do not dwell on the constraints, but introduce some of the most widespread types of algorithms.

The Gerchberg Saxton (GS) algorithm

Already in 1972, the GS algorithm [24] was devised, addressing a long standing problem in quantum mechanics, originally posed by Wolfgang Pauli. The question was how to determine a wave function from measurements of its intensities at two different times. Given the correspondence of our optical problem and quantum mechanics in 2+1 dimensions, this amounts to measuring a wavefield in two different planes, $I_1(x, y)$ and $I_2(x', y')$, respectively, possibly but not necessarily Fourier conjugate planes. The first constraint set consists of all ψ_i that fulfill the wave equation and exhibit intensities $I_1(x, y)$ in the first plane. The second constraint set contains all wavefields ψ_j that

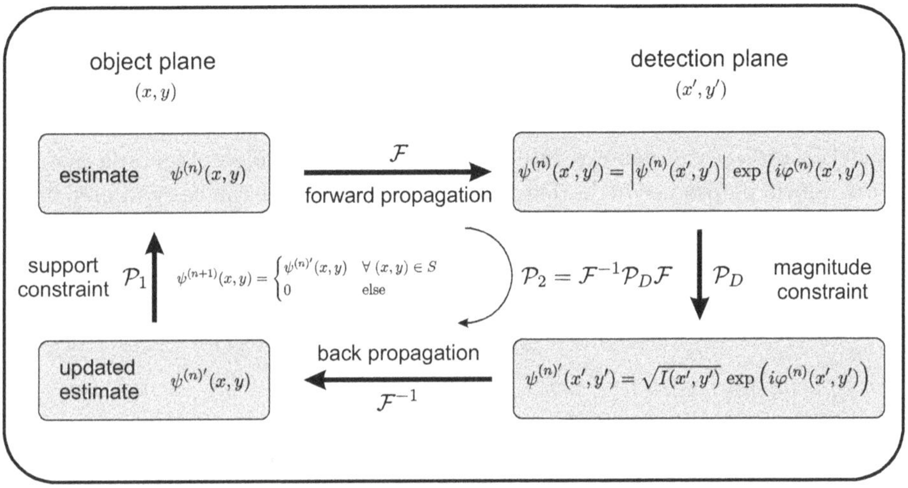

Fig. 6.22: Schematic illustration of the error reduction (ER) algorithm for a far field setting. The algorithm cycles between object and detection plane and projects onto the constraint sets in both planes. In far field diffraction the most commonly used constraint is the finite support S of the object. The error reduction algorithm uses this knowledge in the form of a simple projection, forcing the object to be zero outside S.

fulfill the wave equation and exhibit intensities $I_2(x', y')$ in the second plane. Data compliance is enforced by simple projection onto the data (as detailed below). Consistency with the wave equation is implemented by the propagation between the two planes. The iteration can be initialized by taking the measured intensities I_1 combined with random phases $\varphi_0(x, y) \in [-\pi, \pi[$

$$\psi^{(o)}(x, y) = \sqrt{I_1(x, y)} \cdot \exp[i\varphi_0(x, y)] . \tag{6.145}$$

The projection onto the first set can be written as

$$\mathcal{P}_1[\psi(x, y)] = \sqrt{I_1(x, y)} \cdot \frac{\psi(x, y)}{|\psi(x, y)|} , \tag{6.146}$$

followed by the projection onto the second set formulated as

$$\mathcal{P}_2[\psi(x, y)] = \mathcal{F}^{-1}\left[\sqrt{I_2(x', y')} \cdot \frac{\mathcal{F}[\psi]}{|\mathcal{F}[\psi]|}\right] \tag{6.147}$$

if I_1 and I_2 represent measured intensities in the object and detection plane, respectively, in case of far field imaging. The complete algorithm is then represented by alternating projections onto the two measurements

$$\psi^{(n+1)}(x, y) = \mathcal{P}_1[\mathcal{P}_2[\psi^{(n)}(x, y)]] . \tag{6.148}$$

Note that the Fourier transforms $\mathcal{F}$ and $\mathcal{F}^{-1}$ have to be replaced by the near field propagators $\mathcal{D}_z$ and $\mathcal{D}_{-z}$ for holographic near field imaging (see Equation (6.50) and (6.51)). Both $\mathcal{P}_1$ and $\mathcal{P}_2$ replace only the amplitudes of the reconstructed ψ by the measured amplitudes and keep the phases of the current iterate. For phase objects illuminated by perfect plane waves, the amplitudes in the first plane can be set to one. In this case, the GS algorithm becomes an algorithm for data sets $I(x', y')$ recorded at a single distance. $\psi(x, y)$ denotes the desired wavefield in the object plane, and $\psi(x', y')$ the wavefield propagated to the detection plane. Convergence can be monitored by an error metric measuring the distance between the current iterate $\psi^{(n)}(x', y')$ and the data $I(x', y')$.

$$d^2(|\psi^{(n)}(x', y')|^2) := \frac{1}{N} \sum_{(x',y')} \left(|\psi^{(n)}(x', y')|^2 - I(x', y')\right)^2 , \qquad (6.149)$$

where N denotes the number of detector pixels [25]. Remarks:

- The GS algorithm is a local optimization tool and, therefore, initialization is an important issue.
- It is well suited for retrieving phase objects, since in this case, amplitudes in the first plane can be set to one.
- The GS can show artifacts in the reconstructions, since it projects to the full data, containing sample information and noise.
- For objects that are not of pure phase contrast, the GS projectors can be generalized to account for an object of mixed phase and amplitude contrast, by enforcing a constant ratio of δ and β, i.e., to account for an object of a single material. One can also project only to negative phases and amplitudes < 1, thereby enforcing a very mild range restriction (positive definite electron density). For a large number of iterations, this can also yield surprisingly good results.

The error reduction (ER) algorithm

The error reduction algorithm [19, 20] is a generalization of the GS algorithm by Fienup, who suggested that only one far field measurement $I(x', y')$ and the fact that the object is restricted to a known finite support S in the object plane is sufficient to reconstruct amplitudes and phases in the object plane. Fienup replaced the first constraint set by the set of all functions ψ that fulfill the wave equation and are nonzero only within a known compact support S. The second constraint set (also called the magnitude constraint) is the same as in the Gerchberg–Saxton algorithm. Thus, the corresponding projections are

$$\mathcal{P}_1[\psi(x, y)] = \begin{cases} \psi(x, y) & \forall\,(x, y) \in S \\ 0 & \text{else} \end{cases} \qquad (6.150)$$

and

$$\mathcal{P}_2[\psi(x, y)] = \mathcal{F}^{-1}\left[\sqrt{I(x', y')} \cdot \frac{\mathcal{F}[\psi]}{|\mathcal{F}[\psi]|}\right] \qquad (6.151)$$

for far field imaging. The algorithm is randomly initialized in the object plane. The update rule is equivalent to the GS algorithm

$$\psi^{(n+1)}(x, y) = \mathcal{P}_1[\mathcal{P}_2[\psi^{(n)}(x, y)]] \,, \tag{6.152}$$

see Figure 6.22 for a schematic representation. In practice, the algorithm is used mostly for far field settings. It is known for its simple, fast and robust realization.

The hybrid input output (HIO) algorithm

The HIO algorithm [20] is again a generalization of the error reduction algorithm. The name originated from its background in nonlinear systems theory where the input and output of an operation are linearly combined to form a new iterate. This combination avoids the solution trajectory to depart too rapidly from a given point. It is also the reason for nonlocal behavior. In other words, the solution trajectory can 'hop out' of a local minimum. Here, instead of setting ψ outside the support to zero, the algorithm

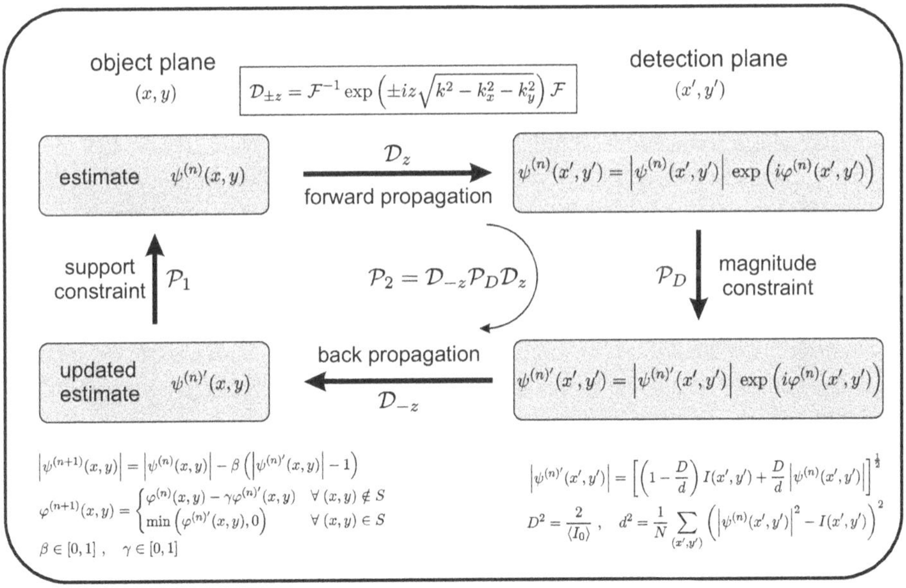

Fig. 6.23: Schematic illustration of the hybrid input output (HIO) algorithm in its modified version for near field propagation imaging [26]. Depending on the optical regime (far field versus near field), not only the propagation operator has to be adapted (from Fourier transformation to Fresnel propagation), but also the prior information. For the far field case, the object function is often modeled in terms of scattering length density, which is zero outside the support S of the object. Contrarily, near field problems are formulated in terms of the exit wave, which is one outside the support. Irrespective of the optical regime, the HIO algorithm is a generalization of the error reduction algorithm, which 'softly' pushes ψ to zero/one outside the support S, avoiding the 'hard' projection of the ER, and thus also avoiding stagnation. Inconsistencies of the measurement can be taken into account by a soft projection [26].

introduces a tunable parameter β to slowly decrease the values of the wavefunction to zero outside the support. The operators can be formulated as

$$\mathcal{P}_1[\psi(x,y)] = \begin{cases} \psi'(x,y) & \forall (x,y) \in S \\ \psi(x,y) - \beta\psi'(x,y) & \text{else,} \end{cases} \tag{6.153}$$

where $\psi'(x,y)$ is the solution of the previous iteration, and

$$\mathcal{P}_2[\psi(x,y)] = \mathcal{F}^{-1}\left[\sqrt{I(x',y')} \cdot \frac{\mathcal{F}[\psi]}{|\mathcal{F}[\psi]|}\right] \tag{6.154}$$

again for the case of far field imaging. The complete algorithm is represented as

$$\psi^{(n+1)}(x,y) = \mathcal{P}_1[\mathcal{P}_2[\psi^{(n)}(x,y)]] \,. \tag{6.155}$$

A schematic illustration is given in Figure 6.23, for the case of a near field propagator (i.e. $\mathcal{F} \rightarrow \mathcal{D}_z$ and $\mathcal{F}^{-1} \rightarrow \mathcal{D}_{-z}$) and with some additional modifications [26]. In practice, one often runs a reconstruction first with the HIO for a nonlocal search, followed by a certain number of ER iterations for local optimization.

One crucial aspect that has not been addressed so far is the support determination. Support as a constraint to enable phasing sounds attractive, but how can we know its precise shape? Indeed, this challenge is well known from far field CDI, where in many cases the support has to be inferred from prior information such as different microscopy methods. More sophisticated solutions, such as the shrink wrap algorithm [51] have been devised to iteratively determine the support from the data, by stepwise reduction. In other words, the constraint is sharpened during the reconstruction. However, this procedure can fail and is difficult to validate if prior information of the sample is not available. Fortunately, in the near field case, holographic reconstruction by a simple Fresnel backpropagation of the intensities already gives a sufficiently precise account of the object's shape and, hence, its support.

By replacing the Fourier operator of far field diffraction $\mathcal{F}$ with the Fresnel near field diffraction $\mathcal{D}_z$ and by modifying the object plane update such that the amplitude is gently pushed towards one and the phase towards zero outside the support, the HIO can be adapted for near field imaging of pure phase objects (mHIO algorithm) [26], yielding superior results in the holographic regime with resolution down to the 20 nm range [5]. Figures 6.24 and 6.25 show illustrative applications.

The relaxed averaged alternating reflection (RAAR) algorithm

The example of the HIO shows that $\mathcal{P}_1$ can be generalized to a linear combination of the input (wavefunction before applying the operator) and the output (wavefunction after application of the operator). This changes the properties of the algorithm significantly and leads to a nonlocal behavior, enabling escape from local minima. However, mathematically speaking, here $\mathcal{P}_1$ is no longer a projector. A better defined nonlocal

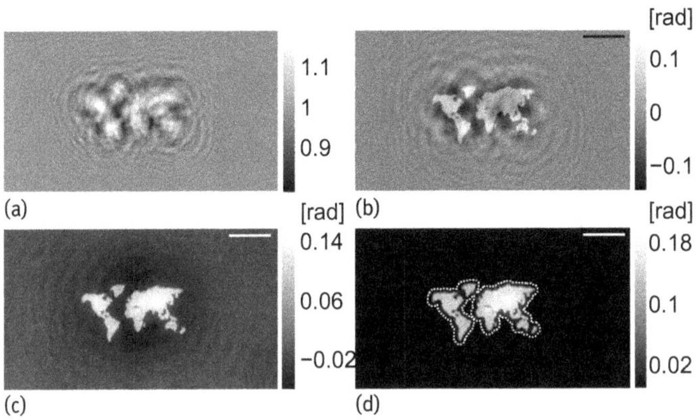

Fig. 6.24: (a) Hologram of a gold test structure recorded at high magnification in a cone beam synchrotron setup [5], along with the corresponding reconstructions by (b) holographic phase reconstruction, (c) phase reconstruction based on the CTF and (d) iterative mHIO phase reconstruction. The support information (*dashed line*) can easily be obtained from the holographic reconstruction. Scale bars 2 μm. From [5].

algorithm with well controlled mathematical properties is a combination of true projections $\mathcal{P}$ and reflections $\mathcal{R}$ from the interface of constraint sets, known as the relaxed averaged alternating reflection (RAAR) algorithm [48]

$$\psi^{(n+1)} = \left(\frac{\beta_n}{2} (\mathcal{R}_1 \mathcal{R}_M + \mathcal{I}) + (1 - \beta_n) \mathcal{P}_M \right) \psi^{(n)} . \tag{6.156}$$

The parameter β_n is a relaxation parameter, $\mathcal{I}$ is the identity operator, $\mathcal{R}_M := 2\mathcal{P}_M - \mathcal{I}$ the reflector taking into account the data (measured intensities) and $\mathcal{R}_1 := 2\mathcal{P}_1 - \mathcal{I}$ the reflector taking into account the constraint set in the object plane. Mathematical statements about convergence of this algorithm can be found in [48]. Furthermore, the mathematics of iterative projection have been addressed in some detail in [49]. Much insight can also be obtained from geometric interpretation in projection and reflection algorithms [50].

In summary, alternating projection algorithms proceed iteratively by projection onto constraint sets, alternating between the object plane and the detection plane. The intermediate wave propagation steps from one plane to the other can be included in the projection operator. For example, propagation to the detector and replacement of amplitudes with the measured data can be written as one projection operator, often called the measurement or modulus constraint. Propagation, however, is instrumental to the performance of the algorithms, since this step mixes and correlates the coefficients of the pixel basis system. However, alternating projection algorithms do not take the mathematical nature of the forward operator directly into account. This is accomplished by iteratively regularized Gauss–Newton (IRGN) methods, which are

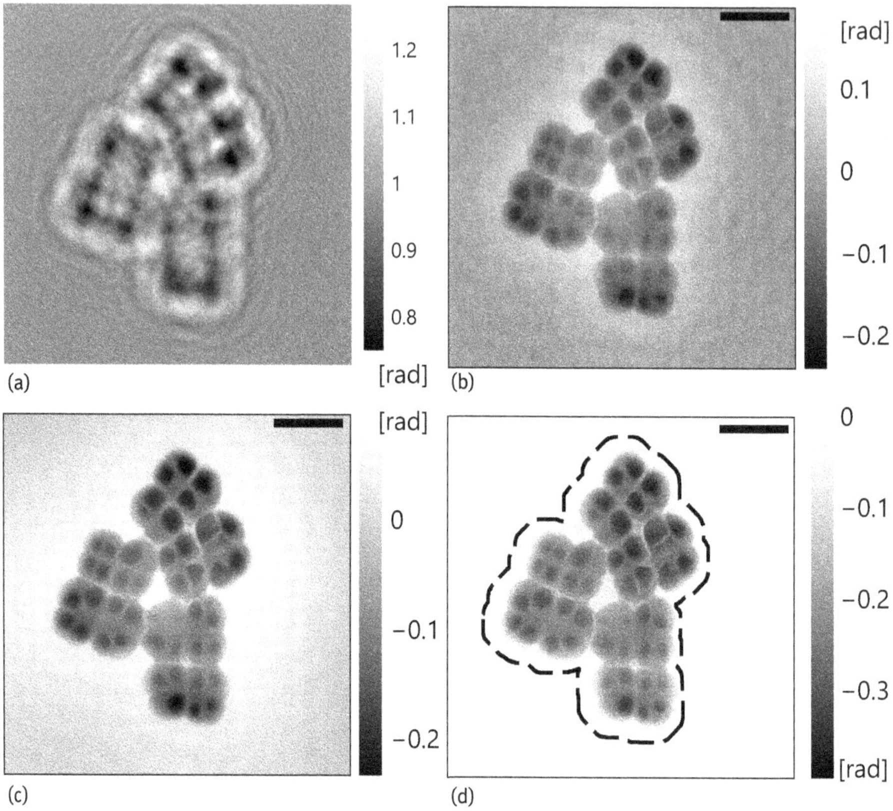

Fig. 6.25: Propagation imaging of *Deinococcus radiodurans* cells [5]. (a) Hologram obtained in a single distance recording at high magnification (cone beam). (b) Phase reconstruction of single distance recording in (a), based on the CTF. (c) CTF phase reconstruction from data recorded at four source-to-object distances (defocus distances). (d) Phase reconstruction utilizing the mHIO scheme. The support information can be obtained from a holographic reconstruction or from (b). The mHIO algorithm clearly yields superior quality. Scale bars 4 μm. From [5].

based on the known Frechet (operator) derivative of the forward operator and have recently been introduced in holographic phase retrieval [53].

6.4 Grating based phase contrast

In general, the aim of phase contrast imaging is to determine the phase shift of the wavefield after propagation through the sample. One way of visualizing the phase is to convert phase variations to measurable intensities by ways of free space propagation.

According to the TIE equation, contrast forms slowly as the beam propagates and is proportional to the second derivative of the phase

$$\Delta I_z(x, y) = \lambda z \nabla^2 \varphi(x, y) \,, \tag{6.157}$$

as discussed above. One may wonder whether contrast can be achieved by a 'short cut', without free space propagation. Indeed, contrast for the object phase $\varphi(x, y)$ can be created or even to some extent amplified by placing suitable optical elements such as diffraction gratings or reflecting Bragg crystals in the beam path behind the object, making the transmission sensitive to wavefront deviations. In particular, this way one can achieve sensitivity to the first spatial derivative of the phase. This has important consequences. Consider the intensity variation (contrast) for a sinusoidal phase variation $\varphi(x) = \varphi_0 \sin(k_x x)$ with spatial frequency k_x, which according to equation (6.157) results in a propagation based contrast proportional to the square of spatial frequency

$$\Delta I = \lambda z k_x^2 \, \sin(k_x x) \,. \tag{6.158}$$

Contrarily, for contrast that is proportional to the first derivative, such as in grating or analyzer based techniques, we would have

$$\Delta I = c k_x \, \sin(k_x x) \,, \tag{6.159}$$

where c is, again, the respective proportionality constant. It follows immediately that the contrast of a first order phase contrast technique is higher than contrast by free propagation for spatial frequencies smaller than

$$k_x^* = \frac{c}{\lambda z} \,, \tag{6.160}$$

while the opposite is true for high spatial frequencies. In practice, free propagation is required for high resolution phase contrast radiography with pixel size in the micron and submicron ranges, and is suitable for preclinical biomedical research. Contrarily, grating based phase contrast is the method of choice for macroscopic imaging, and is suitable for clinical applications. Grating based phase contrast imaging was pioneered mainly by two groups, A. Momose in Japan [57], and independently by a team at the Swiss Light Source, namely C. David, T. Weitkamp and F. Pfeiffer [70, 73, 93]. A major breakthrough was the implementation of grating based imaging at low coherence sources, i.e., the translation from synchrotron research to compact laboratory sources [73] and the invention of dark field contrast by F. Pfeiffer [70]. After a series of technical improvements [14, 58, 70, 73, 93], the extension to tomography [39, 71, 72] and first preclinical tests [29], the technique is now at the point where it could be implemented in clinical practice, provided that the contrast benefits can be sufficiently demonstrated for clinical applications, while keeping the radiation dose at tolerable levels.

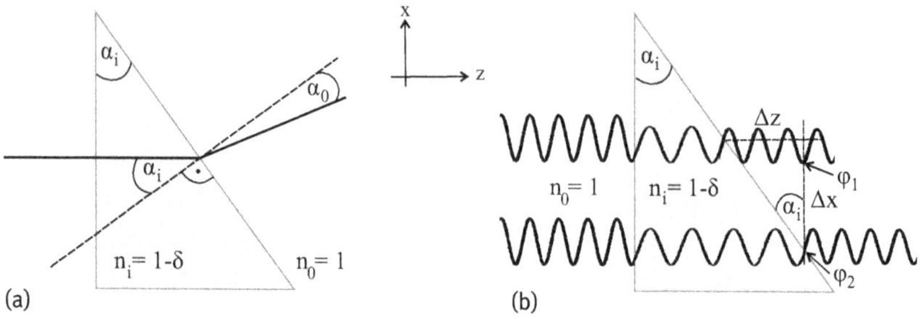

Fig. 6.26: Refraction of a beam impinging on a wedge. The slanted edge at the exit leads to an angular deviation α_r of the exit beam with respect to the optical axis. Two different viewpoints of this situation result in the same α_r: (a) Local refraction from Snell's law, (b) considering the phase shift between two neighboring beams by propagation through a different object thickness. For a more general object, a gradient on the exit phase of the wavefront leads to an angular redirection of the phase front, i.e., a small tilt of a focused beam or of the normal vector of an extended wavefront. This phase gradient and the respective angular deviation are extremely small for typical x-ray refraction indices of matter.

6.4.1 Refraction, wavefront tilts and phase gradient

At the beginning of this chapter, we considered the limits of line integration in radiography by evaluating Snell's law for typical x-ray optical indices, in view of refraction effects. Now we turn this around and want to exploit the angular deviations of the wavefront normal behind an object. We rewrite Snell's law in order to find the angular deviation of the refracted beam with respect to the primary beam for a wedge of homogeneous material with index n_i. To this end, we consider a beam impinging on the vertical side of the wedge, oriented perpendicular to the optical axis, hence entering the medium i without refraction, as sketched in Figure 6.26(a). After propagation in the object, the beam impinges on the slanted edge, undergoing refraction. Next, we want to compute the resulting angular shift α_r of the beam with respect to the optical axis. With subscript 0 denoting the propagation medium, and subscript i denoting the object (wedge), Snell's law is

$$n_i \sin \alpha_i = n_0 \sin \alpha_0 \,, \tag{6.161}$$

with n_i and n_0 the refractive indices of two media, and α_i and α_0 the respective incident and exit angles with respect to the slanted edge. Note that for the geometry given in Figure 6.26, the incidence angle α_i and the wedge angle are identical. In the approximation of small angles, valid in the case of x-ray beams and for air as the usual propagation medium with $n_0 \simeq 1$, this simplifies to

$$\alpha_0 \simeq n_i \alpha_i \,. \tag{6.162}$$

The angular deviation of the refracted beam with respect to the primary beam is then $\alpha_r = \alpha_i - \alpha_0$. For typical values the deviation by refraction is thus on the order of μrad

$$\alpha_r = \alpha_i(1 - n_i) = \alpha_i\,\delta = \alpha_i \underbrace{\frac{\rho_0 \lambda^2 r_0}{2\pi}}_{\substack{3.7\cdot 10^{-7} \\ (\mathrm{H_2O},\, 25\,\mathrm{keV})}} \quad . \tag{6.163}$$

For the same geometry of the slanted wedge, the propagation through a different object thickness results in a phase shift of two neighboring beams given by

$$\Delta\varphi = \varphi_1 - \varphi_2 = k\delta\,\Delta x\,\tan\alpha_i\,. \tag{6.164}$$

In view of a general object with a local interface tilt ('local' wedge angle), we replace the difference quotient by the differential quotient, and within the assumption of small angles obtain

$$\frac{\partial\varphi}{\partial x} \simeq \frac{\Delta\varphi}{\Delta x} = k\delta\alpha_i\,. \tag{6.165}$$

By using this relation to replace $\alpha_i\delta$ in equation (6.163), we can link the angular deviation of the beam (or wavefront) to the phase gradient

$$\alpha_r \simeq \frac{\lambda}{2\pi}\frac{\partial\varphi}{\partial x}\,. \tag{6.166}$$

Thus, propagation through different object thickness, or more generally projected optical path, results in differential phase contrast. If re-directions of the wavefront can be made visible, contrast is formed based on the first derivative of φ. However, the angular differences are extremely small, in the range of nrad to μrad. This requires either highly focused beams and position sensitive detection, as in phase contrast scanning transmission x-ray microscopy (STXM) [35, 60], or detection of wavefront tilts by grating interferometry (also known as Talbot or shearing interferometers). In the following, we consider the optical basics of Talbot interferometry.

6.4.2 Talbot effect and grating based x-ray phase imaging

Talbot effect

Let us first discuss the Talbot effect, which describes selfimaging of a (periodic) field at certain distances, namely at multiples of the so called Talbot distance d_T. To derive this result of Fresnel (near field) diffraction from a periodic object, we consider a wave $\psi_0(x)$ behind a periodic grating with a period p

$$\begin{aligned}\psi_0(x) &= \frac{1}{\sqrt{2\pi}}\int \tilde{\psi_0}(k_x)\,\mathrm{e}^{ik_x x}\,\mathrm{d}k_x \\ &= \frac{1}{\sqrt{2\pi}}\sum_m \tilde{\psi_0}\left(\frac{2\pi m}{p}\right)\mathrm{e}^{2\pi i x m/p}\,\Delta k_x, \qquad m \in \mathbb{Z}\,.\end{aligned} \tag{6.167}$$

Using the free space propagation kernel $\tilde{h} = e^{ikz} e^{izk_x^2/2k}$, the propagated field at a distance z behind the grating can be expressed by

$$\begin{aligned}
\psi(x,z) &= \frac{1}{\sqrt{2\pi}} \int \tilde{\psi_0}(k_x, z)\, e^{ik_x x}\, dk_x \\
&= \frac{1}{\sqrt{2\pi}} \int \tilde{\psi_0}(k_x)\, e^{ikz}\, e^{izk_x^2/2k}\, e^{ik_x x}\, dk_x \\
&= \frac{1}{\sqrt{2\pi}} \sum_m \tilde{\psi_0}\left(\frac{2\pi m}{p}\right) e^{2\pi i x m/p}\, e^{ikz}\, e^{iz(2\pi m/p)^2/2k} \Delta k_x \\
&= \frac{1}{\sqrt{2\pi}} e^{ikz} \sum_m \tilde{\psi_0}\left(\frac{2\pi m}{p}\right) e^{2\pi i x m/p}\, e^{iz(4\pi^2 m^2/(2kp^2))} \Delta k_x\,.
\end{aligned} \tag{6.168}$$

Comparing equation (6.167) and equation (6.168), we can see that the propagated field and the grating's exit field are equal, up to a multiplicative constant e^{ikz}, if the argument of the last exponential in equation (6.168) is a multiple of 2π for all orders m. This is the case for

$$z \cdot \frac{4\pi^2}{k \cdot 2p^2} = n \cdot 2\pi \qquad \text{for } n \in \mathbb{N} \tag{6.169}$$

$$z \cdot \frac{\pi\lambda}{p^2} = n \cdot 2\pi \tag{6.170}$$

$$\Rightarrow\ z = n \cdot \frac{2p^2}{\lambda}\,. \tag{6.171}$$

In other words, the field reproduces itself after propagation by a distance $d_T = \frac{2p^2}{\lambda}$, i.e., the so called Talbot distance. This re-imaging of the wavefield is periodic in d_T. In other words, periodicity of the wavefield in the lateral direction as imposed by the periodic object entails periodicity along the optical axis z. This holds for arbitrary Fresnel number, and (infinitely) extended wavefields. In practical realizations, boundary effects associated with truncation of the wavefield may affect the edge of the detection field. However, these effects are reduced to a single pixel if the detector pixels are larger than the grating periodicity.

Fractional Talbot effect

The periodicity d_T inferred above holds for arbitrary objects functions $o(x)$, as long as they are periodic with $o(x) = o(x+p)$. Inspection of equation (6.168), shows an additional replication of the wavefield in the midplane between two Talbot planes, i.e., after a distance $d_T/2 = \frac{p^2}{\lambda}$. In contrast to the Talbot planes, however, this replicated wavefield is translated along x by $p/2$. Again, this effect is observed for the most general case. Special grating functions $o(x)$ can exhibit additional planes of replication (self imaging) at fractional distances of d_T. This is, for example, the case for a rectangular grating function with binary phase contrast, as is used in most interferometers as the object grating. Such a binary phase grating has three parameters: periodicity

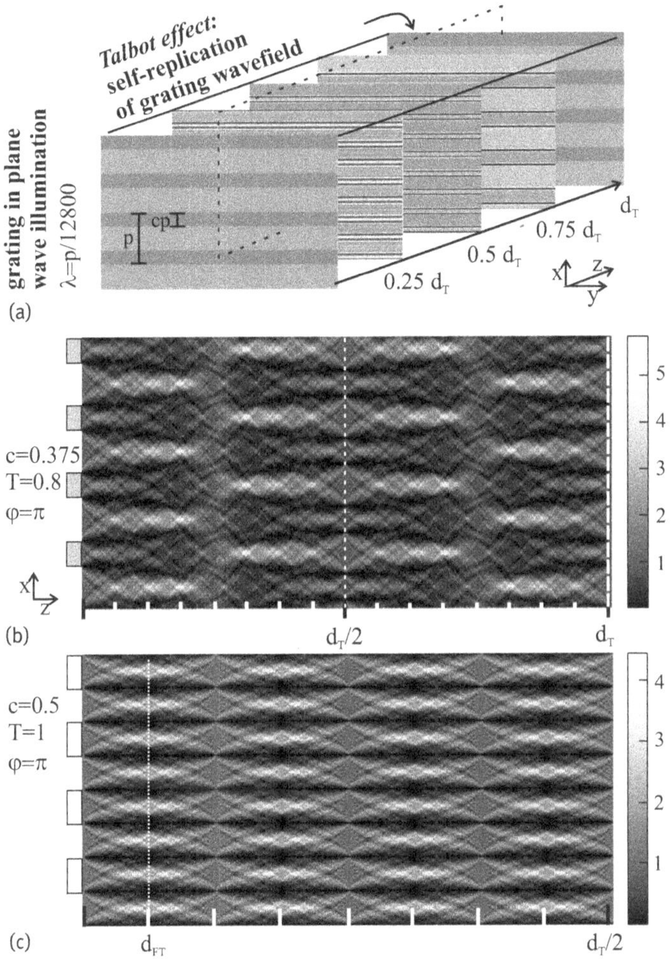

Fig. 6.27: (a) The wavefield behind a periodic object is periodic not only in the lateral planes (perpendicular to the optical axis) but also along the optical axis. This so called Talbot effect is a direct consequence of Fresnel diffraction. Hence, the wavefield behind a grating exhibits a cyclic pattern, with full self replication of the grating's exit field after the so called Talbot distance $z = d_T$. At half the distance $z = d_T/2$, the wavefield also replicates but is shifted by half a grating period $p/2$. (b) The wavefield along the propagation direction z, i.e., the plane formed by the optical axis and the direction of the 1d grating function is sometimes denoted as the Talbot carpet, shown here for a mixed amplitude and phase grating with duty cycle $c = 0.375$. (c) Pure phase grating ($T = 1$) with phase shift π and equal lines and spacing ($c = 0.5$). Interestingly, this special combination with its symmetry yields the wavefield of an amplitude grating at a distance $d = d_T/8$.

p, the duty cycle $0 \leq c \leq 1$, i.e., the fraction of the period 'filled' with phase shifting material, and the phase shift φ. The exit wave behind the phase grating has a homogeneous intensity distribution and so has self image in planes $d_T/2$ and d_T. For interferometry these planes are obviously completely useless. Since the observation of an object is encoded in slight shifts of the perturbed intensity pattern with respect to the unperturbed case, a unit intensity distribution cannot serve for detection. The detection plane is, therefore, chosen such that the periodic phase grating is 'self imaged' to a periodic pattern with strong contrast. This is provided, for example, by an ideal case of a binary phase grating with $\varphi = \pi$ phase shift (matched to the central working photon energy) and duty cycle $c = 0.5$. Plugging this object function into the Fresnel propagation kernel, we observe eight periodic replications (within the period d_{T}) of a rectangular intensity pattern with (ideal) intensity values of one and zero. The 'replicated' wavefield is again a grating function with duty cycle $c = 0.5$ but with the doubled period $p/2$, see the simulated intensity pattern in Figure 6.27. Hence, for each $n \in \mathbb{N}$, there is a plane with maximum intensity contrast

$$d_{FT} = n \cdot \frac{p^2}{8\lambda}\,, \tag{6.172}$$

which can be used to detect slight shifts in the interference pattern, due to refraction in an object.

6.4.3 Talbot effect with object in the beam path

Now suppose that an object is introduced just in front of (or behind) the binary phase grating. The minute angular deviation imparted by phase gradients in the object, as calculated above, leads to a translational shift in the intensity stripes of the fractional Talbot planes. For the simplest case of a wedge the wavefront is tilted by an angle $\alpha_r = \lambda/2\pi\ \partial_x\varphi = \delta \tan\alpha$, where $\alpha_i = \alpha$ is the angle of the wedge (Figure 6.28(c)). For a general object, the local phase gradient can be inferred from the local translational shift of the Talbot self image. In other words, detecting these local shifts with respect to the unperturbed case gives the local phase gradient associated with at a certain lateral location (x, y) of the object, see Figure 6.29. In this regard, it is an advantage that the angular deviations are in the range of nano to microradians. The one-to-one correspondence between detector position and sample position is maintained. However, at large propagation distances, this 'diffraction blurring' could spoil high spatial resolution, and some kind of inversion as in propagation based phase contrast would be required. In practice, the range of pixel sizes between 10 µm to 100 µm typical for grating interferometry is safely treated by a direct pixel-to-pixel correspondence.

Detection of the local phase shift can be performed by a high resolution detector with a pixel size small enough to sample the Talbot image, followed by a subsequent analysis of the local phase shifts. At least one period of the Talbot self image is required to define the local phase shift $\varphi(x, y)$. Hence, several detector pixels yield a single pixel of the phase image, so that suitable cameras need a sufficiently large array of small pixels. In practice, detectors with high efficiencies and large pixel size are

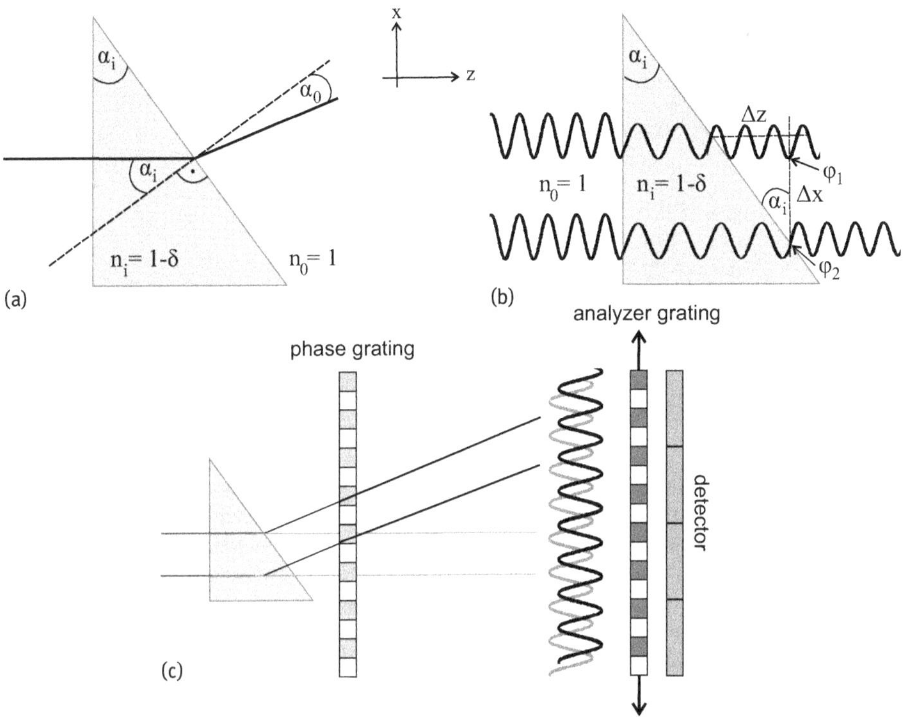

Fig. 6.28: Shift of the Talbot carpet by a wedge like object in the beam path. A homogeneous object with constant phase shift does not change the Talbot interference pattern. Contrarily, a phase gradient such as induced by the wedge in (a) leads to a tilt of the exit wave. As shown before, this tilt can calculated from (a) refraction, or equivalently from (b) the induced difference in phase shift $\Delta\varphi$ for two rays separated by Δx. (c) The tilted wavefront results in a corresponding shift of the interference fringes in the Talbot (self replication) plane. By placing an absorption grating in this plane, the translations are converted to (transmitted) intensity.

used, typically a flat panel detector or a single photon counting pixel detector. The phase shift of the Talbot grating is then detected by placing an absorption grating in front of the detector, with its period matched to the self image. Translating this absorption grating in front of the detector, one records a periodic transmission variation $T(x)$, i.e., a sinusoidal curve directly revealing the phase of the interference pattern. This translational scan is called **phase stepping**, and in practice four translational points are sufficient. Note that for an ideal case, the phase stepping scan would result in a triangular grating function, owing to the convolution of two rectangular gratings. However, finite coherence and, in particular, source size causes blurring (smearing) of the function. Different observables can be generated from the sinusoidal phase stepping curve: a mean intensity revealing the local object transmission, the local phase shift of the interference pattern revealing the phase gradient of the object exit phase and, finally, the amplitude of the sinusoidal oscillation, which we address further be-

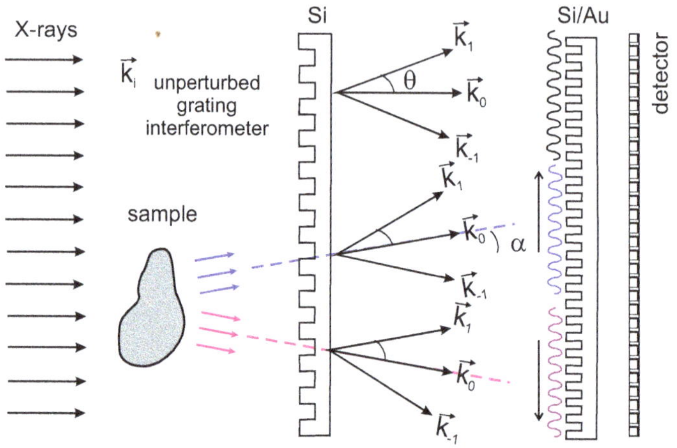

Fig. 6.29: Schematic of phase imaging by grating based x-ray interferometry. A binary phase grating is illuminated by a plane wave. The grating with its different diffraction orders, in particular $m = -1, 0, 1$, acts as a beam splitter, and superposition of the diffracted waves results in the characteristic near field interference pattern of a grating. An absorption (analyzer) grating is placed at a fractional Talbot distance to probe the periodic interference pattern (Talbot self image). Placing an object in the vicinity of the phase grating leads to slight refraction of the beam and hence translational shifts of the self image, resulting in local intensity changes across the analyzer. The angular deviation of the beam due to refraction (phase gradients) can be determined by translating the absorption grating perpendicular to the optical axis and evaluating the intensity changes.

low. Note that for a phase image, the uniaxial phase derivative $\partial_x \varphi(x)$ must first be integrated numerically. In summary, for each pixel (x, y) we have a sinusoidal intensity dependence on the second grating position x_2

$$I_{(x,y)}(x_2) = a_0(x, y) \sum_{m=1}^{\infty} a_m(x, y) \cos\left(\frac{2\pi m x_2}{p_2} + \varphi(x, y)\right) , \qquad (6.173)$$

where p_2 is the period of the analyzer grating, and a_m are the Fourier coefficients to be extracted from the phase stepping scan, e.g., by a fast Fourier transform. The scan is performed first for the empty (reference) beam, then for the sample. The three signals (attenuation, phase, visibility) are then calculated by the ratio of the object over the reference values [6]

$$a_0(x, y) = \frac{a_0^{\mathrm{o}}(x, y)}{a_0^{\mathrm{r}}(x, y)} \qquad \text{transmission} \qquad (6.174)$$

$$\varphi(x, y) = \varphi^{\mathrm{o}}(x, y) - \varphi^{\mathrm{r}}(x, y) \qquad \text{phase of interference stripes} \qquad (6.175)$$

$$V(x, y) = \frac{V^{\mathrm{o}}}{V^{\mathrm{r}}} = \frac{a_0^{\mathrm{r}}(x, y)}{a_0^{\mathrm{o}}(x, y)} \sum_m \frac{a_m^{\mathrm{o}}(x, y)}{a_m^{\mathrm{r}}(x, y)} \qquad \text{visibility/darkfield} \qquad (6.176)$$

where subscript o denotes the object and r the empty beam reference. Initially, only the transmission and phase signals were evaluated, before it was realized by F. PFEIF-

FER that many objects led to a systematic decrease in oscillation amplitude [70]. The corresponding decrease in the visibility of interference stripes has been found to be characteristic for different materials and can, thus, serve as a contrast mechanism. This effect is caused by the fact that the substructure of the object on length scales smaller than the grating period leads to scattering at angles much higher than the refraction angles, effectively smearing out the grating interference pattern.

Dark field contrast
can be quantified by the concept of a so called *linear diffusion coefficient* $\epsilon(x, y, z)$, which accounts for the local property of the sample to scatter radiation with scattering angle θ, modeled for each slice Δz by a Gaussian distribution function

$$P(\theta) = \frac{1}{\sigma\sqrt{2\pi}} \exp\left(-\frac{\theta^2}{2\sigma^2}\right) . \tag{6.177}$$

Subsequent convolution and superposition with a local dark field strength per unit length $\epsilon = \sigma^2/(\Delta z)$ results in a reduction of the visibility signal in the detection plane [6]

$$V(x, y) = \exp\left(-\frac{2\pi^2 d^2}{p^2} \int \epsilon(x, y, z)\, dz\right) , \tag{6.178}$$

where p is the analyzer grating period and d the distance between object and detection plane (analyzer grating). The path integral of the linear diffusion coefficient ϵ in the exponent is analogous to conventional absorption radiography and, hence, has a structure well compatible with tomography, after taking the logarithm of the signal for each projection. While this reductionist interpretation is appealing, one must consider its limitations. The linear integration is valid only when the object thickness is much smaller than the distance d, and strictly speaking is only valid for incoherent scattering. Multiple scattering and large scattering angles, which can spoil the assumed locality (one-to-one correspondence between object and detector pixels), are further complications. Nevertheless, this rather simplistic model has been verified by numerous experiments. Beautiful demonstrations of dark field or scattering contrast have been given for ‘calibration materials’ composed of granular materials with structure sizes below the pixel size, followed by dark field imaging of biological tissue. Figure 6.30 shows a recording for a live mouse.

Coherence, source size and divergence
The above derivation of the Talbot effect is realized assuming a plane wave illumination, i.e., a fully coherent beam in the parallel beam geometry. The Talbot effect also holds for point source (cone beam) imaging, since according to the Fresnel scaling theorem, the point source illumination is equivalent to the parallel beam setup after variable transformation. With the phase grating G1 positioned at distances z_1 from the source, and the analyzer grating G2 at distance $z_1 + z_2$, we have the usual

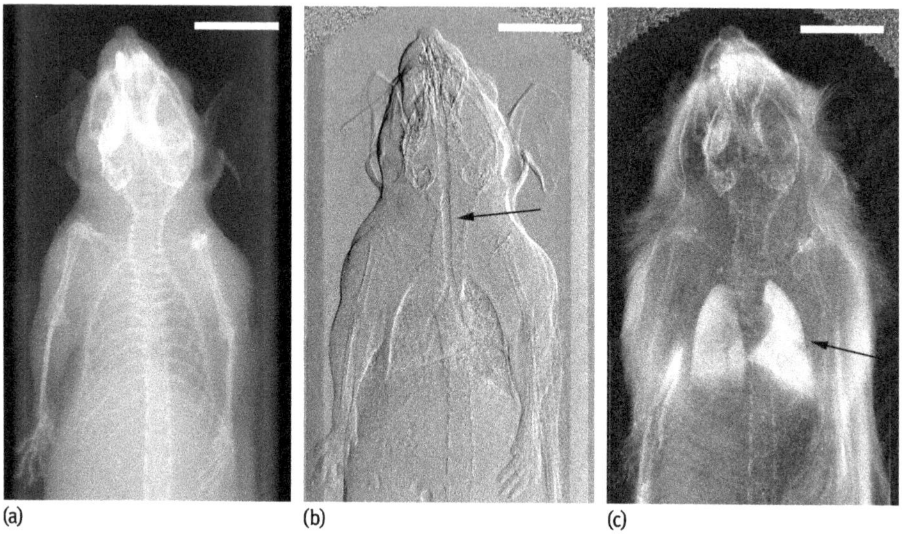

Fig. 6.30: In vivo grating based radiography of a mouse [7]. (a) Conventional absorption contrast. (b) Differential phase contrast. (c) Dark field contrast. All three images are extracted from the same scan of the analyzer grating. The *arrows* mark regions of enhanced contrast, related to the refraction of the trachea (b) and the scattering of the lungs (c). Scale bars: 1 cm. Reprinted by permission from Macmillan Publishers Ltd: Scientific Reports [7], copyright (2013).

geometric magnification $M = (z_1 + z_2)/z_1$ and the effective propagation distance $z_{\text{eff}} = z_1 z_2/(z_1 + z_2)$. Accordingly, the Talbot distance between G1 and G2 in the laboratory frame is

$$d_T \to \tilde{d}_T = \frac{2p^2}{\lambda}\left(\frac{z_1 + z_2}{z_1}\right)^2 . \tag{6.179}$$

The geometric magnification also facilitates the fabrication of the analyzer grating, which is more challenging in view of its aspect ratio, since it works in absorption.

Since laboratory x-ray sources are incoherent, the required partial coherence results only from propagation. Accordingly, the source size s must be sufficiently small, so that the lateral coherence length is larger than the period of the grating $\xi_\perp = \lambda z_1/(2s) \geq p$. The same condition is derived by requiring that the Talbot interference pattern observed at $z_2 = z_1 + d_T$ is not washed out completely by convolution $s z_2/z_1 \leq p_2/2$. For $z_1 = 2\,\text{m}$ and $z_2 = 0.04\,\text{m}$, and a grating period $p = 2\,\mu\text{m}$, a microfocus source with $s \leq 50\,\mu\text{m}$ is, therefore, needed to fulfill the coherence requirements. On the laboratory scale, source sizes are often too large and, in this case, one can place an additional (third) grating in the source plane with $p_0 = p_2 \frac{z_2}{z_1}$ to circumvent the requirements of coherence.

6.4.4 Coded apertures

Phase contrast imaging based on grating interferometry requires a certain degree of coherence, both in the spatial and the temporal domains.[13] Since this is not easily provided by laboratory x-ray sources without costly reduction in flux, the group of A. Olivo at the University College London proposed and developed the use of edge illumination and coded aperture techniques for phase contrast x-ray imaging [62, 65]. As in Talbot interferometry, the beam is structured in the object plane by a periodic array, in this case an absorption mask, resulting in many parallel beamlets. Slight angular changes of the beamlet's direction by refraction in the object is then translated to intensity by a second absorption mask A_2, which partly covers the detector pixels. Since the beamlets do not overlap and interfere, coherence requirements are significantly relaxed. In fact, the coherence length needs to cover only a single beamlet. As sketched in Figure 6.31, the beamlets are defined by the aperture A_1 in front of the object forming an array of beams, each with a width W. Owing to the geometric divergence of most laboratory beams, each beamlet will have broadened to a value $P = W(z_1 + z_2)/z_1$ in the detection plane, by geometric magnification depending on the object to detector distance z_2 and source to object distance z_1. The geometric parameters are chosen such that the width of the beamlet at the detector P is contained within a single detector pixel. A second aperture A_2 in front of the detector is then introduced to place an absorption edge, partly blocking each beamlet. Two measurements are recorded, one with the aperture edge covering the pixel from the left and one from the right side. For each pixel, the corresponding signals I_L and I_R are given by the unblocked intensity contributions of each pixel. The intensity of the partially blocked pixel can then be written as [62]

$$I_{L/R} = \int_{-\infty}^{\infty} dx\, K(x, \pm P/2) I(x) \,, \tag{6.180}$$

with the effective pixel function $K(x, \pm P/2)$, x denoting the pixel position, and $\pm P/2$ the translation of aperture A_2. Finally, the spatial derivative of the phase in the object coordinate ξ is determined from the two detector signals according to

$$\frac{\partial \varphi}{\partial \xi} = c\, \frac{I_L - I_R}{I_R + I_L} \,, \tag{6.181}$$

where the prefactor is $c = P/(2z_2)$ for a perfect point source and $c = \sigma\sqrt{\pi}/(2z_1)$ for an extended (Gaussian) source with width $2\sqrt{\log 2}\sigma$ (FWHM), respectively. It will be interesting to see where this phase contrast technique finds its applications, and how

13 The Talbot effect was first observed for white light, resulting in a peculiar and pretty distribution of colors in the Talbot plane. However, for imaging white light, Talbot interferometry is less well suited, and a certain restriction in bandpass, depending on the Talbot order and grating parameters is required.

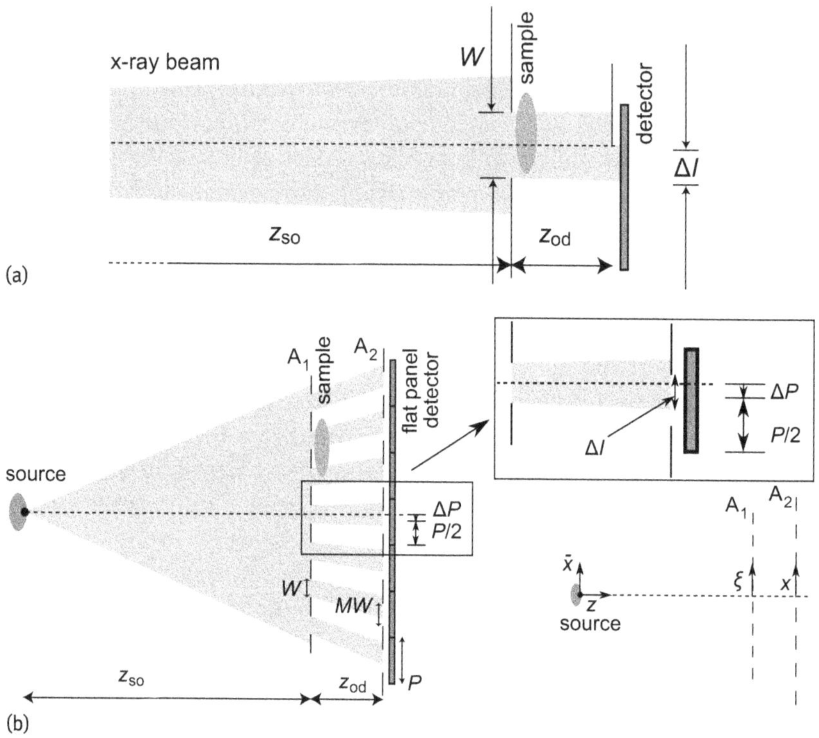

Fig. 6.31: Sketch of the edge illumination (coded aperture) technique for phase contrast imaging. (a) An aperture defines a single beam refracted in a sample. A second aperture blocks part of the pixel in front of the detector, so that minute nanoradian directional shifts of the beamlet become measurable by intensity changes [16]. (b) Generalization of the edge illumination scheme from (a) to a set of multiple beamlets. Reproduced from P.R.T. Munro et al. [61] under license CC BY 4.0, copyright (2013).

these fields of application are delineated with respect to Talbot interferometry. Recording a multitude of separate beamlets could, in principle, also give access to diffraction signals, provided that detectors can resolve the primary beamlet and its small angle scattering for each beamlet separately. Scanning the object through the structured illumination of the many beamlets enables super resolution, i.e., a spatial resolution below the beamlet or pixel size.

Outlook

In conclusion, phase contrast radiography and tomography offer a significant extension of present imaging capabilities, both for clinical diagnostics and for biomedical research, in particular three-dimensional virtual histology. The different applications and modalities are plenty, and the field is still under significant development. Complementing phase contrast, a major step forward was the development of dark field or scattering contrast, as this technique can bring entirely new information to radiographic imaging. Future extensions can be envisioned which employ further properties such as polarization. The current models used to treat both phase and dark field signals are, to some extent, still rudimentary, and further progress can be expected from automatized phase reconstruction, as well as from integration of reconstruction and forward simulations (for example, to treat the scattering contrast more quantitatively). Algebraic reconstruction would offer significant advantages for such a framework, even though it is numerically costly. All-in-one approaches that combine phase retrieval and tomographic reconstruction into one step [78] offer further potential for improvements, at least for analytical high resolution biomedical research.

In view of conceptualizing x-ray dark field contrast, it is appropriate to consider it as an effective scattering contrast. In an effective manner, it sums up contributions from different real structure effects in the sample on length scales that are cut off by the pixel size. These can, for example, be incoherent scatter events or refraction on length scales too small to show up in the differential phase. Compared to diffraction contrast in STXM experiments, the scattering signal cannot be resolved in momentum space, unless one systematically varies the grating period. At the same time, however, the major advantage of dark field imaging is to reconcile an intrinsic sensitivity to small structural features with macroscopic low dose x-ray imaging, by virtue of converting scattering into a contrast mode. Further work should also be directed at validating to which extent the neglect of second order phase contrast (by free propagation) in the current treatment of grating based phase contrast imaging is justified. Ultimately, it will be necessary to conceive a unified description that takes into account first and second order phase contrast.

Interestingly, grating based imaging is the first phase contrast technique that can be complemented and enhanced with dark field contrast, but it will probably not be the only one. Some interesting initial demonstrations of how scattering contrast can also be exploited in the setting of propagation based phase contrast have been given in [98, 99]. An alternative would be to record data that is redundant in view of coherent diffraction, for example, by recording in multiple planes, and to use the 'misfit' or inconsistency in the pure propagation phase contrast model to trace down incoherent scattering contributions.

Recommended references for further reading

The textbook by Paganin [69] is without doubt the most important reference for this chapter, regarding coherent x-ray optics in breadth and depth, along with a number of relevant monographs [2, 21, 25, 41, 45, 76], published in the *Göttingen series in X-ray Physics* and reporting PhD dissertations, which have been supervised by one of the authors (T. S.). Further, we recommend the following textbooks, from which we have learned a lot and have taken inspiration: the textbook by Lauterborn and Kurz [46] for an introduction to the Kirchhoff–Fresnel integrals and to spatial and longitudinal coherence, Goodman's textbook [28] in particular for the discussion of boundary conditions in the diffraction integrals and for sampling criteria, the textbook by Saleh and Teich [82] in particular for the PWE and its solutions, Wolf's textbook on coherence [96] as a general reference, and more specifically for coherence in x-ray imaging the review by Nugent [63], and Margaritondo's textbook on application of synchrotron radiation for biomedical research [54].

References

[1] J. Als-Nielsen and D. McMorrow. *Elements of Modern X-ray Physics.* John Wiley & Sons, 2nd edition, 2011.

[2] M. Bartels. *Cone-beam x-ray phase contrast tomography of biological samples: Optimization of contrast, resolution and field of view.* PhD thesis, Universität Göttingen, 2013.

[3] M. Bartels, V. H. Hernandez, M. Krenkel, T. Moser, and T. Salditt. Phase contrast tomography of the mouse cochlea at microfocus x-ray sources. *Appl. Phys. Lett.*, 103(8):083703, 2013.

[4] M. Bartels, M. Krenkel, P. Cloetens, W. Möbius, and T. Salditt. Myelinated mouse nerves studied by x-ray phase contrast zoom tomography. *Journal of Structural Biology*, 192:561–568, 2015.

[5] M. Bartels, M. Krenkel, J. Haber, R. N. Wilke, and T. Salditt. X-ray holographic imaging of hydrated biological cells in solution. *Phys. Rev. Lett.*, 114:048103, 2015.

[6] M. Bech. *X-ray imaging with a grating interferometer.* PhD thesis, Faculty of Science, University of Copenhagen, 2009.

[7] M. Bech, A. Tapfer, A. Velroyen, A. Yaroshenko, B. Pauwels, J. Hostens, P. Bruyndonckx, A. Sasov, and F. Pfeiffer. In-vivo dark-field and phase-contrast x-ray imaging. *Sci. Rep.*, 3:3209, 2013.

[8] M. N. Boone, W. Devulder, M. Dierick, L. Brabant, E. Pauwels, and L. Van Hoorebeke. Comparison of two single-image phase-retrieval algorithms for in-line x-ray phase-contrast imaging. *J. Opt. Soc. Am. A*, 29(12):2667–2672, 2012.

[9] E. O. Brigham. *The Fast Fourier Transform and its Applications.* Prentice-Hall, 1974.

[10] A. V. Bronnikov. Theory of quantitative phase-contrast computed tomography. *J. Opt. Soc. Am. A*, 19(3):472–480, 2002.

[11] A. Burvall, U. Lundström, P. A. C. Takman, D. H. Larsson, and H. M. Hertz. Phase retrieval in X-ray phase-contrast imaging suitable for tomography. *Opt. Express*, 19(11):10359–10376, 2011.

[12] T. Butz. *Fouriertransformation für Fußgänger.* Vieweg+Teubner Verlag / Springer Fachmedien Wiesbaden GmbH, Wiesbaden, 2012.

[13] P. Cloetens, W. Ludwig, J. Baruchel, J.-P. Guigay, P. Pernot-Rejmankova, M. Salome-Pateyron, M. Schlenker, J.-Y. Buffiere, E. Maire, and G. Peix. Hard x-ray phase imaging using simple propagation of a coherent synchrotron radiation beam. *J. Phys. D: Appl. Phys.*, 32(10A):A145–A151, 1999.

[14] C. David, B. Nohammer, H. H. Solak, and E. Ziegler. Differential x-ray phase contrast imaging using a shearing interferometer. *Appl. Phys. Lett.*, 81(17):3287–3289, 2002.

[15] T. J. Davis. Dynamical X-ray diffraction from imperfect crystals: a solution based on the Fokker–Planck equation. *Acta Crystallogr. A*, 50(2):224–231, 1994.

[16] P. C. Diemoz, M. Endrizzi, C. E. Zapata, Z. D. Pešić, C. Rau, A. Bravin, I. K. Robinson, and A. Olivo. X-ray phase-contrast imaging with nanoradian angular resolution. *Phys. Rev. Lett.*, 110:138105, 2013.

[17] F. Döring, A. L. Robisch, C. Eberl, M. Osterhoff, A. Ruhlandt, T. Liese, F. Schlenkrich, S. Hoffmann, M. Bartels, T. Salditt, and H.-U. Krebs. Sub-5 nm hard x-ray point focusing by a combined Kirkpatrick-Baez mirror and multilayer zone plate. *Opt. Express*, 21(16):19311–19323, 2013.

[18] C. Eberl, F. Döring, T. Liese, F. Schlenkrich, B. Roos, M. Hahn, T. Hoinkes, A. Rauschenbeutel, M. Osterhoff, T. Salditt, and H.-U. Krebs. Fabrication of laser deposited high-quality multilayer zone plates for hard x-ray nanofocusing. *Appl. Surf. Sci.*, 307(0):638–644, 2014.

[19] J. R. Fienup. Reconstruction of an object from the modulus of its Fourier transform. *Opt. Lett.*, 3(1):27–29, 1978.

[20] J. R. Fienup. Phase retrieval algorithms: a comparison. *Appl. Opt.*, 21(15):2758–2769, 1982.

[21] C. Fuhse. *X-ray waveguides and waveguide-based lensless imaging*. PhD thesis, University of Göttingen, 2006.

[22] C. Fuhse and T. Salditt. Finite-difference field calculations for two-dimensionally confined x-ray waveguides. *Appl. Opt.*, 45(19):4603–4608, 2006.

[23] G. Geloni, E. Saldin, L. Samoylova, E. Schneidmiller, H. Sinn, T. Tschentscher, and M. Yurkov. Coherence properties of the European XFEL. *New Journal of Physics*, 12(3):035021, 2010.

[24] R. W. Gerchberg and W. O. Saxton. A Practical Algorithm for the Determination of Phase from Image and Diffraction Plane Pictures. *Optik*, 35(2):237–246, 1972.

[25] K. Giewekemeyer. *A study on new approaches in coherent x-ray microscopy of biological specimens*. PhD thesis, Universität Göttingen, 2011.

[26] K. Giewekemeyer, S. P. Krüger, S. Kalbfleisch, M. Bartels, C. Beta, and T. Salditt. X-ray propagation microscopy of biological cells using waveguides as a quasipoint source. *Phys. Rev. A*, 83(2):023804, 2011.

[27] K. Giewekemeyer, P. Thibault, S. Kalbfleisch, A. Beerlink, C. M. Kewish, M. Dierolf, F. Pfeiffer, and T. Salditt. Quantitative biological imaging by ptychographic x-ray diffraction microscopy. *Proc. Nat. Acad. Sci. USA*, 107(2):529–534, 2010.

[28] J. W. Goodman. *Introduction to Fourier Optics*. Roberts & Company: Englewood, Colorado, 2005.

[29] L. B. Gromann, D. Bequé, K. Scherer, K. Willer, L. Birnbacher, M. Willner, J. Herzen, S. Grandl, K. Hellerhoff, J. I. Sperl, F. Pfeiffer, and C. Cozzini. Low-dose, phase-contrast mammography with high signal-to-noise ratio. *Biomed. Opt. Express*, 7(2):381–391, 2016.

[30] A. Groso, R. Abela, and M. Stampanoni. Implementation of a fast method for high resolution phase contrast tomography. *Opt. Express*, 14(18):8103–8110, 2006.

[31] A. Groso, M. Stampanoni, R. Abela, P. Schneider, S. Linga, and R. Müller. Phase contrast tomography: An alternative approach. *Appl. Phys. Lett.*, 88:214104, 2006.

[32] T. E. Gureyev, T. J. Davis, A. Pogany, S. C. Mayo, and S. W. Wilkins. Optical Phase Retrieval by Use of First Born- and Rytov-Type Approximations. *Appl. Opt.*, 43(12):2418–2430, 2004.

[33] T. E. Gureyev, C. Raven, A. Snigirev, I. Snigireva, and S. W. Wilkins. Hard x-ray quantitative non-interferometric phase-contrast microscopy. *J. Phys. D: Appl. Phys.*, 32(5):563–567, 1999.

[34] J. Hagemann, A.-L. Robisch, D. R. Luke, C. Homann, T. Hohage, P. Cloetens, H. Suhonen, and T. Salditt. Reconstruction of wave front and object for inline holography from a set of detection planes. *Opt. Express*, 22(10):11552–11569, 2014.

[35] C. Holzner, M. Feser, S. Vogt, B. Hornberger, S. B. Baines, and C. Jacobsen. Zernike phase contrast in scanning microscopy with X-rays. *Nature Physics*, 6:883–887, 2010.

[36] C. Homann, T. Hohage, J. Hagemann, A.-L. Robisch, and T. Salditt. Validity of the empty-beam correction in near-field imaging. *Phys. Rev. A*, 91:013821, 2015.

[37] The MathWorks Inc. Matlab. Natick, Massachusetts, United States.
[38] J. D. Schmidt. *Numerical simulation of optical wave propagation with examples in MATLAB*. Society of Photo-Optical Instrumentation Engineers (SPIE), Bellingham, Washington, USA, 2010
[39] T. Köhler and F. Noo. Comment on "Region-of-Interest Tomography for Grating-Based X-ray Differential Phase-Contrast Imaging". *Phys. Rev. Lett.*, 102:039801, 2009.
[40] Y. V. Kopylov, A. V. Popov, and A. V. Vinogradov. Application of the parabolic wave equation to X-ray diffraction optics. *Optics Communications*, 118(5–6):619–636, 1995.
[41] M. Krenkel. *Cone-beam x-ray phase contrast tomography for the observation of single cells in whole organs*. PhD thesis, Universität Göttingen, 2015.
[42] M. Krenkel, M. Bartels, and T. Salditt. Transport of intensity phase reconstruction to solve the twin image problem in holographic x-ray imaging. *Opt. Express*, 21(2):2220–2235, 2013.
[43] M. Krenkel, A. Markus, M. Bartels, C. Dullin, F. Alves, and T. Salditt. Phase-contrast zoom tomography reveals precise locations of macrophages in mouse lungs. *Sci. Rep.*, 5:09973, 2015.
[44] M. Krenkel, M. Töpperwien, F. Alves, and T. Salditt. Three-dimensional single-cell imaging with x-ray waveguides in the holographic regime. *Acta Cryst. A*, 73(4):282–292, 2017.
[45] S. P. Krüger. *Optimization of waveguide optics for lensless x-ray imaging*. PhD thesis, Universität Göttingen, 2010.
[46] W. Lauterborn and T. Kurz. *Coherent optics: fundamentals and applications; with 73 problems and complete solutions*. Springer, 2003.
[47] M. A. Leontovich and V. A. Fock. Solution of propagation of electromagnetic waves along the earth's surface by the method of parabolic equations. *J. Phys. USSR*, 10(1):13–23, 1946.
[48] D. R. Luke. Relaxed averaged alternating reflections for diffraction imaging. *Inverse Problems*, 21(1):37, 2005.
[49] D. R. Luke, J. V. Burke, and R. G. Lyon. Optical wavefront reconstruction: Theory and numerical methods. *SIAM review*, 44(2):169–224, 2002.
[50] S. Marchesini. Invited Article: A unified evaluation of iterative projection algorithms for phase retrieval. *Rev. Sci. Instrum.*, 78(1):011301, 2007.
[51] S. Marchesini, H. He, H. N. Chapman, S. P. Hau-Riege, A. Noy, M. R. Howells, U. Weierstall, and J. C. H. Spence. X-ray image reconstruction from a diffraction pattern alone. *Phys. Rev. B*, 68(14):140101, 2003.
[52] D. Marcuse. *Theory of dielectric optical waveguides*. Elsevier, 2013.
[53] S. Maretzke, M. Bartels, M. Krenkel, T. Salditt, and T. Hohage. Regularized Newton methods for x-ray phase contrast and general imaging problems. *Opt. Express*, 24(6):6490–6506, 2016.
[54] G. Margaritondo. *Elements of Synchrotron Light for Biology, Chemistry, and Medical Research*. Oxford Univ. Press, 2002.
[55] J. Miao, P. Charalambous, J. Kirz, and D. Sayre. Extending the methodology of X-ray crystallography to allow imaging of micrometre-sized non-crystalline specimens. *Nature*, 400(6742):342–344, 1999.
[56] G. Mie. Beiträge zur Optik trüber Medien, speziell kolloidaler Metallösungen. *Annalen der Physik*, 330(3):377–445, 1908.
[57] A. Momose. Demonstration of phase-contrast x-ray computed tomography using an x-ray interferometer. *Nucl. Instrum. Methods Phys. Res., Sect. A*, 352(3):622–628, 1995.
[58] A. Momose, S. Kawamoto, I. Koyama, Y. Hamaishi, K. Takai, and Y. Suzuki. Demonstration of X-ray Talbot Interferometry. *Jpn. J. Appl. Phys.*, 42(7B):L866, 2003.
[59] J. Moosmann, R. Hofmann, and T. Baumbach. Single-distance phase retrieval at large phase shifts. *Opt. Express*, 19(13):12066–12073, 2011.
[60] G. R. Morrison and B. Niemann. Differential phase contrast x-ray microscopy. In Jürgen Thieme, editor, *X-ray microscopy and spectromicroscopy, Würzburg, 1996*, pages I – 85–94, Berlin, 1998. Springer.

[61] P. R. T. Munro, M. Endrizzi, P. C. Diemoz, C. K. Hagen, M. B. Szafraniec, T. P. Millard, C. E. Zapata, R. D. Speller, and A. Olivo. Medicine, material science and security: the versatility of the coded-aperture approach. *Philosophical Transactions of the Royal Society of London A: Mathematical, Physical and Engineering Sciences*, 372(2010), 2014.

[62] P. R. T. Munro, K. Ignatyev, R. D. Speller, and A. Olivo. Phase and absorption retrieval using incoherent x-ray sources. *Proc. Nat. Acad. Sci. USA*, 109(35):13922–13927, 2012.

[63] K. A. Nugent. Coherent methods in the X-ray sciences. *Adv. Phys.*, 59(1):1–99, 2010.

[64] A. Olivo, F. Arfelli, D. Dreossi, R. Longo, R. H. Menk, S. Pani, P. Poropat, L. Rigon, F. Zanconati, and E. Castelli. Preliminary study on extremely small angle x-ray scatter imaging with synchrotron radiation. *Phys. Med. Biol.*, 47(3):469, 2002.

[65] A. Olivo and R. Speller. A coded-aperture technique allowing x-ray phase contrast imaging with conventional sources. *Appl. Phys. Lett.*, 91(7):074106, 2007.

[66] M. Osterhoff and T. Salditt. Coherence filtering of x-ray waveguides: analytical and numerical approach. *New Journal of Physics*, 13(10):103026, 2011.

[67] D. Paganin, S. C. Mayo, T. E. Gureyev, P. R. Miller, and S. W. Wilkins. Simultaneous phase and amplitude extraction from a single defocused image of a homogeneous object. *J. Microsc.*, 206(Pt 1):33–40, 2002.

[68] D. Paganin and K. A. Nugent. Noninterferometric Phase Imaging with Partially Coherent Light. *Phys. Rev. Lett.*, 80:2586–2589, 1998.

[69] D. M. Paganin. *Coherent X-Ray Optics*. New York: Oxford University Press, 2006.

[70] F. Pfeiffer, M. Bech, O. Bunk, P. Kraft, E. F. Eikenberry, Ch. Brönnimann, C. Grünzweig, and C. David. Hard-x-ray dark-field imaging using a grating interferometer. *Nature Materials*, 7:134–137, 2008.

[71] F. Pfeiffer, C. David, O. Bunk, T. Donath, M. Bech, G. Le Duc, A. Bravin, and P. Cloetens. Region-of-interest tomography for grating-based x-ray differential phase-contrast imaging. *Phys. Rev. Lett.*, 101(16):168101, 2008.

[72] F. Pfeiffer, C. Kottler, O. Bunk, and C. David. Hard X-Ray Phase Tomography with Low-Brilliance Sources. *Phys. Rev. Lett.*, 98:108105, 2007.

[73] F. Pfeiffer, T. Weitkamp, O. Bunk, and C. David. Phase retrieval and differential phase-contrast imaging with low-brilliance x-ray sources. *Nature Physics*, 2(4):258–261, 2006.

[74] A. Pogany, D. Gao, and S. W. Wilkins. Contrast and resolution in imaging with a microfocus x-ray source. *Rev. Sci. Instrum.*, 68(7):2774–2782, 1997.

[75] W. H. Press, S. A. Teukolsky, W. T. Vetterling, and B. P. Flannery. *Numerical recipes: the art of scientific computing*. Cambridge University Press, 3rd edition, 2007.

[76] A.-L. Robisch. *Phase retrieval for object and probe in the optical near-field*. PhD thesis, University of Göttingen, 2015.

[77] A.-L. Robisch, K. Kröger, A. Rack, and T. Salditt. Near-field ptychography using lateral and longitudinal shifts. *New J. Phys.*, 17(7):073033, 2015.

[78] A. Ruhlandt, M. Krenkel, M. Bartels, and T. Salditt. Three-dimensional phase retrieval in propagation-based phase-contrast imaging. *Phys. Rev. A*, 89:033847, 2014.

[79] A. Ruhlandt and T. Salditt. Three-dimensional propagation in near-field tomographic X-ray phase retrieval. *Acta Crystallogr. A*, 72(2), 2016.

[80] T. Salditt, S. Kalbfleisch, M. Osterhoff, S. P. Krüger, M. Bartels, K. Giewekemeyer, H. Neubauer, and M. Sprung. Partially coherent nano-focused x-ray radiation characterized by Talbot interferometry. *Opt. Express*, 19(10):9656–9675, 2011.

[81] T. Salditt, S. P. Krüger, C. Fuhse, and C. Bahtz. High-Transmission Planar X-Ray Waveguides. *Phys. Rev. Lett.*, 100(18):184801–184804, 2008.

[82] B. E. A. Saleh and M. C. Teich. *Fundamentals of Photonics*. Wiley, 1991.

[83] M. Slaney, A. C. Kak, and L. E. Larsen. Limitations of imaging with first-order diffraction tomography. *IEEE Transactions on Microwave Theory and Techniques*, 32(8):860–874, 1984.
[84] Y. Sung and G. Barbastathis. Rytov approximation for x-ray phase imaging. *Opt. Express*, 21(3):2674–2682, 2013.
[85] M. R. Teague. Deterministic phase retrieval: a Green's function solution. *J. Opt. Soc. Am.*, 73(11):1434–1441, 1983.
[86] P. Thibault, M. Dierolf, O. Bunk, A. Menzel, and F. Pfeiffer. Probe retrieval in ptychographic coherent diffractive imaging. *Ultramicroscopy*, 109(4):338–343, 2009.
[87] M. Töpperwien, M. Krenkel, F. Quade, and T. Salditt. Laboratory-based x-ray phase-contrast tomography enables 3d virtual histology. *Proc. SPIE*, 9964, 2016.
[88] M. Töpperwien, M. Krenkel, D. Vincenz, F. Stöber, A.M. Oelschlegel, J. Goldschmidt, and T. Salditt. Three-dimensional mouse brain cytoarchitecture revealed by laboratory-based x-ray phase-contrast tomography. *Sci. Rep.*, 7:42847, 2017.
[89] Mareike Töpperwien, Martin Krenkel, Kristin Müller, and Tim Salditt. Phase-contrast tomography of neuronal tissues: from laboratory-to high resolution synchrotron ct. In *SPIE Optical Engineering+ Applications*, pages 99670T–99670T. International Society for Optics and Photonics, 2016.
[90] L. Turner, B. Dhal, J. Hayes, A. Mancuso, K. Nugent, D. Paterson, R. Scholten, C. Tran, and A. Peele. X-ray phase imaging: Demonstration of extended conditions for homogeneous objects. *Opt. Express*, 12(13):2960–2965, 2004.
[91] D. Voelz. *Computational Fourier Optics: A MATLAB tutorial*, volume TT89 of *Tutorial Texts in Optical Engineering*. SPIE – the International Society for Optical Engineering, 2011.
[92] D. G. Voelz and M. C. Roggemann. Digital simulation of scalar optical diffraction: revisiting chirp function sampling criteria and consequences. *Appl. Opt.*, 48(32):6132–6142, 2009.
[93] T. Weitkamp, A. Diaz, C. David, F. Pfeiffer, M. Stampanoni, P. Cloetens, and E. Ziegler. X-ray phase imaging with a grating interferometer. *Opt. Express*, 13(16):6296–6304, 2005.
[94] R. N. Wilke. Zur Phasenrekonstruktion von X-Feldern mit gekrümmten Phasenfronten. Master's thesis, Universitaet Goettingen, 2010.
[95] Y. De Witte, M. Boone, J. Vlassenbroeck, M. Dierick, and L. Van Hoorebeke. Bronnikov-aided correction for x-ray computed tomography. *J. Opt. Soc. Am. A*, 26(4):890–894, 2009.
[96] E. Wolf. *Introduction to the theory of coherence and polarization of light.* Cambridge University Press, 2007.
[97] S. Zabler, P. Cloetens, J.-P. Guigay, J. Baruchel, and M. Schlenker. Optimization of phase contrast imaging using hard x rays. *Rev. Sci. Instrum.*, 76(7):073705, 2005.
[98] I. Zanette, S. Lang, A. Rack, M. Dominietto, M. Langer, F. Pfeiffer, T. Weitkamp, and B. Müller. Holotomography versus x-ray grating interferometry: A comparative study. *Appl. Phys. Lett.*, 103(24), 2013.
[99] I. Zanette, M.-C. Zdora, T. Zhou, A. Burvall, D.H. Larsson, P. Thibault, H.M. Hertz, and F. Pfeiffer. X-ray microtomography using correlation of near-field speckles for material characterization. *Proc. Nat. Acad. Sci.*, 112(41):12569–12573, 2015.

Symbols and abbreviations used in Chapter 6

a	structure size
A	atomic mass number (nucleon number)
$A(\underline{r})$	envelope function

α_{c}	critical angle for total external reflection of x-rays
$\underline{B}$	magnetic induction
BAC	Bronnikov aided correction
β	imaginary part of the refractive index
CDI	coherent diffractive imaging
CTF	contrast transfer function
$\underline{D}$	electric displacement
$\mathcal{D}_{\Delta}$	free space diffraction operator
$\mathcal{D}_{\Delta}^{\mathrm{F}}$	Fresnel diffraction operator
DFT	discrete Fourier transform
δ	deviation of the real part of the refractive index from unity
E	photon energy
$\underline{E}$	electric field
ER	error reduction
ϵ_0	electric permittivity of vacuum
ϵ_{r}	relative permittivity of a medium
$\epsilon(x, y, z)$	linear diffusion coefficient
F	Fresnel number
FEL	free electron laser
FOV	field of view
FWHM	full width at half maximum
f	atomic form factor
f'	real part of atomic dispersion correction
f''	imaginary part of atomic dispersion correction
$\mathcal{F}, \mathcal{F}^{-1}$	continuous and inverse continuous Fourier transform
GS	Gerchberg–Saxton
Γ	autocorrelation function in object plane
$\Gamma_{1,2}(t_1, t_2)$	mutual coherence function
$\gamma_{1,2}(t_1, t_2)$	normalized mutual coherence function
$\underline{H}$	magnetic field
HIO	hybrid input/output
$\hbar$	reduced Planck's constant
iDFT	inverse discrete Fourier transform
I	intensity
IRGN	iteratively regularized Gauss–Newton
$\underline{k} = (k_x, k_y, k_z)^{\mathrm{T}},\ \lvert k\rvert = \frac{2\pi}{\lambda}$	wavevector and wavenumber
$\underline{k}_{\perp}$	spatial frequencies of image in units of inverse unit length
λ	x-ray wavelength
M	geometric magnification
MBA	modified Bronnikov algorithm

μ	linear attenuation coefficient
μ_0	magnetic permeability of vacuum
μ_r	relative permeability of a medium
$n = 1 - \delta + i\beta$	x-ray refractive index
N_A	Avogadro's number
∇	nabla operator
$o(x, y)$	object transmission function
$O(\underline{r})$	scattering potential
PWE	paraxial wave equation
$\mathcal{P}$	projection operator
φ	phase
$\underline{q}$	scattering vector or momentum transfer
$\underline{q}_\perp$	spatial frequencies of image in inverse pixel units
r_0	classical electron radius (Thomson scattering length)
$\underline{r}_\perp = (x, y)^T$	lateral coordinates
R	reflection coefficient
RAAR	relaxed averaged alternating reflection
$\mathcal{R}$	reflection operator
ρ	mass density
ρ_a	number density of atoms
ρ_e	electron density
STXM	scanning transmission x-ray microscopy
t	time
TIE	transport-of-intensity equation
τ	object transmission function
$u(x_1, y_1, 0)$	unpropagated wavefield in the object plane $z = 0$
$u(x_2, y_2, z)$	propagated wavefield at distance z
$W_{1,2}(\omega)$	cross spectral density
ω	angular frequency
$\xi_\perp$	lateral coherence length
$\xi_\parallel$	longitudinal coherence length
χ	normalized, unitless spatial frequency
Z	atomic charge number (proton number)
z_{eff}	effective propagation distance

7 Object reconstruction: nonideal conditions and noise

We started the book with an introduction to image processing, presenting very general elements of signal theory, such as Fourier filtering or the description of image formation based on the theory of linear systems. In the last chapter, we can now go further and address the inverse process – reconstruction of an object from image data – also on a general level. In contrast to previous chapters on tomography, nuclear imaging and phase contrast radiography, we do not present specific solutions to a given reconstruction problem, but rather the general mathematical concepts of so called **inverse problems**. These advanced imaging reconstruction methods are particularly useful for reconstruction problems under nonideal conditions and noise, and can significantly enhance the specific approaches discussed before, for example, tomographic reconstruction or phase retrieval.

7.1 Inverse problems

Consider an object $\underline{f}$ which is imaged, for example, in radiography or tomography. If the imaging system can be described by a (linear) operator A, the first task is to predict (or to simulate) the image $\underline{g}$

$$\underline{g} = A\underline{f} \,. \tag{7.1}$$

This is denoted as the direct problem. It typically already involves significant physical modeling and approximations. Once we have confidence that a suitable description of the imaging process is found in the form of an operator A, we can go further and address the inverse problem, i.e., reconstruction of the object $\underline{f}$ from the data $\underline{g}$, rather than just predicting the data for a known object. In more general terms, we do not want to calculate the effect for a given cause, but infer the cause for a given effect. Unfortunately, direct problems are often accompanied by a loss of information. For example, several different objects could cause the same image. The solution of an inverse problem would, hence, imply information gain [2]. The need for additional information is a characteristic feature of *ill posed* inverse problems. An extensive treatment of inverse problems in imaging can be found in the textbook of M. Bertero and P. Boccacci [2], and in [1].

7.1.1 Regularization techniques for inverse problems

Problems of the form of Equation (7.1) can be characterized as well posed or ill posed. Well posed problems are characterized by existence, uniqueness and continuity of the

DOI 10.1515/9783110426694-007

solution with respect to the data. Contrarily, an ill posed problem is defined by one or several of the following properties:

1. The problem does not have a solution: $\underline{g} \notin \text{range}(A)$ (nonexistence).
2. The problem has multiple solutions: the kernel of A contains more elements than just 0, i.e., $\text{Ker}(A) \neq \{0\}$ (nonuniqueness).
3. Two "distant" objects $\underline{f}_0$ and $\underline{f}_1$ have very similar images $\underline{g}_0$ and $\underline{g}_1$. This may happen, for instance, when the inverse of A does not exist, but also if it exists but is not continuous. Due to such a discontinuity with respect to the data, small changes in the recorded data due to noise will lead to unreasonable solutions and conclusions on the original object [2] (noncontinuity).

These issues can be addressed and removed step by step. Concerning (1), we ask that an admissible $\underline{f}$ fulfill

$$\underline{f} = \text{argmin}_{\underline{f}'} \left| A\underline{f}' - \underline{g} \right| , \tag{7.2}$$

i.e., we seek for a solution $\underline{f}$ for which $A\underline{f}$ is closest to $\underline{g}$. However, there may still be infinitely many such solutions if $\text{Ker}(A) \neq \{0\}$. To address this issue (2), we require $\underline{f}$ to be of minimal norm $|\underline{f}|$. It follows that $\underline{f}$ does not have any components in $\text{Ker}(A)$. By these two steps, we have compensated for nonexistence and nonuniqueness, leading to a unique solution $\underline{f}^+$. The corresponding matrix A^+ defined implicitly by

$$\underline{f}^+ = A^+ \underline{g} \tag{7.3}$$

is called the Moore–Penrose inverse or pseudo inverse of A. The pseudo inverse has the following properties [1, 7]

$$AA^+A = A \tag{7.4}$$

$$A^+AA^+ = A^+ \tag{7.5}$$

$$(AA^+)^* = AA^+ \tag{7.6}$$

$$(A^+A)^* = A^+A \, . \tag{7.7}$$

For illustration purposes, we consider the pseudo inverse for the special case of A Hermitian[1]. As a Hermitian matrix A can be brought to diagonal form by a unitary operator U[2] and a diagonal matrix D

$$D = \begin{pmatrix} \lambda_1 & & \\ & \cdots & \\ & & \lambda_k \end{pmatrix} , \tag{7.8}$$

as

$$A = U^\dagger D U , \tag{7.9}$$

1 An operator A is called Hermitian if $A = (A^*)^{\mathrm{T}} = A^\dagger$.

2 An operator U is unitary if $U^\dagger = U^{-1}$.

and the pseudo inverse can be formulated in terms of the pseudo-inverse D^+ of D [1]

$$A^+ = (U^\dagger D U)^+ = U^+ D^+ U^{\dagger +} = U^{-1} D^+ U = U^\dagger D^+ U\,. \tag{7.10}$$

D^+ can be written as

$$D^+ = \begin{pmatrix} \lambda_1^{-1} & & \\ & \dots & \\ & & \lambda_k^{-1} \end{pmatrix}, \tag{7.11}$$

where we set $\lambda_k^{-1} := 0$ for all zero eigenvalues $\lambda_k = 0$. If all eigenvalues are nonzero, the pseudoinverse becomes the inverse $D^+ = D^{-1}$. The following argument illustrates how the pseudoinverse guarantees uniqueness of the solution $\underline{f}^+$. Due to the properties of D^+, $\underline{f}^+ = A^+\underline{g}$ does not have any components in Ker(A) because if $\lambda_k = 0$, then the component of $\underline{f}^+$ along the kth eigendirection is set to zero by the corresponding entry of 0 in D^+. Furthermore, $\left|A\underline{f}^+ - \underline{g}\right| = \left|AA^+\underline{g} - \underline{g}\right|$ is minimal, since in the difference $AA^+\underline{g} - \underline{g}$, all components where $\lambda_k \neq 0$ are removed and the remaining components can not be removed, since they are in the kernel of A. Finally, no other vector than $\underline{f}^+$ is better suited, since $\underline{f}^+$ is already of minimum norm because it has no components in the kernel of A.

We still must address the last issue: small changes in $\underline{g}$ can lead to large changes in $\underline{f}$. We again illustrate this behavior with a Hermitian matrix: If $\lambda_k \neq 0$ is small, λ_k^{-1} is large and enhances errors in the corresponding sub-space. This becomes particularly pronounced for matrices of infinite dimension with, for example, $\lambda_k > 0$ and $\lambda_k \to 0$ for $k \to \infty$. Here A^+ is not continuous, such that even for a very small noise contribution $\epsilon\Delta\underline{g}$ (with $\epsilon \to 0$) we may have

$$\lim_{\varepsilon\to 0} A^+(\underline{g}_0 + \varepsilon\Delta\underline{g}) \neq A^+\underline{g}_0\,. \tag{7.12}$$

Here, $\Delta\underline{g}$ denotes noise inherent in the data and $\epsilon \geq 0$ its strength. Without loss of generality we can assume that $|\Delta\underline{g}| \leq 1$. To reestablish continuity, we introduce a regularization strategy [4]. The basic idea of regularization is to replace the operator A^+ by a family of more well behaved ones. Which member of the family is actually used depends on a so called regularization parameter γ, which is a function of the noise strength, i.e., $\gamma = \gamma(\varepsilon)$. For noise free data, the regularization parameter should be zero. Therefore, we demand that the function $\gamma(\varepsilon)$ satisfy

$$\gamma(\varepsilon) \to 0 \;\text{ for }\; \varepsilon \to 0\,. \tag{7.13}$$

For regularization, we seek an operator T_γ that is continuous for $\gamma > 0$ and replaces A^+ in the presence of noise. We would like it to have the properties

$$T_\gamma\left(\underline{g}\right) \xrightarrow{\gamma\to 0} A^+\underline{g} \tag{7.14}$$

for all $\underline{g}$ and

$$||T_{\gamma(\varepsilon)}|| \cdot \varepsilon \xrightarrow{\varepsilon\to 0} 0 \tag{7.15}$$

for the operator norm of $T_{\gamma(\varepsilon)}$[3], because then it follows that

$$\left|T_{\gamma(\varepsilon)}(\underline{g}_0 + \varepsilon\Delta\underline{g}) - A^+\underline{g}_0\right| \leq \left|T_{\gamma(\varepsilon)}\varepsilon\Delta\underline{g}\right| + \left|T_{\gamma(\varepsilon)}\underline{g}_0 - A^+\underline{g}_0\right| \tag{7.16}$$

$$\leq ||T_{\gamma(\varepsilon)}|| \cdot |\Delta\underline{g}|\varepsilon + \left|T_{\gamma(\varepsilon)}\underline{g}_0 - A^+\underline{g}_0\right| , \tag{7.17}$$

and with Equations (7.14), (7.13) and (7.15), we have for the limit $\varepsilon \to 0$

$$||T_{\gamma(\varepsilon)}|| \cdot |\Delta\underline{g}|\varepsilon + \left|T_{\gamma(\varepsilon)}\underline{g}_0 - A^+\underline{g}_0\right| \xrightarrow{\varepsilon\to 0} 0 . \tag{7.18}$$

Hence, instead of calculating $\underline{f}^+ = A^+(\underline{g}_0 + \varepsilon\Delta\underline{g})$, which for $\varepsilon \to 0$ does not necessarily go to $A^+\underline{g}_0$, one can compute $\underline{f}^+ = T_{\gamma(\varepsilon)}(\underline{g}_0 + \varepsilon\Delta\underline{g})$, which for $\varepsilon \to 0$ becomes the inversion $A^+\underline{g}_0$ of the noise free data. Under the condition that we know the error ε, we have a reasonable inverse $\underline{f}^+$, which controls the error with respect to the inversion of the noise free data. The task is then to determine a suitable $T_{\gamma(\varepsilon)}$ and $\gamma(\varepsilon)$. One possible solution is the so called **Tikhonov regularization**

$$T_\gamma = (A^\dagger A + \gamma I)^{-1} A^\dagger , \tag{7.19}$$

where I is the identity matrix. Because $A^\dagger A$ is Hermitian, it has real and, in fact, non-negative eigenvalues. The latter is due to the fact that for any complex vector $\underline{v}$ of suitable dimension, the scalar product $(\underline{v}, A^\dagger A\underline{v}) = (A\underline{v}, A\underline{v}) = |A\underline{v}|^2 \geq 0$, i.e., $A^\dagger A$ is positive semidefinite. It follows that $A^\dagger A + \gamma I$ can be inverted for all $\gamma > 0$. One can show that $T_\gamma \underline{g} \xrightarrow{\gamma\to 0} A^+\underline{g}$ for all $\underline{g}$. If the inverse of A exists, i.e., $A^{-1} = A^+$, this follows from direct insertion

$$(A^\dagger A + \gamma \cdot 1)^{-1} A^\dagger = \left[A^\dagger A\left(1 + \gamma(A^\dagger A)^{-1}\right)\right]^{-1} A^\dagger \tag{7.20}$$

$$= \left(1 + \gamma(A^\dagger A)^{-1}\right)^{-1} (A^\dagger A)^{-1} A^\dagger \tag{7.21}$$

$$= \left(1 - \gamma(A^\dagger A)^{-1} + \gamma^2(A^\dagger A)^{-2} + \ldots\right) A^{-1} \tag{7.22}$$

$$\xrightarrow{\gamma\to 0} A^{-1} = A^+ . \tag{7.23}$$

Of course, the interesting point is that this also holds for operators A, which cannot be inverted.

Evaluation of $T_\gamma \underline{g}$ involves calculation of matrix products and matrix inverses, which may be prohibitive in high dimensions. Therefore, we now show that Tikhonov regularization can also be formulated as a minimization problem

$$\underline{f} = \operatorname{argmin}_{\underline{f}'} \left(|A\underline{f}' - \underline{g}|^2 + \gamma|\underline{f}'|^2\right) , \tag{7.24}$$

3 The operator norm of a linear operator $A : X \to Y$ over the two normed vector spaces X and Y can be defined as $||A|| = \sup_{\underline{v}\in X\setminus \underline{0}} \frac{|A\underline{v}|_Y}{|\underline{v}|_X}$.

with the discrepancy functional $|A\underline{f}' - \underline{g}|^2$ and a penalty $\gamma|\underline{f}'|^2$ containing the L_2 norm of the trial function $\underline{f}'$. By calculating the derivative of $|A\underline{f}' - \underline{g}|^2 + \gamma|\underline{f}'|^2$ with respect to $\underline{f}'$ and setting the result to zero, we obtain

$$0 = \frac{\partial}{\partial f'_k}\left[|A\underline{f}' - \underline{g}|^2 + \gamma|\underline{f}'|^2\right] \tag{7.25}$$

$$0 = 2\sum_i A^{\mathrm{T}}_{ki}(\sum_j A_{ij}f'_j - g_i) + 2\gamma f'_k \tag{7.26}$$

Or, more generally, for complex A,

$$0 = A^\dagger(A\underline{f}' - g) + \gamma\underline{f}' \,. \tag{7.27}$$

Rearranging the last equation gives

$$A^\dagger\underline{g} = (A^\dagger A\underline{f}' + \gamma\underline{f}') \,, \tag{7.28}$$

which is solved by

$$\underline{f}' = (A^\dagger A + \gamma I)^{-1}A^\dagger\underline{g} = T_\gamma\underline{g} \,. \tag{7.29}$$

We can see that the solution of the minimization problem coincides with application of the Tikhonov operator T_γ to the data $\underline{g}$. In other words, we could also use a numerical optimization procedure to find the solution to the regularized problem, which may offer advantages for large matrices. Importantly, the optimization approach can also be used for nonlinear operators.

7.1.2 Maximum likelihood approach

Noisy data means that we have to deal with random events, which are described by probability functions. The object itself (or, more generally, a physical observable μ_0) is assumed to be deterministic, but the measurement process or measurement channel is assumed to be probabilistic. The measured data X is just one realization of the underlying probability function $P(X|\mu = \mu_0)$ for given and fixed μ_0 [2]. The task is to determine the object μ_0 that is most likely to have resulted in the measured data X.

For example, X could denote the number of photons counted in one pixel, with a measurement process governed by the Poisson distribution

$$P(X|\mu) = \frac{\mu^X}{X!}\,e^{-\mu} \,, \tag{7.30}$$

and μ_0 the concentration of the radioactive tracer material in the respective tissue, which we want to reconstruct. To this end, a so called “estimator” is required, i.e., a function $\hat{\mu} = \hat{\mu}(X)$ that provides a value for μ_0 given one or several independent values of X. One such an estimator is the “maximum likelihood estimator” (ML). The idea is to find that value of the parameter μ which maximizes the probability of having obtained X. With P given, the only unknown parameter is the object μ itself. When we thus regard P as a function of μ for fixed X, this function is called "likelihood." We find the optimal value of μ by maximizing $P(X|\mu)$:

$$\mu = \text{argmax}\Big(P(X|\mu)\Big) \quad \rightarrow \quad 0 = \frac{\partial}{\partial\mu} P(X|\mu)\,. \tag{7.31}$$

Equivalently, in practice, one usually seeks the minimizer of the negative log likelihood $-\log P(X|\mu)$ and solves $0 = \frac{\partial}{\partial\mu}\log P(X|\mu)$, which has the same solution as $0 = \frac{\partial}{\partial\mu}P(X|\mu)$. Again taking the example of a Poisson random variable, the log likelihood becomes

$$\log\left[P(X|\mu)\right] = \log\left[\frac{\mu^X}{X!}\,e^{-\mu}\right] = -\mu + X\log\mu - \log X!\,, \tag{7.32}$$

and minimization of the negative log likelihood results in

$$0 = \frac{\partial}{\partial\mu}\Big(-\mu + X\log\mu - \log X!\Big) \tag{7.33}$$

$$= -1 + \frac{X}{\mu} \quad \Rightarrow \hat{\mu}(X) = X\,. \tag{7.34}$$

Hence, the log likelihood is maximized for a given realization X when the parameter μ coincides with the measured X. Furthermore, in this simple case, the estimator $\hat{\mu}$ is unbiased, i.e., the expectation value of $\hat{\mu}$ is the same as the true value μ_0

$$\langle\hat{\mu}\rangle = \langle X\rangle = \mu_0\,, \tag{7.35}$$

although this does not need to be the case in general.

This can be generalized to the case of having more than one measurement $X_1, \ldots, X_n$. The joint likelihood of $(X_1, \ldots, X_n)$ is given by

$$P(X_1, \ldots, X_n|\mu) = \prod_{i=1}^{n} P(X_i|\mu)\,. \tag{7.36}$$

Minimizing the corresponding negative log likelihood results in

$$0 = \frac{\partial}{\partial\mu}\log\prod_{i=1}^{n}P(X_i|\mu) = \sum_{i=1}^{n}\frac{\partial}{\partial\mu}\log P(X_i|\mu) = \sum_{i=1}^{n}\left(-1 + \frac{X_i}{\mu}\right) \quad \Rightarrow \hat{\mu} = \frac{1}{n}\sum_{i=1}^{n}X_i\,, \tag{7.37}$$

which shows that the natural tendency to take averages of multiple measurements is, in the case of Poisson statistics at least, very much justified because it is identical to the ML estimator.

Consider slightly more complicated configurations, for example, the Radon transform, where we have to deal with several measurements X_i and several parameters μ_i. In the case of SPECT, the μ_i are the line integrals through a radioactive tracer density in the patient's body

$$\mu(\underline{n}_\theta, s) = \int d^2x\,\delta(\underline{x}\cdot\underline{n}_\theta - s)\,f(\underline{x})\,. \tag{7.38}$$

Discretizing for simplicity, this can be written in discrete form with the approximated Radon integral kernel A_{ij} as

$$\mu_i = \sum_j A_{ij} f_j \,. \tag{7.39}$$

Here, we would like to obtain an estimate of $\underline{f}$ (the discretized object of interest) by maximizing the likelihood

$$P(X_1, X_2, \ldots \mid \underbrace{\mu_i = \sum_j A_{ij} f_j}_{\text{Radon transform}}, i = 1, 2, \ldots) \tag{7.40}$$

for having measured the noisy signals X_i. Since the X_i are mutually independent, their joint probability distribution factorizes, and we can write

$$\begin{aligned} 0 &= \frac{\partial}{\partial f_k} \log \prod_{i=1}^{n} P\left(X_i | \mu_i = \sum_j A_{ij} f_j \right) && (7.41)\\ &= \sum_{i=1}^{n} \frac{\partial}{\partial f_k} (-\mu_i + X_i \log \mu_i - \log X_i!) && (7.42)\\ &= \sum_{i=1}^{n} \left(-A_{ik} + \frac{X_i}{\sum_j A_{ij} f_j} A_{ik} \right) && (7.43)\\ &= \sum_{i=1}^{n} \left(-1 + \frac{X_i}{\sum_j A_{ij} f_j} \right) A_{ik} \,. && (7.44) \end{aligned}$$

If A_{ij} is invertible, it follows for the ML estimator $\hat{\underline{f}}$ that

$$\begin{aligned} 0 &= -1 + \frac{X_i}{\sum_j A_{ij} f_j} && (7.45)\\ \Leftrightarrow X_i &= \sum_j A_{ij} f_j && (7.46)\\ \Leftrightarrow \hat{\underline{f}} &= A^{-1} X \,. && (7.47) \end{aligned}$$

Hence, the 'simple' inversion of the noisy data is equal to the ML estimator.

In the more general case that A is not invertible, the ML estimator for Poisson distributed random data must be obtained by solving the optimization problem

$$\hat{\underline{f}} = \operatorname{argmin} \left[\sum_i \left(\sum_j A_{ij} f_j - X_i \log \sum_j A_{ij} f_j \right) \right] . \tag{7.48}$$

Apart from Poisson distributions, the ML approach can be applied to all kinds of probability functions $P(X|\mu)$, including the simple case of Gaussian noise. Of course, the functionals to be minimized will look different from those in the Poisson case.

7.1.3 Bayesian inference techniques

In the preceding section, we assumed the object to be deterministic and we did not impose any constraints on the object or physical function to be reconstructed. Yet, in many problems we may have some prior information about the object, in addition to the measured data X. For example, considering the distribution μ of a tracer material in a patient's body, we know that the sum over all μ_i cannot exceed the dose administered to the patient beforehand

$$\sum_i \mu_i \leq \text{administered Dose} . \tag{7.49}$$

However, this is not the only a priori information available. One can also consider the object μ itself to be a realization of a stochastic process with known probability distribution $Q(\mu)$. Using probabilistic instead of deterministic prior information is the underlying idea of Bayesian methods compared to ML techniques [2]. For example, let us assume that $0 \leq \mu \leq \tilde{\mu}$ is uniformly distributed. The measured signal X still follows the Poisson distribution with parameter μ. Now consider the perfectly reasonable case that $X = \lceil \tilde{\mu} + 1 \rceil$ is measured. The ML method would suggest that the most likely object $\hat{\mu}$ underlying the data X is $\hat{\mu} = X = \lceil \tilde{\mu} + 1 \rceil$. However, this solution is unreasonable, since we know that μ is not allowed to exceed the value of $\tilde{\mu}$. What then is a better estimator for μ in this case? The answer is given by Bayes's theorem. Let X and μ be two correlated random variables. The marginal probability of finding μ is $Q(\mu)$. The conditional probability of finding X – provided that μ has already been determined – is $P(X|\mu)$. The joint probability of $P(X, \mu)$ is then given by the product of $Q(\mu)$ and $P(X|\mu)$

$$P(X, \mu) = P(X|\mu) \cdot Q(\mu) . \tag{7.50}$$

On the other hand, the joint probability $P(X, \mu)$ for the combination of μ and X can equivalently be expressed as the product of the probability $P(X) = \int d\mu\, P(X, \mu)$ to measure X and the joint probability $P(\mu|X)$ to obtain the value of μ for given X

$$P(X, \mu) = P(X) \cdot P(\mu|X) . \tag{7.51}$$

From this, **Bayes's theorem** follows, which allows us to calculate the conditional probability $P(\mu|X)$ as

$$P(\mu|X) = \frac{P(X|\mu) \cdot Q(\mu)}{P(X)} . \tag{7.52}$$

The benefit of this theorem is that we gain a statement about the a posteriori probability of μ, provided that we have measured X. Since we know both $Q(\mu)$ and $P(X|\mu)$, $P(X) = \int d\mu\, P(X, \mu) = \int d\mu P(X|\mu) \cdot Q(\mu)$ can be calculated. Hence, knowing the a priori probability $Q(\mu)$ we can estimate μ by maximizing the a posteriori probability $P(\mu|X)$ or equivalently by minimizing $-\log P(\mu|X)$

$$\hat{\mu} = \operatorname{argmin} \left(-\log P(X|\mu) - \log Q(\mu) + \log P(X) \right) . \tag{7.53}$$

Since $P(X)$ does not depend on μ, it is even sufficient to evaluate

$$\hat{\mu} = \operatorname{argmin}\left(-\log P(X|\mu) - \log Q(\mu)\right) . \tag{7.54}$$

This technique is called the maximum a posteriori (MAP) or Bayesian inference method. For the example above, $Q(\mu)$ can be expressed as

$$Q(\mu) = \frac{\Theta(\mu)\Theta(\tilde{\mu} - \mu)}{\tilde{\mu}} , \tag{7.55}$$

where $\Theta(\mu)$ is the Heaviside step function[4]. Here, the Bayesian inference method results in

$$\begin{aligned} \hat{\mu} &= \operatorname{argmin}\left[-\log\left(\frac{\Theta(\mu)\Theta(\tilde{\mu} - \mu)}{\tilde{\mu}} e^{-\mu} \frac{\mu^X}{X!}\right)\right] & (7.56)\\ &= \operatorname{argmin}\left[-\log\left(\Theta(\mu)\Theta(\tilde{\mu} - \mu)\right) + \mu - X \log \mu\right] . & (7.57) \end{aligned}$$

If $0 \leq X \leq \tilde{\mu}$, the minimum is found at $\hat{\mu} = X$, which coincides with the $\hat{\mu}$ obtained by the ML strategy. If, however, $X > \tilde{\mu}$, the minimum is found at $\hat{\mu} = \tilde{\mu}$.
In the limiting case of uniformly distributed μ, i.e., $Q(\mu) = \text{const.}$, meaning that no further information on μ is available, the Bayesian inference method

$$\hat{\mu} = \operatorname{argmin}\left(-P(X|\mu)\right) \tag{7.58}$$

is the same as the ML method.

Similarly to ML approach, the Bayesian inference technique can be generalized to more sophisticated problems. The main challenge is to find a suitable model for the a priori probability $Q(\mu)$, since this requires already quite detailed knowledge about the object to be reconstructed. Again, this points to a fundamental aspect of treating noisy data: Noise causes a loss of information and the reconstruction problems are ill posed. Hence, additional knowledge $Q(y)$ has to be invoked to compensate for the information deficit.

We close this section with a remark about the connection between the methods introduced. At first, we saw that in order to solve a deterministic inverse problem, the Moore–Penrose inverse alone is not a good method, since it gives qualitatively wrong results in the presence of small (deterministic) perturbations if the problem is ill posed. In order to overcome this difficulty, Tikhonov regularization was introduced. This method is based purely on analytic convergence properties of operators

4 The Heaviside step function is defined as

$$\Theta(\mu) = \begin{cases} 1 & \text{if } \mu \geq 0 \\ 0 & \text{else.} \end{cases}$$

when the strength of the perturbation goes to 0. The recipe to apply this method can be formulated as a minimization problem of a sum of a quadratic data discrepancy term plus a quadratic penalty term. Our next approach was to consider ML estimation for stochastic inverse problems. Again, we saw that this can be expressed as a minimization problem, but not necessarily over a quadratic data discrepancy term, but rather one that is adapted to the noise properties. However, we saw that this approach may be identical to a simple inversion of the measured data, which is (just like the Moore–Penrose inverse) of no use if the problem is ill posed. Lastly, we introduced MAP estimation, which is again expressed as a minimization problem over a data fidelity term plus a regularization term. This method is capable of exploiting a priori information and can, thus, overcome the ill posed nature of the problem, just like Tikhonov regularization. It should be kept in mind, however, that despite their deceptively similar form, the methods apply to conceptually very different situations. In the case of the Moore–Penrose inverse and Tikhonov regularization, we have a deterministic setting in which the perturbation is formally assumed to go to 0 and where we assume nothing more than a few analytic properties of the operators involved. In contrast, in the case of ML and MAP estimation, we have a stochastic setting with a finite amount of noise with known distribution and available (physical) prior information about the problem at hand.

7.2 Reconstruction of two-dimensional images

The reconstruction of objects in 3d space (like the radioactive tracer distribution in the case of SPECT) is only one example of inverse problems. Next, we focus on 2d images, where similar procedures can be used to retrieve the respective objects. Here, we concentrate on denoising and sharpening algorithms.

Let us consider a 2d planar image $\underline{X}$ of an object $\underline{f}$ described as[5]

$$X_i = f_i + \xi_i \,, \tag{7.59}$$

where ξ_i represents Gaussian white noise with standard deviation σ

$$P(\xi_i) = \frac{1}{\sqrt{2\pi}\sigma} e^{-\frac{\xi_i^2}{2\sigma^2}} \,. \tag{7.60}$$

Importantly, the noise is uncorrelated for all ξ_i. Furthermore, due to the PSF $\underline{\mathcal{P}}$ of the optical system, each image is blurred, see Figure 7.1. This blurring can be modeled by a convolution of the original object $\underline{f}$ with $\underline{\mathcal{P}}$

$$X_i = \left(\underline{\mathcal{P}} * \underline{f}\right)_i + \xi_i \,. \tag{7.61}$$

5 Here 2d images and objects are formulated as one $1D$ array, where each element is a pixel of the image/the object. The 2d image $X(x, y)$ or object $f(x, y)$ is recovered by rearranging the elements of the 1d array in two dimensions.

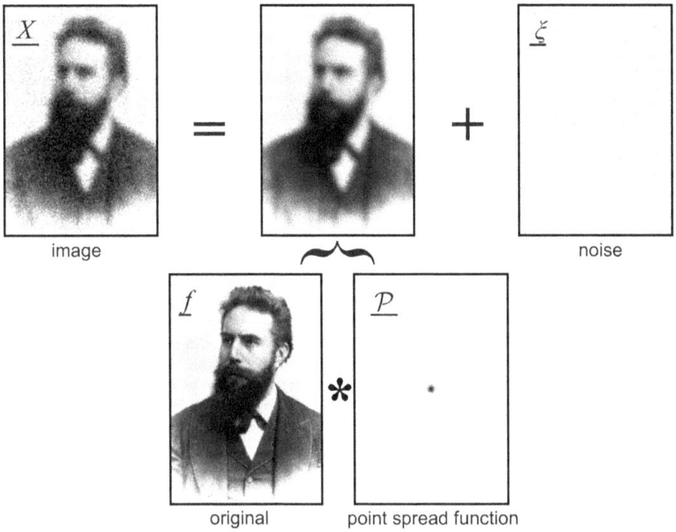

Fig. 7.1: Illustration of equation (7.61). The image is modeled as the superposition of the blurred original and noise. The purpose of image reconstruction is to sharpen and denoise the image in order to retrieve the original object.

$\mathcal{P}(\underline{y})f(\underline{x}-\underline{y})$ is the component of f which is mapped from position $(\underline{x}-\underline{y})$ to position $\underline{x}$. Without loss of generality, we can choose

$$\sum_i \mathcal{P}_i = 1 \,, \tag{7.62}$$

since otherwise the total intensity of $\underline{X}$ is reduced with respect to the total intensity of $\underline{f}$. Next, we address how we can reconstruct objects from such blurred and noisy images.

7.2.1 Images as Markov random fields

To begin, we recall the Bayesian inference method

$$\underline{f}_s = \operatorname{argmin}\left(-\log Q(\underline{f}) - \log P(\underline{X}|\underline{f})\right) \,, \tag{7.63}$$

described in the last section. It is of general applicability and can be used for planar images as well as for (tomographic) 3d images, provided that the original objects (noise free and not blurred) can be modeled as an ensemble of random variables. With the formulation for an N pixel image according to equation (7.61), and the noise model $P(\xi_i)$ (see equation (7.60)), and given that $\underline{\xi}$ and $\underline{\mathcal{P}} * \underline{f}$ are well known, the conditional probability $P(\underline{X}|\underline{f})$ becomes

$$P(\underline{X}|\underline{f}) = \prod_{i=1}^{N} P(X_i|f_i) = \prod_{i} P(\xi_i|f_i) = \prod_{i=1}^{N} P\left(\underbrace{Xi - (\underline{\mathcal{P}} * \underline{f})_i}_{=\xi_i} \Big| f_i \right)$$
$$= \prod_{i=1}^{N} \frac{1}{\sqrt{2\pi}\sigma} \exp\left[-\frac{\left(X_i - (\underline{\mathcal{P}} * \underline{f})_i\right)^2}{2\sigma^2} \right]$$
$$= \left(\frac{1}{\sqrt{2\pi}\sigma} \right)^N \exp\left[-\sum_{i=1}^{N} \frac{\left(X_i - (\underline{\mathcal{P}} * \underline{f})_i\right)^2}{2\sigma^2} \right]$$
$$= \left(\frac{1}{\sqrt{2\pi}\sigma} \right)^N \exp\left[-\frac{|(\underline{X} - (\underline{\mathcal{P}} * \underline{f}))|^2}{2\sigma^2} \right] . \tag{7.64}$$

Hence, the logarithmic part in equation (7.63) is

$$\log(P(\underline{X}|\underline{f})) = \log\left[\left(\frac{1}{\sqrt{2\pi}\sigma} \right)^N \right] - \frac{|(\underline{X} - (\underline{\mathcal{P}} * \underline{f}))|^2}{2\sigma^2} , \tag{7.65}$$

where only the last term is relevant for minimizing equation (7.63).

For the evaluation of equation (7.63), we still need a suitable description $Q(\underline{f})$ for the object $\underline{f}$. To this end, we invoke a stochastic process known as the Markov random field (MRF), proposed in 1984 by S. Geman and D. Geman as a general framework to model $Q(\underline{f})$ in imaging problems [3]. An object consisting of N intensity values can be considered as the realization of a random process $\underline{f}$ with elements f_i. The probability of a particular element f_i to adopt the value ϕ_i shall only depend on the values ϕ_j of neighboring elements of f_i. Hence, the MRF is a vector of random numbers with the following property

$$Q\left(f_i = \varphi_i \mid f_j = \varphi_j, j \neq i, j \in \{1, \dots, N\}\right) = Q\left(f_i = \varphi_i \mid f_j = \varphi_j, j \in G_i\right) , \tag{7.66}$$

where G_i is the local neighborhood of site i. Equation (7.66) states that the intensity value of f_i only depends on the intensity values of its surrounding pixels. Such a neighborhood G_i can be interpreted in a broad sense: nearest and next nearest neighbors may also contribute to G_i. The important point is that G_i is small compared to the number of pixels N of the entire object. Hence, the underlying idea of Markov random field models is that the global structure of an object is determined by its local, physical properties. This implies that the object is smooth except for a few discontinuities such as edges. Next, we have to address the question of how the conditional probability distributions $Q\left(f_i = \varphi_i|f_j = \varphi_j, j \in G_i\right)$ determine the joint probability distribution $Q(f_1, f_2, \dots f_N) = Q(\underline{f})$. To formulate the single $Q\left(f_i = \varphi_i|f_j = \varphi_j, j \in G_i\right)$, one has to identify the rules that predict the intensity value of a pixel for given intensities of neighboring pixels. Such general rules could be obtained by analyzing a huge amount of images, followed by the reconstruction of the joint probability distribution $Q(\underline{f})$. Unfortunately, this turns out to be very cumbersome and not realizable in practice.

Luckily there exists a different way to find the joint probability distribution of the object. In 1971, J. Hammersley and P. Clifford stated in an unpublished note that $Q(\underline{f})$ can be expressed by a Gibbs distribution, since any conditional distribution $Q\left(f_i = \varphi_i | f_j = \varphi_j, j \in G_i\right)$ associated with a Markov random field has a joint distribution function $Q(\underline{f})$, which can be related to a Gibbs distribution [5] as

$$Q(\underline{f}) = \frac{1}{Z} \exp\left(-\beta H(\underline{f})\right) , \tag{7.67}$$

where Z is a normalizing constant and $H(\underline{f})$ is defined as

$$H(\underline{f}) = \sum_{\text{cliques } C} V_C\left(\{f_i | i \in C\}\right) , \tag{7.68}$$

where a clique is a collection of pixels, as discussed further below. Now, $Q(\underline{f})$ is a Gibbs distribution specified by an energy functional $H(\underline{f})$, which is the summation of the "potential energies" V_C of the individual cliques C. The normalizing factor Z is the summation or integration of $\exp\left(-\beta H(\underline{f})\right)$ over all possible realizations of $\underline{f}$. Gibbs distributions are well known from statistical physics, where the parameter β is the inverse temperature, H is a Hamilton function and Z the corresponding partition function.

Next, we illustrate the concept of cliques. A neighborhood G_i can be subdivided into collections of pixels, so called cliques, which have to fulfill the following properties. First, a clique includes the pixel of interest f_i and, second, any two members of a clique are neighbors. Figure 7.2(a) illustrates the eight-point neighborhood of the pixel colored in gray, while (b) depicts cliques that can be defined within this eight-point neighborhood.

While $Q(\underline{f})$ is now easily constructed, the question remains how to model the different V_C. Indeed, the choice of V_C depends on the problem. Typical examples of V_C for a clique consisting of two elements are

$$V_C(f_i - f_j) = (f_i - f_j)^2 , \tag{7.69}$$

$$V_C(f_i - f_j) = \frac{1}{1 + (f_i - f_j)^\alpha} , \tag{7.70}$$

$$V_C(f_i - f_j) = \log(\cosh(f_i - f_j)) , \tag{7.71}$$

which are illustrated in Figure 7.3.

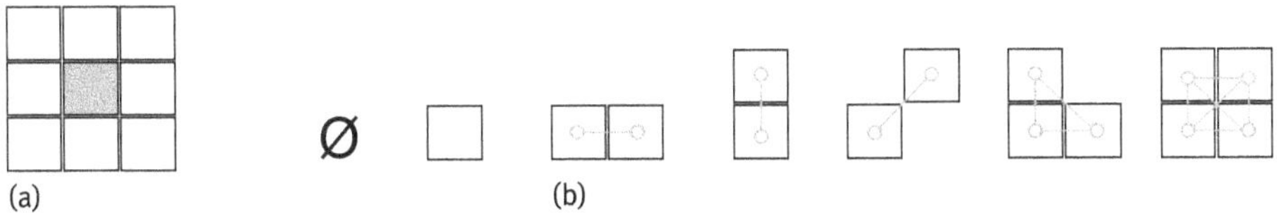

Fig. 7.2: (a) An eight-point neighborhood of the gray pixel. (b) Cliques that can be defined within the eight-point neighborhood.

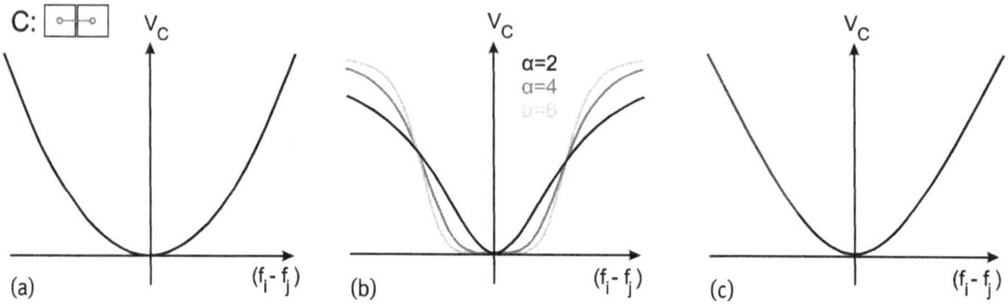

Fig. 7.3: Different definitions of the potential V_C of a clique consisting of two pixels. (a) $V_C(f_i - f_j) = (f_i - f_j)^2$, (b) $V_C(f_i - f_j) = \frac{1}{1+(f_i-f_j)^\alpha}$, (c) $V_C(f_i - f_j) = \log(\cosh(f_i - f_j))$.

In general, it is not recommendable to use the quadratic potential given in equation (7.69), since it suppresses edges too strongly. Hence, due to the constraint imposed by V_C the image will contain less noise, but edges will appear washed out. To preserve edges, one chooses V_C to be quadratic for small intensity deviations of neighboring pixels and to be less steep for values of $|f_i - f_j|$ exceeding a defined threshold. Given appropriate models for V_C and with the joint probability distribution $Q(\underline{f})$ formulated as a Gibbs distribution, $Q(\underline{f})$ can be replaced in the Bayesian inference method by equation (7.67)

$$\underline{f}_s = \operatorname{argmin}\left(-\log Q(\underline{f}) - \log P(\underline{X}|\underline{f})\right) \tag{7.72}$$

$$= \operatorname{argmin}\left(\beta H(\underline{f}) - \log P(\underline{X}|\underline{f})\right) \tag{7.73}$$

$$= \operatorname{argmin}\left(\beta \sum_{\text{cliques } C} V_C(\{f_i | i \in C\}) - \log P(\underline{X}|\underline{f})\right), \tag{7.74}$$

and finally with equation (7.65)

$$\underline{f}_s = \operatorname{argmin}\left(\underbrace{\beta \sum_{\text{cliques } C} V_C(\{f_i | i \in C\})}_{\text{de-noising}} + \underbrace{\frac{|(\underline{X} - (\mathcal{P} * \underline{f}))|^2}{2\sigma^2}}_{\text{sharpening}}\right). \tag{7.75}$$

When calculating $\underline{f}_s$ by equation (7.75), we have simultaneously achieved denoising and deblurring of the image $\underline{X}$ by considering a priori information on the object (contained in $Q(\underline{f})$) and on the imaging system (contained in $P(\underline{X}|\underline{f})$). Furthermore, in equation (7.75) the meaning of the parameter β becomes clear. It determines the ratio between sharpening and denoising. To evaluate equation (7.75), different algorithms exist, yet minimization is still not trivial, since the functional (7.75) may contain many local minima, which can lead to stagnation of the optimization. To this end, methods known from statistical physics, such as simulated annealing and parallel tempering, can provide suitable ways out.

Finally, an appropriate value for β has to be chosen. As an illustrative example, we demonstrate a proper choice for a Gaussian PSF given by

$$\mathcal{P}(x, y) \approx \exp\left(-\frac{x^2 + y^2}{2l^2}\right) . \tag{7.76}$$

The parameter l is the length scale, over which single pixel values are smeared out. In other words, finer object features than l are washed in the image. The convolution $\mathcal{P}(x, y) * f(x, y)$ corresponds to a multiplication in Fourier space

$$\mathcal{P}(x, y) * f(x, y) = \mathcal{F}^{-1}\left(\tilde{\mathcal{P}}(k_x, k_y) \cdot \tilde{f}(k_x, k_y)\right) , \tag{7.77}$$

where

$$\tilde{\mathcal{P}}(k_x, k_y) \approx \exp\left(-\frac{l^2 \cdot \left(k_x^2 + k_y^2\right)}{2}\right) . \tag{7.78}$$

Now we assume that the original object $f(x, y)$ may be disturbed by periodic, noisy structures of length scales l_0. For $l_0 \leq l$, these disturbances are not visible in $|X(x, y) - \mathcal{P}(x, y) * f(x, y)|$, see Figure 7.4. Hence, particular frequencies can have been manipulated without significant effect in $|X(x, y) - \mathcal{P}(x, y) * f(x, y)|$. For example, one can think of a function $\tilde{\tilde{f}}(k_x, k_y)$ such that

$$\tilde{\tilde{f}}(k_x, k_y) = \begin{cases} 0 & |k| \neq \frac{1}{l_0} \\ \text{random number} & |k| \approx \frac{1}{l_0}, \quad \text{where } l_0 \approx l , \end{cases} \tag{7.79}$$

which can be added to $\tilde{f}(k_x, k_y)$. Once the corrupted $\mathcal{F}^{-1}\left[\tilde{f}(k_x, k_y) + \tilde{\tilde{f}}(k_x, k_y)\right](x, y)$ is convolved with $\mathcal{P}(x, y)$ with the characteristic feature size $l \approx l_0$, the measured signal resulting from the noise free $f(x, y)$ and the signal resulting from the corrupted object

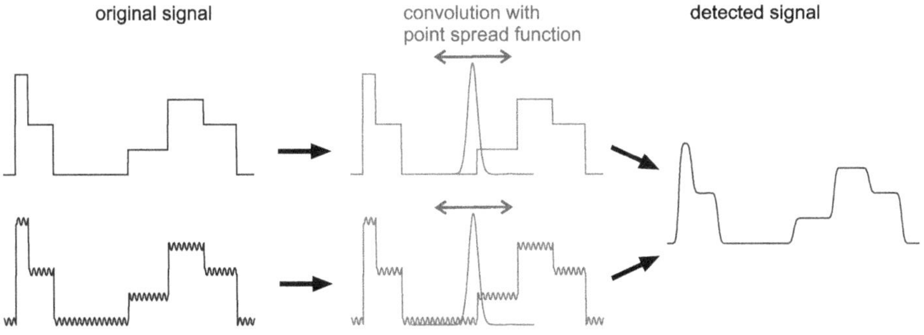

Fig. 7.4: The PSF washes out features below a characteristic structure size. The measurement of a noise free signal and the measurement of a signal deteriorated by additional high frequencies cannot be distinguished. The proper parameter β in optimization is inversely proportional to this length scale (in pixel units).

can no longer be distinguished, see Figure 7.4. This effect, which can lead to reconstruction of objects that are corrupted by high frequency artifacts, is undesirable and has to be suppressed by a suitable regularization technique. Such a regularizing effect can be introduced by the potential V_C and, in particular, it can be tuned by the parameter β

$$\beta \cdot V_C(\underline{f} + \underline{\hat{f}}) \approx \beta \cdot V_C(\underline{f}) + \beta \underline{\hat{f}} \cdot \nabla V_C(\underline{f}) + \dots \quad . \tag{7.80}$$

Recall that $V_C(\underline{f})$ only depends on those elements f_i with $i \in C$. This implies that also $\underline{\hat{f}} \cdot \nabla V_C(\underline{f})$ is only affected by those f_i with $i \in C$. Because the f_i are neighbors, they are separated by the length of 1 pixel. Since $\tilde{\hat{f}}(\underline{k}) \neq 0$ for $|k| \approx l_0^{-1} \approx l^{-1}$, the component $\hat{f}_i$ varies on scales of l with respect to its neighboring pixels. In order for the second component in the expansion given in equation (7.80) to significantly contribute to V_C, β has to fulfill

$$\beta \cdot l^{-1} \approx \mathcal{O}(1) \, . \tag{7.81}$$

In other words, the length scale for denoising as controlled by the parameter β should be comparable to the length scale l on which the convolution with the PSF washes out the image.

7.2.2 Image deconvolution by the Wiener filter

The Wiener filter is an entirely different and, in fact, more common approach than MRF based optimization to sharpen and denoise images. In this case, the problem of finding appropriate parameter settings is circumvented because these are estimated from the content of the image itself. However, the statistical properties have to obey more stringent conditions, such as translational invariance. We still consider the object as a stochastic variable and define the correlation $C_{\underline{f}\underline{g}}$ of two (object) functions $\underline{f}$ and $\underline{g}$ as

$$C_{\underline{f}\underline{g}}(i,j) = \mathbb{E}\left[f_i g_j\right] \,, \tag{7.82}$$

where $\mathbb{E}$ denotes the expectation value. We assume that the autocorrelation depends only on the distance between pixels

$$C_{\underline{f}\underline{f}}(i,j) = C_{\underline{f}\underline{f}}(i-j) \,, \tag{7.83}$$

in other words, we require statistical invariance under translations, which is actually quite a strong restriction on the object. The imaging process is again modeled as

$$\underline{X} = \left(\underline{\mathcal{P}} * \underline{f}\right) + \underline{\xi} \,, \tag{7.84}$$

with the PSF $\underline{\mathcal{P}}$ and independent Gaussian noise $\underline{\xi}$ with variance σ^2. The idea underlying the Wiener filter is to recover the original object $\underline{f}$ by convolution of the image $\underline{X}$

with a function $\underline{h}$. To this end, $\underline{h}$ has to be chosen such that the average discrepancy between $\underline{h} * \underline{X}$ and $\underline{f}$ is minimal

$$\begin{aligned}\mathbb{E}\left[\left(\underline{h} * \underline{X} - \underline{f}\right)^2\right] &= \mathbb{E}\left[\underline{Z}^2\right] - 2 \cdot \mathbb{E}\left[\underline{Z}\underline{f}\right] + \mathbb{E}\left[\underline{f}^2\right] \\ &= \sum_i \left(C_{\underline{Z}\underline{Z}}(i,i) - 2C_{\underline{Z}\underline{f}}(i,i) + C_{\underline{f}\underline{f}}(i,i)\right) ,\end{aligned} \tag{7.85}$$

where single components Z_i of $\underline{Z}$ are

$$Z_i = \sum_j h_j X_{i-j} . \tag{7.86}$$

We first rewrite the correlations in equation (7.85) by using the properties of the expectation value

$$\begin{aligned}C_{\underline{Z}\underline{Z}}(i,i) = C_{\underline{Z}\underline{Z}}(0) &= \mathbb{E}\left[\sum_j h_j X_{i-j} \sum_k h_k X_{i-k}\right] \\ &= \sum_j h_j \sum_k h_k \mathbb{E}\left[X_{i-j} X_{i-k}\right] \\ &= \sum_j h_j \sum_k h_k C_{\underline{X}\underline{X}}(j-k) ,\end{aligned} \tag{7.87}$$

and

$$\begin{aligned}C_{\underline{Z}\underline{f}}(i,i) = C_{\underline{Z}\underline{f}}(0) &= \mathbb{E}\left[\sum_j h_j X_{i-j} f_i\right] \\ &= \sum_j h_j \mathbb{E}\left[X_{i-j} f_i\right] \\ &= \sum_j h_j C_{\underline{f}\underline{X}}(0) .\end{aligned} \tag{7.88}$$

The results obtained in equation (7.87) and equation (7.88) are plugged in the discrepancy functional given in equation (7.85). We minimize

$$\hat{h} = \operatorname{argmin}_h \left[\sum_j h_j \sum_k h_k C_{\underline{X}\underline{X}}(j-k) - 2 \cdot \sum_j h_j C_{\underline{f}\underline{X}}(j) + C_{\underline{f}\underline{f}}(0)\right] \tag{7.89}$$

by setting the derivative of the discrepancy functional with respect to all h_j to zero

$$\begin{aligned}0 &= \frac{\partial}{\partial h_j}\left[\sum_{j'} h_{j'} \sum_k h_k C_{\underline{X}\underline{X}}(j'-k) - 2 \cdot \sum_{j'} h_{j'} C_{\underline{f}\underline{X}}(j') + C_{\underline{f}\underline{f}}(0,0)\right] \\ &= 2 \cdot \sum_k h_k \, C_{\underline{X}\underline{X}}(j-k) - 2 \cdot C_{\underline{f}\underline{X}}(j) ,\end{aligned} \tag{7.90}$$

where we used that $C_{\underline{X}\underline{X}}$ is symmetric. Recognizing the convolution between $\underline{h}$ and $\underline{X}$ in the last step and applying the convolution theorem

$$\sum_k h_k\, C_{\underline{X}\underline{X}}(j-k) = \mathcal{F}^{-1}\left[\tilde{h}_k \cdot \tilde{C}_{\underline{X}\underline{X}}(k)\right](j) \tag{7.91}$$

we find for the best filter $\underline{h}$

$$\underline{h} = \mathcal{F}^{-1}\left[\frac{\tilde{C}_{\underline{f}\underline{X}}}{\tilde{C}_{\underline{X}\underline{X}}}\right]. \tag{7.92}$$

It remains to compute $\tilde{C}_{\underline{X}\underline{f}}$ and $\tilde{C}_{\underline{X}\underline{X}}$. For $\tilde{C}_{\underline{X}\underline{f}}$ we find (with arbitrary i)

$$\begin{aligned}
\tilde{C}_{\underline{f}\underline{X}}(k) &= \mathcal{F}\left[C_{\underline{f}\underline{X}}(i, i-j)\right](k) \\
&= \frac{1}{\sqrt{N}}\sum_j C_{\underline{f}\underline{X}}(i, i-j)\exp\left(-i\frac{2\pi}{N}jk\right) \\
&= \frac{1}{\sqrt{N}}\sum_j \mathbb{E}\left[f_i\cdot\left(\left(\underline{\mathcal{P}}*\underline{f}\right)_{i-j}+\xi_{i-j}\right)\right]\exp\left(-i\frac{2\pi}{N}jk\right) \\
&= \frac{1}{\sqrt{N}}\sum_j \mathbb{E}\left[f_i\left(\sum_l \mathcal{P}_l\cdot f_{i-j-l}+\xi_{i-j}\right)\right]\exp\left(-i\frac{2\pi}{N}jk\right) \\
&= \frac{1}{\sqrt{N}}\sum_j \exp\left(-i\frac{2\pi}{N}jk\right)\sum_l \mathcal{P}_l\cdot\mathbb{E}\left[f_i\cdot f_{i-j-l}\right] \\
&= \frac{1}{\sqrt{N}}\sum_l \mathcal{P}_l\exp\left(i\frac{2\pi}{N}lk\right)\sum_j \exp\left(-i\frac{2\pi}{N}(j+l)\,k\right)\cdot C_{\underline{f}\underline{f}}(j+l) \\
&= \sqrt{N}\tilde{\underline{\mathcal{P}}}^*(k)\cdot\tilde{C}_{\underline{f}\underline{f}}(k)\,,
\end{aligned} \tag{7.93}$$

where we used the facts that the expectation value $\mathbb{E}\left[\underline{\xi}\right]$ is zero and that $\mathcal{P}$ is real such that $\tilde{\mathcal{P}}(-k) = \tilde{\mathcal{P}}^*(k)$. In an analogous way we can derive

$$\tilde{C}_{\underline{X}\underline{X}} = |\tilde{\underline{\mathcal{P}}}|^2\cdot\tilde{C}_{\underline{f}\underline{f}} + \sigma^2\,. \tag{7.94}$$

Hence, the most suitable filter is

$$\underline{h} = \mathcal{F}^{-1}\left[\frac{\tilde{C}_{\underline{f}\underline{X}}}{\tilde{C}_{\underline{X}\underline{X}}}\right] = \mathcal{F}^{-1}\left[\frac{\tilde{\underline{\mathcal{P}}}^*}{|\tilde{\underline{\mathcal{P}}}|^2+\frac{\sigma^2}{\tilde{C}_{\underline{f}\underline{f}}}}\right]. \tag{7.95}$$

Importantly, the filter depends on the $C_{\underline{f}\underline{f}}$ which has to be inferred from the data, e.g., according to equation (7.94). In the case when there is no blurring, i.e. in the case that for each position i the PSF is a delta distribution $\mathcal{P}_i = \delta_i$, we find

$$\tilde{C}_{\underline{X}\underline{X}} = \tilde{C}_{(\underline{f}+\underline{\xi})\underline{X}} = \tilde{C}_{\underline{f}\underline{X}} + \tilde{C}_{\underline{\xi}\underline{X}} = \tilde{C}_{\underline{f}\underline{X}} + \tilde{C}_{\underline{\xi}\underline{f}} + \tilde{C}_{\underline{\xi}\underline{\xi}} = \tilde{C}_{\underline{f}\underline{X}} + \sigma^2 \tag{7.96}$$

and from equation (7.92) we find that the ideal filter is

$$\underline{h} = \mathcal{F}^{-1}\left[\frac{\tilde{C}_{\underline{X}\underline{X}}-\sigma^2}{\tilde{C}_{\underline{X}\underline{X}}}\right]. \tag{7.97}$$

The correlation $\tilde{C}_{\underline{X}\,\underline{X}}$ can be estimated directly from the data $\underline{X}$ by

$$\tilde{C}_{\underline{X}\,\underline{X}}(j) \approx \frac{1}{V} \sum_i X_i X_{i-j} \,, \tag{7.98}$$

where V is a normalization constant. However, as soon as the PSF is nontrivial, one has to find suitable approaches to determine $\tilde{C}_{\underline{f}\,\underline{X}}$.

Comparing the Wiener filter

$$\underline{h} = \mathcal{F}^{-1} \left[\frac{\tilde{C}_{\underline{f}\,\underline{X}}}{\tilde{C}_{\underline{X}\,\underline{X}}} \right] = \frac{\tilde{\underline{\mathcal{P}}}^*}{|\underline{\mathcal{P}}|\, \tilde{C}^2_{\underline{X}\,\underline{X}} + \frac{\sigma^2}{\tilde{C}_{\underline{f}\,\underline{f}}}} \,, \tag{7.99}$$

and the Tikhonov regularization technique

$$T_\gamma = (A^\dagger A + \gamma I)^{-1} A^\dagger \,, \tag{7.100}$$

which was discussed before, one notices obvious similarities. Indeed, the Wiener filter is a different formulation of Tikhonov regularization with the advantage that γ can be directly obtained from the data. We can see that the Wiener filter solves the problem $\underline{X} = \underline{\mathcal{P}} * \underline{f} + \underline{\xi}$ in an adequate way. However, it requires the statistical properties of the object $\underline{f}$ to be invariant under translations. Unfortunately, this is only a suitable representation of $\underline{f}$ in a few cases.

7.2.3 Total variation denoising

We close the chapter with a brief discussion of total variation (TV). Minimization with respect to a regularization term containing the TV norm is a concept that is widely used in image denoising. Consider possible mathematical spaces to which an object may belong. Depending on the problem, different choices are suitable. Examples for spaces with decreasing number of elements are: (i) the space of distributions, (ii) the space of L^p functions, for which $\int \mathrm{d}^n x \, |f|^p < \infty$, or (iii) the space of functions with bounded variation. The last example is the space of functions f, which have a bounded TV, i.e., the integral

$$\mathrm{TV}(f) := \int \mathrm{d}^n x \, |\nabla f| \tag{7.101}$$

is finite. The TV integral corresponds to the L^1 norm of the gradient of f. It is obviously well defined for differentiable functions, but it can also be generalized to noncontinuous functions. One useful property of the TV norm is that it tolerates edges, because the derivative of an edge, i.e., a step function, results in a delta distribution and the integration over a delta distribution is finite. Contrarily, functions with rapid oscillations are characterized by large values of the TV norm. Thus, in order to denoise an

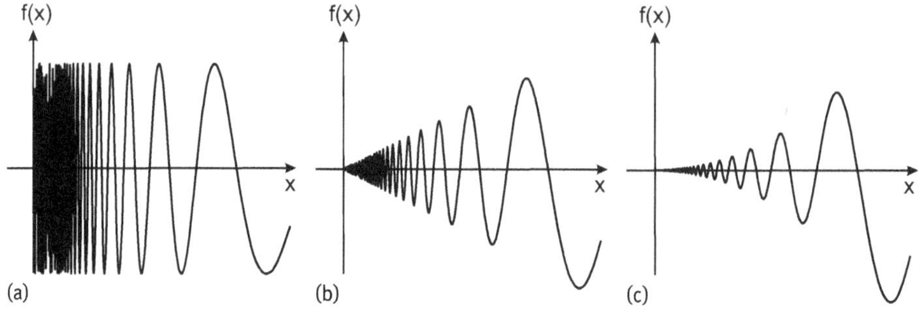

Fig. 7.5: (a,b) Examples of functions that are not elements of the space of bounded variation. Contrarily, example (c) is of bounded variation.

image $\underline{X}$ while preserving edges, it is reasonable to find a function $\underline{f}_s$ that is close to the data and has minimum TV [6].

Examples of functions that are not of bounded variations are

$$f(x) = \sin\left(\frac{1}{x}\right) \quad \text{and} \quad f(x) = x \cdot \sin\left(\frac{1}{x}\right),$$

since for $x \to 0$ they oscillate increasingly rapidly, see Figure 7.5(a–b). Contrarily, the function

$$f(x) = x^2 \cdot \sin\left(\frac{1}{x}\right)$$

is an example of a bounded variation, because x^2 is strong enough to suppress the oscillating behavior of $\sin\left(\frac{1}{x}\right)$ for $x \to 0$, see Figure 7.5(c).

An intrinsic property of the TV norm is the fact that it preserves edges while it suppresses rapid oscillations. Therefore, it can be used as a means to remove noise without compromising image sharpness. To achieve denoising and sharpening, the TV norm of the object has to be minimized. If an image is modeled as in equation (7.61), disturbed by Gaussian noise (equation (7.60)) and smeared out by the PSF $\underline{\mathcal{P}}$, we have to minimize the likelihood

$$P(\underline{X}|\underline{f}) = \frac{|\underline{X} - \underline{P} * \underline{f}|^2}{\sigma^2} + \lambda \cdot TV(\underline{f}), \tag{7.102}$$

where λ is a parameter that controls the weight of the TV norm in the functional $P(\underline{X}|\underline{f})$. Compared to the MRF methods (with Gibbs free energy), TV denoising shows good preservation of edges. However, it tends to induce spots. This becomes intuitively clear when considering the example of the two functions $f(x)$ and $g(x)$ defined by

$$f(x) = cx \quad \text{for} \quad 0 \le x \le L$$
$$g(x) = c \cdot \Theta(x - x_0) \quad \text{for} \quad 0 \le x \le L,$$

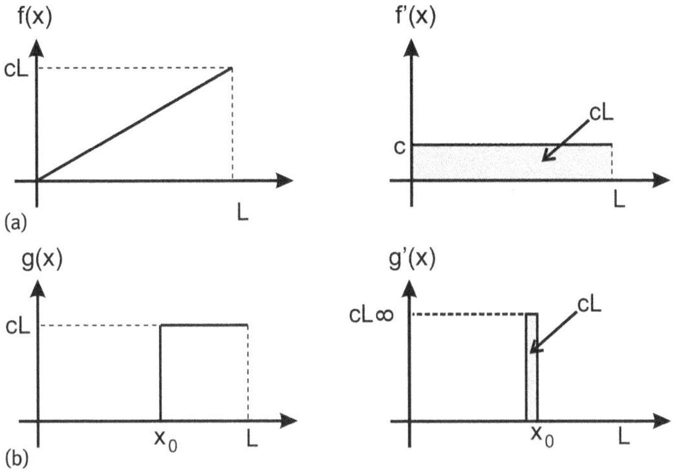

Fig. 7.6: The TV norm of a linear function (a) is identical to the TV norm of an edge (b).

which exhibit identical TV norms, see Figure 7.6, namely

$$TV(f(x)) = \int_0^L dx\, c = Lc$$

$$TV(g(x)) = cL \cdot \int_0^L dx\, \delta(x - x_0) = cL\,.$$

Hence, the TV norm is not a criterion that can help to distinguish between linear functions and edges. We see that edges are not suppressed with respect to a smooth slope. In the presence of noise, the shape that is more likely to represent the noise is favored and, in general, this will not be the smooth function.

References

[1] H. H. Barrett and K. J. Myers. *Foundations of Image Science*. Wiley, 2003.

[2] M. Bertero and P. Boccacci. *Introduction to inverse problems in imaging*. IoP Publishing, 1998.

[3] S. Geman and D. Geman. Stochastic Relaxation, Gibbs Distributions, and the Bayesian Restoration of Images. *IEEE Transactions on Pattern Analysis and Machine Intelligence*, PAMI-6(6):721–741, 1984.

[4] W. C. Karl. Regularization in image restoration and reconstruction. *Handbook of Image and Video Processing*, pages 183–202, 2005.

[5] J. D. Lafferty, A. McCallum, and F. C. N. Pereira. Conditional Random Fields: Probabilistic Models for Segmenting and Labeling Sequence Data. In *Proceedings of the Eighteenth International Conference on Machine Learning*, ICML '01, pages 282–289, San Francisco, CA, USA, 2001. Morgan Kaufmann Publishers Inc.

[6] L. I. Rudin, S. Osher, and E. Fatemi. Nonlinear total variation based noise removal algorithms. *Physica D: Nonlinear Phenomena*, 60(1):259–268, 1992.
[7] H. Schwarz. *Mehrfachregelungen. Grundlagen einer Systemtheorie: Zweiter Band.* Springer-Verlag, 2013.

Symbols and abbreviations used in Chapter 7

A	(linear) operator describing an imaging system
A^+	Moore–Penrose inverse (pseudoinverse) of A
$C_{\underline{f}\,\underline{g}}$	correlation of two functions
D	diagonal matrix
ε	noise strength
$\underline{f}$	object function
$\underline{f}'$	trial function
$\hat{\underline{f}}$	maximum likelihood estimator of $\underline{f}$
$\underline{g}$	image/data
$\Delta\underline{g}$	noise in the image/data
$\gamma(\varepsilon)$	regularization parameter (as a function of noise strength)
$\underline{h}$	filter
λ_i	eigenvalue
ML	maximum likelihood
MAP	maximum a posteriori
MRF	Markov random field
μ	parameter of Poisson distribution
$\hat{\mu}$	estimator for the parameter μ
$P(\mu, X)$	joint probability to obtain μ and X
$P(\mu\|X)$	conditional probability to obtain μ for given X
$\underline{\mathcal{P}}$	point spread function of an optical system
T_γ	Tikhonov operator (with regularization parameter γ)
TV	total variation
Θ	Heaviside step function
U	unitary operator
X	measured data or random variable
$\underline{X}$	2d planar image
ξ_i	Gaussian white noise in pixel i

Index

www.ingramcontent.com/pod-product-compliance
Lightning Source LLC
Chambersburg PA
CBHW080905220326
41598CB00034B/5485